Zsófia Lendek, Thierry Marie Guerra, Robert Babuška, and Bart De Schutter

Stability Analysis and Nonlinear Observer Design Using Takagi-Sugeno Fuzzy Models

Studies in Fuzziness and Soft Computing, Volume 262

Editor-in-Chief

Prof. Janusz Kacprzyk
Systems Research Institute
Polish Academy of Sciences
ul. Newelska 6
01-447 Warsaw
Poland
E-mail: kacprzyk@ibspan.waw.pl

Further volumes of this series can be found on our homepage: springer.com

Vol. 247. Michael Glykas
Fuzzy Cognitive Maps, 2010
ISBN 978-3-642-03219-6

Vol. 248. Bing-Yuan Cao
Optimal Models and Methods with Fuzzy Quantities, 2010
ISBN 978-3-642-10710-8

Vol. 249. Bernadette Bouchon-Meunier, Luis Magdalena, Manuel Ojeda-Aciego, José-Luis Verdegay, Ronald R. Yager (Eds.)
Foundations of Reasoning under Uncertainty, 2010
ISBN 978-3-642-10726-9

Vol. 250. Xiaoxia Huang
Portfolio Analysis, 2010
ISBN 978-3-642-11213-3

Vol. 251. George A. Anastassiou
Fuzzy Mathematics: Approximation Theory, 2010
ISBN 978-3-642-11219-5

Vol. 252. Cengiz Kahraman, Mesut Yavuz (Eds.)
Production Engineering and Management under Fuzziness, 2010
ISBN 978-3-642-12051-0

Vol. 253. Badredine Arfi
Linguistic Fuzzy Logic Methods in Social Sciences, 2010
ISBN 978-3-642-13342-8

Vol. 254. Weldon A. Lodwick, Janusz Kacprzyk (Eds.)
Fuzzy Optimization, 2010
ISBN 978-3-642-13934-5

Vol. 255. Zongmin Ma, Li Yan (Eds.)
Soft Computing in XML Data Management, 2010
ISBN 978-3-642-14009-9

Vol. 256. Robert Jeansoulin, Odile Papini, Henri Prade, and Steven Schockaert (Eds.)
Methods for Handling Imperfect Spatial Information, 2010
ISBN 978-3-642-14754-8

Vol. 257. Salvatore Greco, Ricardo Alberto Marques Pereira, Massimo Squillante, Ronald R. Yager, and Janusz Kacprzyk (Eds.)
*Preferences and Decisions,*2010
ISBN 978-3-642-15975-6

Vol. 258. Jorge Casillas and Francisco José Martínez López
Marketing Intelligent Systems Using Soft Computing, 2010
ISBN 978-3-642-15605-2

Vol. 259. Alexander Gegov
Fuzzy Networks for Complex Systems, 2010
ISBN 978-3-642-15599-4

Vol. 260. Jordi Recasens
Indistinguishability Operators, 2010
ISBN 978-3-642-16221-3

Vol. 261. Chris Cornelis, Glad Deschrijver, Mike Nachtegael, Steven Schockaert, and Yun Shi (Eds.)
35 Years of Fuzzy Set Theory, 2010
ISBN 978-3-642-16628-0

Vol. 262. Zsófia Lendek, Thierry Marie Guerra, Robert Babuška, and Bart De Schutter
Stability Analysis and Nonlinear Observer Design Using Takagi-Sugeno Fuzzy Models, 2010
ISBN 978-3-642-16775-1

Zsófia Lendek, Thierry Marie Guerra,
Robert Babuška, and Bart De Schutter

Stability Analysis and Nonlinear Observer Design Using Takagi-Sugeno Fuzzy Models

Springer

Authors

Dr. Zsófia Lendek
Delft Center for Systems and Control
Delft University of Technology
Mekelweg 2
2628 CD Delft
The Netherlands

Prof. Dr. Bart De Schutter
Delft Center for Systems and Control
Delft University of Technology
Mekelweg 2
2628 CD Delft
The Netherlands

Prof. Dr. Thierry Marie Guerra
Université de Valenciennes et du
Hainaut-Cambrésis
LAMIH CNRS FRE 3304
Le Mont Houy BP311
59313 Valenciennes Cedex 9
France

Prof. Dr. Robert Babuška
Delft Center for Systems and Control
Delft University of Technology
Mekelweg 2
2628 CD Delft
The Netherlands

ISBN 978-3-642-26567-9 ISBN 978-3-642-16776-8 (eBook)

DOI 10.1007/978-3-642-16776-8

Studies in Fuzziness and Soft Computing ISSN 1434-9922

Typeset & Cover Design: Scientific Publishing Services Pvt. Ltd., Chennai, India.

Printed on acid-free paper

9 8 7 6 5 4 3 2 1

springer.com

Preface

Many problems in decision making, monitoring, fault detection, and control rely on the knowledge of state variables and time-varying parameters that are not directly measured by sensors. In such situations, observers, or estimators, can be employed that use the measured input and output signals along with a dynamic model of the system in order to estimate the unknown states or parameters. An essential requirement in designing an observer is to guarantee the convergence of the estimates to the true values or at least to a small neighborhood of the true values. For linear models, a wide array of estimation techniques are available, such as the Kalman filter and its variants. However, no general method exists for the design of estimators for nonlinear systems. The design and tuning of a nonlinear observer is generally complicated and involves large computational costs.

This book provides a range of methods and tools to design observers for nonlinear systems represented by a special type of a dynamic nonlinear model – the Takagi–Sugeno (TS) fuzzy model. The TS model is a convex combination of affine linear models. This structure facilitates stability analysis and observer design by using effective algorithms based on Lyapunov functions and linear matrix inequalities. TS models are known to be universal approximators and, in addition, a broad class of nonlinear systems can be exactly represented as a TS system.

In the fuzzy systems literature, observer design is typically considered as a dual problem to control design, and as such it has not received much attention yet. This books aims at filling this gap by addressing observer design for TS systems in its own right, with a special attention to large-scale, decentralized systems. To this end, three particular structures of large-scale TS models are considered: cascaded systems, distributed systems, and systems affected by unknown disturbances. The reader will find in-depth theoretical analysis accompanied by illustrative examples and simulations of real-world systems. Stability analysis of TS models is also addressed in detail.

The intended audience are graduate students and researchers both from academia and industry. For newcomers to the field, the book provides a concise introduction dynamic TS fuzzy models along with two methods to construct TS models for a given nonlinear system.

While this monograph focusses mainly on the theory and methodology of state and parameter estimation in nonlinear distributed dynamic systems, the methods presented can readily be used in applications such as industrial processes, traffic systems, environmental systems, energy and water distribution networks, and so on.

Supplementary information relevant to this book is available at the website:

http://www.dcsc.tudelft.nl/fuzzybook/

Comments, suggestions, or questions concerning the book or the website are welcome. Interested readers are encouraged to get in touch with the authors using the contact information on the website.

We thank Janusz Kacprzyk, the series editor, for giving us the opportunity to publish our book with Springer, and the editorial and production team at Springer, especially Thomas Ditzinger, for their valuable help. We gratefully acknowledge the financial support of the BSIK-ICIS project Interactive Collaborative Information Systems (grant no. BSIK03024 of Senter, Ministry of Economic Affairs of the Netherlands), as well as of the International Campus on Safety and Intermodality in Transportation, the Nord-Pas-de-Calais Region, the European Community, the Regional Delegation for Research and Technology, the French Ministry of Higher Education and Research.

Finally, we thank our families for their continual understanding, patience, and support.

July 2010

Zsófia Lendek
Thierry Marie Guerra
Robert Babuška
Bart De Schutter

Contents

1 Introduction 1
1.1 Observer Design for TS Fuzzy Systems 1
1.2 Outline 3

2 Takagi-Sugeno Fuzzy Models 5
2.1 TS Fuzzy Models 5
2.2 Dynamic TS Fuzzy Models 7
2.3 Constructing TS Models 11
2.3.1 The Sector Nonlinearity Approach 11
2.3.2 Linearization 17
2.4 Summary 24

3 Stability Analysis of TS Fuzzy Systems 25
3.1 Introduction 25
3.2 Preliminaries 25
3.2.1 Notation 25
3.2.2 Linear Matrix Inequalities 26
3.3 Stability Analysis of TS Systems 30
3.3.1 Quadratic Stability 30
3.3.2 D-Stability 31
3.3.3 Leaving the Quadratic Stability Framework 33
3.4 State Feedback Stabilization 36
3.4.1 H_∞ Attenuation 38
3.4.2 Robust Control 41
3.5 Output Feedback Stabilization 45
3.6 Input-to-State Stability 46
3.7 Summary 48

4 Observers for TS Fuzzy Systems 49
4.1 Observer Design for TS Systems 49
4.2 Observer Design: Measured Scheduling Vector 52
4.3 Observer Design: Estimated Scheduling Vector 62
4.4 Observer-Based Stabilization 68
4.5 Summary 71

5 Cascaded TS Systems and Observers 73
5.1 Introduction 73
5.2 Stability of Cascaded Dynamic Systems 74
5.2.1 Cascaded Dynamic Systems 74
5.2.2 Partitioning a Nonlinear System 76
5.2.3 Stability of Cascaded Systems 80
5.3 Cascaded TS Fuzzy Systems 82
5.3.1 Stability Analysis of Cascaded TS Systems 82
5.3.2 Convergence Rate of Cascaded Systems 88
5.4 Cascaded TS Fuzzy Observers 90
5.4.1 Measured Scheduling Vector 92
5.4.2 Estimated Scheduling Vector 96
5.5 Summary 102

6 Distributed TS Systems and Observers 103
6.1 Introduction 103
6.2 Distributed Stability Analysis of TS Systems 106
6.2.1 Parallel Stability Analysis 106
6.2.2 Sequential Stability Analysis 114
6.3 Distributed Observer Design for TS Systems 125
6.3.1 General Framework 125
6.3.2 Sequential Design: Measured Scheduling Vector 130
6.3.3 Sequential Design: Estimated Scheduling Vector 138
6.4 Summary 147

7 Adaptive Observers for TS Systems 149
7.1 Introduction 149
7.2 Unknown Input Estimation 151
7.3 Estimation of Unknown Polynomial Inputs 158
7.3.1 Measured Scheduling Vector 158
7.3.2 Estimated Scheduling Vector 166
7.4 Estimation of Unmodelled Dynamics 174
7.4.1 Measured Scheduling Vector 174
7.4.2 Estimated Scheduling Vector 179
7.5 Summary 182

Glossary 183

References 185

Index 195

Acronyms

TS	Takagi-Sugeno
LMI	Linear Matrix Inequality
ISS	Input-to-State Stable
GAS	Globally Asymptotically Stable
UGAS	Uniformly Globally Asymptotically Stable
PDC	Parallel Distributed Compensation
SISO	Single-Input Single-Output
MIMO	Multi-Input Multi-Output
FP	Feasibility Problem
EVP	Eigenvalue Problem
GEVP	Generalized Eigenvalue Problem

Chapter 1
Introduction

1.1 Observer Design for TS Fuzzy Systems

In order to understand how a system works, one needs to have information on certain important quantities associated with the system. Many problems in decision making, monitoring, and control require the knowledge of the variables, i.e., states and parameters of the system involved. In practical situations, measuring all these variables may not be possible due to technical or economical reasons. Therefore, the *estimation* of states and parameters in dynamic systems is an important prerequisite for safe and economical operation. Hence, estimation is an integral part in applications such as process monitoring, fault detection, and process optimization. Moreover, any state feedback control design requires the knowledge of state variables.

Observers in general use the input and output signals of a system, together with a system's model. They generate an estimate of the system's state, which may then be further employed, in control, monitoring, fault detection, etc. Observers were first proposed and developed by Luenberger in the sixties (Luenberger, 1966). Since the early developments, observers for linear and nonlinear systems with both known and unknown inputs have been developed (Saif and Guan, 1992; Ruiz Vargas and Hemerly, 2001; Bergsten et al., 2001; Welch and Bishop, 2002; Huang and Dey, 2005; Hyun et al., 2006; Besançon, 2006; Priscoli et al., 2006).

In this book, dynamic systems are modeled in the state space framework, using a state transition model, which describes the evolution of the states over time, and a measurement model, which relates the measurement to the states. The mathematical description of the system used is

$$\begin{aligned}\dot{\boldsymbol{x}}(t) &= \boldsymbol{f}(\boldsymbol{x}(t), \boldsymbol{u}(t), \boldsymbol{\theta}(t))\\ \boldsymbol{y}(t) &= \boldsymbol{h}(\boldsymbol{x}(t), \boldsymbol{u}(t), \boldsymbol{\zeta}(t))\end{aligned} \tag{1.1}$$

where $\boldsymbol{f}$ is the state transition function, describing the evolution of the states over time, $\boldsymbol{h}$ is the measurement function, relating the measurements to the states, $\boldsymbol{x}$ is

Zs. Lendek et al.: Stability and Observer Design Using TS Models, STUDFUZZ 262, pp. 1–4.
springerlink.com

the vector of the state variables, $\boldsymbol{u}$ is the vector of the input or control variables, $\boldsymbol{\theta}$ and $\boldsymbol{\zeta}$ are (unknown/uncertain) parameters, and $\boldsymbol{y}$ denotes the measurement vector.

Given the state space model (1.1), the problem of state estimation arises as soon as the measured output does not coincide with the whole state, i.e., $\boldsymbol{y} \neq \boldsymbol{x}$. Unlike for linear systems, there is no systematic procedure to design a state observer for a given nonlinear model. The problem becomes more difficult when some parameters in the model are not exactly known.

In order to design observers, in this book we represent nonlinear systems by Takagi-Sugeno (TS) fuzzy models of the form

$$\begin{aligned}\dot{\boldsymbol{x}}(t) &= \sum_{i=1}^{m} w_i(\boldsymbol{z}(t))(A_i\boldsymbol{x}(t) + B_i\boldsymbol{u}(t) + a_i) \\ \boldsymbol{y}(t) &= \sum_{i=1}^{m} w_i(\boldsymbol{z}(t))(C_i\boldsymbol{x}(t) + c_i)\end{aligned} \tag{1.2}$$

where m is the number of local models, A_i, B_i, C_i, are the matrices and a_i and c_i are the biases of the ith local model, $\boldsymbol{z}(t)$ is the vector of the scheduling variables, which may depend on the states, inputs, measurements, or other exogenous variables, and $w_i(\boldsymbol{z})(t)$, $i = 1, 2, \ldots, m$ are normalized membership functions, i.e., $w_i(\boldsymbol{z}(t)) \geq 0$ and $\sum_{i=1}^{m} w_i(\boldsymbol{z}(t)) = 1, \forall t \in \mathbb{R}$. Note that throughout the remainder of the book, the explicit time-dependence of the variables is omitted.

Such a model presents several advantages. The TS model is a universal approximator (Fantuzzi and Rovatti, 1996), and many nonlinear systems can be exactly represented in a compact set of state variables as TS systems (Ohtake et al., 2001). Moreover, (1.2) is the convex combination of local affine models, which facilitates the stability analysis and controller and observer design for such systems. In addition, many already available stability and design conditions for TS system can be formulated as linear matrix inequalities (Boyd et al., 1994; Scherer and Weiland, 2005; Tanaka and Wang, 1997; Tanaka et al., 1998), for which efficient algorithms exist that test their feasibility.

This book presents *stability analysis* and *observer design* methods for nonlinear systems, represented by TS fuzzy models. For a large-scale or time-varying system, the analysis of the system and the design and tuning of an observer may be complicated and involve large computational costs. Therefore, to decrease the computational costs, before analyzing the system or designing an observer for it, we consider the structure of the system. Three classes of TS fuzzy systems are investigated: systems in a cascaded form, distributed systems, and systems affected by unknown disturbances.

Looking at the research concerning control and observer design for TS systems in the last decades, one can see that while control design for TS systems has gained an increased interest, observer design is generally considered the dual problem of controller design, and is therefore assumed to be a side-issue. Even in output-feedback control, where observers are frequently used, the design usually relies on the separation principle (Jiang, 2000; Uang and Chen, 2000; Jiang et al., 2001; Tseng, 2008;

Guelton et al., 2009; Huang et al., 2009), which, since nonlinear systems are concerned, is valid only in restricted cases and under strong assumptions on the model. However, as argued in the beginning of this section, estimation of the unmeasured variables in dynamic systems is an essential part in process monitoring, fault detection, process optimization, and control. This books therefore aims to address this issue.

1.2 Outline

The book is organized into 7 chapters, as follows.

Chapter 2 is used to introduce the necessary notations and background. In particular, the dynamic Takagi-Sugeno fuzzy system that is used further on is introduced, together with two methods that can be used to construct TS models based on a given nonlinear system of the form (1.1). The first method presented, the sector nonlinearity approach, can be used to obtain an exact fuzzy representation of the nonlinear system considered, in a compact set of the state-space. By using the second method, Taylor series expansion in several operating points, an approximate model is obtained.

Chapter 3 reviews methods and algorithms that can be used to analyze the stability of TS fuzzy systems. These methods are in general derived from analysis using a Lyapunov function, and are therefore stated as sufficient conditions. These conditions are formulated such that their feasibility can be verified by solving linear matrix inequalities (LMIs).

Chapter 4 introduces observers and reviews methods for designing observers for TS systems, and briefly describes observer-based control. Since the design methods actually rely on determining the observer gains such that the resulting estimation error dynamics are stable, the methods presented in this chapter can be seen as extensions of the methods in Chapter 3.

Chapter 5 presents techniques for the stability analysis and observer design of a special type of distributed systems, *cascaded systems*. An important class of distributed systems can be represented as a cascade of subsystems. For general nonlinear systems, the stability of the individual subsystems does not imply the stability of the cascaded system. In this chapter, results are presented for the cascade of TS fuzzy models, as the stability analysis of a cascaded TS system may be performed by analyzing the individual subsystems. The cascaded approach is also described for observer design. A stable observer can be obtained by designing observers independently for the subsystems. Moreover, we show that the cascaded design does not lead to a loss of performance in the terms of the estimation error decay-rate.

Chapter 6 concerns *general distributed* systems. Many physical systems, such as power systems, material processing systems, and communication and transportation networks are composed of interconnected lower-dimensional subsystems. In many cases, large-scale systems are not cascaded, but distributed, i.e., the influence among the subsystems is not in one way only. In Chapter 6 we consider such systems, where each subsystem is represented by a TS fuzzy model and we present results for the stability analysis and observer design of distributed TS systems. In addition, we also consider systems whose structure is not fixed, and where subsystems may be added to or removed from the system.

Finally, Chapter 7 considers TS systems that *change over time* or that are influenced by *unknown inputs*, for which adaptive observers can be designed. Such observers simultaneously estimate the states and unknown inputs or parameters of a system. The design of observers in the presence of unknown inputs is an important problem, since in many cases not all inputs are known, and the unknown inputs may represent effects of actuator or plant component failures. The observer is designed based on the known part of the fuzzy model. These observers guarantee either the asymptotic convergence to zero or the boundedness of the estimation error.

Chapter 2
Takagi-Sugeno Fuzzy Models

In this chapter we first introduce the continuous-time Takagi-Sugeno (TS) fuzzy systems that are employed throughout the book. In the second part of the chapter, we present methods to construct TS models that represent or approximate a nonlinear dynamic system starting from a given model of this system.

2.1 TS Fuzzy Models

Traditionally, the class of linear, time-invariant systems has dominated the systems and control field (Kailath, 1980; Franklin et al., 1990; Åström and Wittenmark, 1990). Thanks to their linearity and time-invariance these systems are easy to analyze and well-established methods and algorithms exist to design observers and controllers for them. The disadvantage of such models is that they fail to describe nonlinear systems globally. An accurate approximation of a nonlinear system can only be expected in the vicinity of an equilibrium point.

In this book, we consider continuous-time dynamic TS fuzzy systems (Takagi and Sugeno, 1985). These systems, as used in this book, are mathematical models of a special form, with the property that they are able to exactly represent or to approximate to an arbitrary degree of accuracy a large class of nonlinear systems in a compact st of the state space.

The TS fuzzy model, originally proposed by Takagi and Sugeno (1985), consists of an *if-then* rule base. The rule antecedents partition a subset of the model variables into fuzzy sets. The consequent of each rule is a simple functional expression. The ith rule is described as

Model rule i:
If z_1 *is* Z_1^i *and ... and* z_p *is* Z_p^i *then* $\boldsymbol{y} = \mathbf{F}_i(\boldsymbol{z})$

where the vector $\boldsymbol{z}$ has p components, z_j, $j = 1, 2, \ldots, p$, and stands for the vector of antecedent variables; these variables are also called *scheduling variables*, as their values determine the degree to which rules are active. The sets Z_j^i, $j = 1, 2, \ldots, p$, $i = 1, 2, \ldots, m$, where m is the number of rules, are the antecedent fuzzy sets. The

Zs. Lendek et al.: Stability and Observer Design Using TS Models, STUDFUZZ 262, pp. 5–24.
springerlink.com

value of a scheduling variable z_j belongs to a fuzzy set Z_j^i with a truth value given by the membership function $\omega_{ij} : \mathbb{R} \to [0, 1]$. The truth value for an entire rule is determined based on the individual premise variables, using a conjunction operator (Kruse et al., 1994) such as the *minimum*

$$\varphi_i(\boldsymbol{z}) = \min_j \{\omega_{ij}(z_j)\}$$

or the *algebraic product*

$$\varphi_i(\boldsymbol{z}) = \prod_{j=1}^{p} \omega_{ij}(z_j) \tag{2.1}$$

The obtained truth value is normalized

$$w_i(\boldsymbol{z}) = \frac{\varphi_i(\boldsymbol{z})}{\sum_{j=1}^{m} \varphi_j(\boldsymbol{z})} \tag{2.2}$$

assuming that $\sum_{j=1}^{m} \varphi_j(\boldsymbol{z}) \neq 0$, i.e., that for any allowed combination of the scheduling variables at least one rule has a truth value greater than zero. In what follows, the expression $w_i(\boldsymbol{z})$ is referred to as the normalized membership function.

The output of a rule i is the value given by the consequent vector function $\mathbf{F}_i$, and usually depends on the scheduling variables; $\boldsymbol{y}$ is the output of the model, computed as the weighted combination of the output of the rules. Using $w_i(\boldsymbol{z})$, the output of the model is expressed as a function of $\boldsymbol{z}$ as

$$\boldsymbol{y} = \sum_{i=1}^{m} w_i(\boldsymbol{z}) \mathbf{F}_i(\boldsymbol{z}) \tag{2.3}$$

In general, the consequents of the rules (the functions $\mathbf{F}_i$) may also depend on exogenous variables, i.e., on variables that do not appear in the scheduling vector. In such a case, the output of the fuzzy model is given as

$$\boldsymbol{y} = \sum_{i=1}^{m} w_i(\boldsymbol{z}) \mathbf{F}_i(\boldsymbol{z}, \boldsymbol{\theta})$$

where $\boldsymbol{\theta}$ denotes the vector of exogenous variables and p_θ denotes the number of these variables. This model is a fuzzy model, since each rule can be rewritten as

Model rule i:
If z_1 *is* Z_1^i *and* ... *and* z_p *is* Z_p^i *and*
and θ_1 *is* $Z_{\theta,1}^i$ *and* ... *and* θ_{p_θ} *is* Z_{θ,p_θ}^i *then* $\boldsymbol{y} = \mathbf{F}_i(\boldsymbol{z}, \boldsymbol{\theta})$

where the antecedent fuzzy sets $Z_{\theta,j}^i$, $j = 1, 2, \ldots, p_\theta$, $i = 1, 2, \ldots, m$ are determined such that they cover the whole space where the variables θ_j are defined, i.e., the membership functions $\omega_{\theta,ij}(\theta_j) = 1$, $j = 1, 2, \ldots, p_\theta$, $i = 1, 2, \ldots, m$.

Example 2.1. Consider the function $y(x, a) = \text{abs}(x) + ax^2$, with $x \in \mathbb{R}$, and $a \in [-1, 1]$. This function can be expressed as the two-rule fuzzy model

Model rule 1:
If x *is* Z_1^1 *and* a *is* A_1 *then* $y = -x + ax^2$

Model rule 2:
If x *is* Z_1^2 *and* a *is* A_2 *then* $y = x + ax^2$

where $A_1 = A_2 = A = [-1, 1]$, Z_1^1 denotes the set of negative real numbers and Z_1^2 denotes the set of non-negative real numbers. In the description above, we have two scheduling variables, x and a. However, we also have $A_1 = A_2$, and the truth value of a *is* A is always 1. To simplify the rules, the above model is written as

Model rule 1:
If x *is* Z_1^1 *then* $y = -x + ax^2$

Model rule 2:
If x *is* Z_1^2 *then* $y = x + ax^2$

i.e., only x is explicitly given as a scheduling variable, although the consequent functions also depend on a. □

2.2 Dynamic TS Fuzzy Models

In this book we consider TS models that represent nonlinear dynamic systems. Therefore let a dynamic system be given as

$$\begin{aligned} \dot{\boldsymbol{x}} &= \boldsymbol{f}(\boldsymbol{x}, \boldsymbol{u}, \boldsymbol{\theta}) \\ \boldsymbol{y} &= \boldsymbol{h}(\boldsymbol{x}, \boldsymbol{\zeta}) \end{aligned} \tag{2.4}$$

where $\boldsymbol{f}$ and $\boldsymbol{h}$ are smooth nonlinear functions, with $\boldsymbol{f}$ representing the state model and with $\boldsymbol{h}$ representing the measurement model, $\boldsymbol{x} \in \mathbb{R}^{n_x}$ is the state vector, $\boldsymbol{u} \in \mathbb{R}^{n_u}$ is the input vector, $\boldsymbol{y} \in \mathbb{R}^{n_y}$ is the measurement vector, and $\boldsymbol{\theta}$ and $\boldsymbol{\zeta}$ represent vectors of constant parameters or other exogenous variables that act on the system. A TS fuzzy system that represents or approximates the nonlinear system (2.4) is expressed as a set of m fuzzy rules of the following form

Model rule i:
If z_1 *is* Z_1^i *and ... and* z_p *is* Z_p^i *then*

$$\begin{aligned} \dot{\boldsymbol{x}} &= \widehat{\boldsymbol{f}}_i(\boldsymbol{x}, \boldsymbol{u}, \boldsymbol{\theta}) \\ \boldsymbol{y} &= \widehat{\boldsymbol{h}}_i(\boldsymbol{x}, \boldsymbol{\zeta}) \end{aligned}$$

where z_j, $j = 1, 2, \ldots, p$, represent the scheduling variables, and $\widehat{\boldsymbol{f}}_i$ and $\widehat{\boldsymbol{h}}_i$ are the consequent functions of the ith rule. The scheduling variables are usually chosen as a subset of the state, input, output, or other exogenous variables in the system, or they are functions of the states, inputs, outputs, or exogenous variables.

The membership functions $\omega_{ij}(z_j)$ are chosen such that their truth values are in $[0, 1]$, and for any allowed value of $\boldsymbol{z}$ at least one of the rules is active. Then, the truth values of the rules are computed using (2.1), and normalized.

Using (2.3), the rules are combined into

$$\begin{aligned}\dot{\boldsymbol{x}} &= \frac{\sum_{i=1}^{m} \varphi_i(\boldsymbol{z}) \widehat{\boldsymbol{f}}_i(\boldsymbol{x}, \boldsymbol{u}, \boldsymbol{\theta})}{\sum_{i=1}^{m} \varphi_i(\boldsymbol{z})} = \sum_{i=1}^{m} w_i(\boldsymbol{z}) \widehat{\boldsymbol{f}}_i(\boldsymbol{x}, \boldsymbol{u}, \boldsymbol{\theta}) \\ \boldsymbol{y} &= \frac{\sum_{i=1}^{m} \varphi_i(\boldsymbol{z}) \widehat{\boldsymbol{h}}_i(\boldsymbol{x}, \boldsymbol{\zeta})}{\sum_{i=1}^{m} \varphi_i(\boldsymbol{z})} = \sum_{i=1}^{m} w_i(\boldsymbol{z}) \widehat{\boldsymbol{h}}_i(\boldsymbol{x}, \boldsymbol{\zeta})\end{aligned}$$

The consequent functions $\widehat{\boldsymbol{f}}_i$ and $\widehat{\boldsymbol{h}}_i$ are usually less complex than the original nonlinear functions $\boldsymbol{f}$ and $\boldsymbol{h}$, and are in general chosen as constant, linear, or affine functions. Since these consequents are typically valid only locally, i.e., where the value of the corresponding normalized membership function is nonzero, in the sequel they will also be referred to as "local models".

In this book, we use TS fuzzy systems with linear or affine local models. Therefore, the rules have the following form[1]

Model rule i:
If z_1 *is* Z_1^i *and ... and* z_p *is* Z_p^i *then*

$$\begin{aligned}\dot{\boldsymbol{x}} &= A_i \boldsymbol{x} + B_i \boldsymbol{u} \\ \boldsymbol{y} &= C_i \boldsymbol{x}\end{aligned}$$

for linear models, and

Model rule i:
If z_1 *is* Z_1^i *and ... and* z_p *is* Z_p^i *then*

$$\begin{aligned}\dot{\boldsymbol{x}} &= A_i \boldsymbol{x} + B_i \boldsymbol{u} + a_i \\ \boldsymbol{y} &= C_i \boldsymbol{x} + c_i\end{aligned}$$

for affine models. In the expressions above, A_i, B_i, C_i are the matrices and a_i, c_i are the biases of the ith local model. The final outputs of the TS system are computed as

[1] Note that $\dot{\boldsymbol{x}}$ and $\boldsymbol{y}$ in the consequent parts are interpreted as linguistic variables, and the output of each rule is given only by the expressions $A_i \boldsymbol{x} + B_i \boldsymbol{u}$ and $C_i \boldsymbol{x}$. The notation

$$\begin{aligned}\dot{\boldsymbol{x}} &= A_i \boldsymbol{x} + B_i \boldsymbol{u} \\ \boldsymbol{y} &= C_i \boldsymbol{x}\end{aligned}$$

for the consequent part is common in the literature, and is therefore also used in this book.

$$\begin{aligned}\dot{\boldsymbol{x}} &= \sum_{i=1}^{m} w_i(\boldsymbol{z})(A_i\boldsymbol{x} + B_i\boldsymbol{u}) \\ \boldsymbol{y} &= \sum_{i=1}^{m} w_i(\boldsymbol{z})C_i\boldsymbol{x}\end{aligned} \tag{2.5}$$

for models with linear consequents and

$$\begin{aligned}\dot{\boldsymbol{x}} &= \sum_{i=1}^{m} w_i(\boldsymbol{z})(A_i\boldsymbol{x} + B_i\boldsymbol{u} + a_i) \\ \boldsymbol{y} &= \sum_{i=1}^{m} w_i(\boldsymbol{z})(C_i\boldsymbol{x} + c_i)\end{aligned} \tag{2.6}$$

for models with affine consequents. In the sequel, we only use the notation (2.5) for linear and (2.6) for affine TS models, respectively, with the understanding that the membership functions $w_i(\boldsymbol{z})$, $i = 1, 2, \ldots, m$ are normalized, i.e., $w_i(\boldsymbol{z}) \geq 0$ and $\sum_{i=1}^{m} w_i(\boldsymbol{z}) = 1$. Thanks to the normalized membership functions, the linear (affine) dynamic TS model is in fact a convex combination of local linear (affine) models. This property facilitates the stability analysis of the fuzzy system (see Chapter 3).

Example 2.2. Consider the nonlinear dynamic system

$$\begin{aligned}\dot{x}_1 &= -x_1 + x_1x_2 \qquad y = x_1 \\ \dot{x}_2 &= x_1 - 3x_2\end{aligned} \tag{2.7}$$

with $x_1, x_2 \in [-1, 1]$. This system can be exactly represented (using the sector nonlinearity approach, see Section 2.3.1) by the following TS fuzzy system with linear consequents

Model rule 1:
If z_1 *is around* -1 *then*

$$\dot{\boldsymbol{x}} = \begin{pmatrix} -2 & 0 \\ 1 & -3 \end{pmatrix} \boldsymbol{x}$$

$$y = x_1$$

Model rule 2:
If z_1 *is around* 1 *then*

$$\dot{\boldsymbol{x}} = \begin{pmatrix} 0 & 0 \\ 1 & -3 \end{pmatrix} \boldsymbol{x}$$

$$y = x_1$$

In the model above, the scheduling variable z_1 is chosen as x_2, the fuzzy sets are Z_1^1 = 'around -1', Z_1^2 = 'around 1', and the corresponding membership functions are $\omega_{11} = (1 - z_1)/2$ and $\omega_{21} = (1 + z_1)/2$, respectively. It can be easily seen that with these membership functions, we have

$$\frac{1-x_2}{2}\begin{pmatrix}-2 & 0\\ 1 & -3\end{pmatrix}\begin{pmatrix}x_1\\ x_2\end{pmatrix}+\frac{1+x_2}{2}\begin{pmatrix}0 & 0\\ 1 & -3\end{pmatrix}\begin{pmatrix}x_1\\ x_2\end{pmatrix}=\begin{pmatrix}-x_1+x_1x_2\\ x_1-3x_2\end{pmatrix}$$
$$\frac{1-x_2}{2}x_1+\frac{1+x_2}{2}x_1=x_1=y$$

i.e., the fuzzy model is an exact representation of the nonlinear system (2.7) in the compact set $S = \{x_1, x_2 \in [-1, 1]\}$.

Consider now the nonlinear system

$$\begin{aligned}\dot{x}_1 &= -x_1 + x_1x_2 + 1 \qquad y = x_1\\ \dot{x}_2 &= x_1 - 2x_2 - 1\end{aligned}$$

with $x_1, x_2 \in [-1, 1]$. This system can be approximated by a TS system with linear consequents, or can be exactly represented, similarly to (2.7), by the following TS fuzzy system with affine consequents

Model rule 1:
If z_1 *is around* -1 *then*

$$\dot{\boldsymbol{x}} = \begin{pmatrix}-2 & 0\\ 1 & -3\end{pmatrix}\boldsymbol{x} + \begin{pmatrix}1\\ -1\end{pmatrix}$$
$$y = x_1$$

Model rule 2:
If z_1 *is around* 1 *then*

$$\dot{\boldsymbol{x}} = \begin{pmatrix}0 & 0\\ 1 & -3\end{pmatrix}\boldsymbol{x} + \begin{pmatrix}1\\ -1\end{pmatrix}$$
$$y = x_1$$

where the scheduling variable is $z_1 = x_2$, and the antecedent fuzzy sets and the corresponding membership functions are defined as Z_1^1 = 'around -1', Z_1^2 = 'around 1', and $\omega_{11} = (1 - z_1)/2$, $\omega_{21} = (1 + z_1)/2$, respectively. □

The variables of TS fuzzy models are usually defined on a compact set. On such a compact set, models of the form (2.5) and (2.6) have been proven to be able to approximate any nonlinear function to an arbitrary degree of accuracy (Wang and Mendel, 1992; Kosko, 1994; Ying, 1994; Fantuzzi and Rovatti, 1996).

For stability analysis, in general, TS fuzzy models with linear consequents are used. Exploiting the fact that in the case of linear consequents, all the local models have the same equilibrium point, zero, Lyapunov stability analysis can naturally be employed. The stability analysis of affine TS models is in general more involved. Results for the stability analysis of both linear and affine TS models are presented in Chapter 3.

For observer design, in this book, we employ affine TS models, since a TS model with affine consequents can represent a larger class of nonlinear systems than those with linear consequents.

Regarding the form of the TS fuzzy models, two differences between the models used for observer and controller design have to be mentioned. The first difference is that affine local models are rarely used for controller design. This is because, using most current control design methods, the affine terms have to be compensated for in each rule, that is possible only in special cases. However, for observer design, affine local models do not present a problem.

The second difference is that while in fuzzy control design it is assumed that the membership functions do *not* depend on the control input, so as to avoid having to solve implicit equations, in observer design this does not represent a problem. For observer design, the input is considered as a known (measured) variable, and therefore the membership functions may depend on it.

2.3 Constructing TS Models

Two main approaches can be used to obtain TS fuzzy models: 1) identifying the model using measured or simulated data and 2) analytic construction of a TS model that exactly represents or approximates a given nonlinear dynamic system.

Of the two classes above, identification has so far only been applied to the construction of discrete-time TS models. Since in this book we consider continuous-time TS systems, methods for the identification of TS systems are not presented, but the interested reader is referred to (Driankov et al., 1993; Abonyi et al., 2002; Babuška et al., 2002; Johansen and Babuška, 2003; Kukolj and Levi, 2004; Kaymak and van den Berg, 2004; Angelov and Filev, 2004a).

Several methods exist that construct a fuzzy representation or an approximation of a given nonlinear system. Among these, the sector nonlinearity approach (Ohtake et al., 2001) can be employed to obtain a TS model that is an exact fuzzy representation of a given nonlinear system. Using the method described in Chapter 14 of (Tanaka and Wang, 2001) a TS fuzzy model can be constructed such that both the nonlinear system and its derivative are approximated. Other methods that approximate a given nonlinear system are dynamic linearization (Johansen et al., 2000), which is in fact a Taylor series expansion in several operating points, or the method developed by Kiriakidis (2007).

In this section, two of the above methods are presented in detail: 1) the sector nonlinearity approach and 2) linearization.

2.3.1 *The Sector Nonlinearity Approach*

The sector nonlinearity approach has first been described by Ohtake et al. (2001). This approach is one of the most frequently used approaches for constructing TS models for fuzzy control design, as it can obtain an exact fuzzy representation of a given nonlinear system in a compact set of the state space.

The method has originally been developed for nonlinear systems of the form

$$\begin{aligned} \dot{\boldsymbol{x}} &= \boldsymbol{f}^{\mathrm{m}}(\boldsymbol{x},\boldsymbol{u})\boldsymbol{x} + \boldsymbol{g}^{\mathrm{m}}(\boldsymbol{x},\boldsymbol{u})\boldsymbol{u} \\ \boldsymbol{y} &= \boldsymbol{h}^{\mathrm{m}}(\boldsymbol{x},\boldsymbol{u})\boldsymbol{x} \end{aligned} \tag{2.8}$$

In the expression above, $\boldsymbol{f}^{\mathrm{m}}$, $\boldsymbol{g}^{\mathrm{m}}$, and $\boldsymbol{h}^{\mathrm{m}}$ are smooth nonlinear matrix functions, $\boldsymbol{x} \in \mathbb{R}^{n_x}$ is the state vector, $\boldsymbol{u} \in \mathbb{R}^{n_u}$ is the input vector, and $\boldsymbol{y} \in \mathbb{R}^{n_y}$ the measurement vector. The elements of the matrix functions $\boldsymbol{f}^{\mathrm{m}}$, $\boldsymbol{g}^{\mathrm{m}}$, and $\boldsymbol{h}^{\mathrm{m}}$, are assumed to be bounded. Furthermore, in general, all variables are assumed to be defined on a compact set.

However, since in this book we consider also affine TS models, the sector nonlinearity approach is presented for more general nonlinear systems of the form

$$\begin{aligned} \dot{\boldsymbol{x}} &= \boldsymbol{f}^{\mathrm{m}}(\boldsymbol{x}, \boldsymbol{u})\boldsymbol{x} + \boldsymbol{g}^{\mathrm{m}}(\boldsymbol{x}, \boldsymbol{u})\boldsymbol{u} + \boldsymbol{a}(\boldsymbol{x}, \boldsymbol{u}) \\ \boldsymbol{y} &= \boldsymbol{h}^{\mathrm{m}}(\boldsymbol{x}, \boldsymbol{u})\boldsymbol{x} + \boldsymbol{c}(\boldsymbol{x}, \boldsymbol{u}) \end{aligned} \tag{2.9}$$

with the same assumptions on $\boldsymbol{f}^{\mathrm{m}}$, $\boldsymbol{g}^{\mathrm{m}}$, and $\boldsymbol{h}^{\mathrm{m}}$, and the variables as for (2.8), and, furthermore, with $\boldsymbol{a}$ and $\boldsymbol{c}$ smooth nonlinear vector functions, with all their elements bounded. Note that (2.9) is more general than (2.8), which is commonly used to obtain TS fuzzy models with linear consequents. In fact, any nonlinear dynamic system can be written in the form (2.9). However, since most methods for stability analysis of TS models concern models with linear consequents, to facilitate the analysis and the design, whenever possible, a representation of the form (2.8), i.e., without the affine terms is preferred.

With the assumptions above, the terms of the matrix functions $\boldsymbol{f}^{\mathrm{m}}$, $\boldsymbol{g}^{\mathrm{m}}$, and $\boldsymbol{h}^{\mathrm{m}}$, and of the vector functions $\boldsymbol{a}$ and $\boldsymbol{c}$ are either constants or bounded.

The scheduling variables are chosen as $z_j(\cdot) \in [\underline{\mathrm{nl}}_j, \overline{\mathrm{nl}}_j]$, $j = 1, 2, \ldots, p$, where z_j denote the non-constant terms in $\boldsymbol{f}^{\mathrm{m}}$, $\boldsymbol{g}^{\mathrm{m}}$, $\boldsymbol{h}^{\mathrm{m}}$, $\boldsymbol{a}$, and $\boldsymbol{c}$, and $\underline{\mathrm{nl}}_j$ and $\overline{\mathrm{nl}}_j$ are the minimum and maximum[2], respectively, of z_j. Then, for each z_j, two weighting functions can be constructed as

$$\eta_0^j(\cdot) = \frac{\overline{\mathrm{nl}}_j - z_j(\cdot)}{\overline{\mathrm{nl}}_j - \underline{\mathrm{nl}}_j} \quad \eta_1^j(\cdot) = 1 - \eta_0^j(\cdot) \quad j = 1, 2, \ldots, p$$

These two weighting functions are normalized, i.e., $\eta_0^j(\cdot) \geq 0$, $\eta_1^j(\cdot) \geq 0$, and $\eta_0^j + \eta_1^j = 1$, for any value of z_j. Moreover, z_j can be expressed as $z_j = \underline{\mathrm{nl}}_j \eta_0^j(z_j) + \overline{\mathrm{nl}}_j \eta_1^j(z_j)$, i.e., the weighted sum of the two extrema.

The fuzzy sets corresponding to both weighting functions are defined on $[\underline{\mathrm{nl}}_j, \overline{\mathrm{nl}}_j]$, i.e., the domain where z_j takes its values. These fuzzy sets are denoted in the sequel by $\bar{Z}_0^j$ and $\bar{Z}_1^j$.

The rules of the TS system are constructed such that all the terms z_j, $j = 1, 2, \ldots, p$, are taken into account, i.e., the rules have the form

Model rule i:
If z_1 *is* Z_1^i *and ... and* z_p *is* Z_p^i *then*

$$\begin{aligned} \dot{\boldsymbol{x}} &= A_i \boldsymbol{x} + B_i \boldsymbol{u} + a_i \\ \boldsymbol{y} &= C_i \boldsymbol{x} + c_i \end{aligned}$$

[2] A differentiable function defined on a compact set attains its minimum and maximum.

where Z_j^i, $i = 1, 2, \ldots, m$, $j = 1, 2, \ldots, p$, can be either $\bar{Z}_0^j$ or $\bar{Z}_1^j$. Consequently, the TS system consists of $m = 2^p$ rules.

The membership function of rule i is computed as the product of the weighting functions that correspond to the fuzzy sets in the rule, i.e.,

$$w_i(\boldsymbol{z}) = \prod_{j=1}^{p} \omega_{ij}(z_j) \tag{2.10}$$

where $\omega_{ij}(z_j)$ is either $\eta_0^j(z_j)$ or $\eta_1^j(z_j)$, depending on which weighting function is used in the rule. Thanks to the construction of the weighting functions, the resulting membership functions are normal, i.e., $w_i(\boldsymbol{z}) \geq 0$ and $\sum_{i=1}^{m} w_i(\boldsymbol{z}) = 1$.

The matrices A_i, B_i, C_i, and the vectors a_i and c_i are constructed by substituting the elements corresponding to the weighting functions used in rule i, i.e., $\underline{\mathrm{nl}}_j$ for η_0^j, and $\overline{\mathrm{nl}}_j$ for η_1^j, respectively, into the matrix and vector functions $\boldsymbol{f}^{\mathrm{m}}$, $\boldsymbol{g}^{\mathrm{m}}$, $\boldsymbol{h}^{\mathrm{m}}$, $\boldsymbol{a}$, and $\boldsymbol{c}$.

Then, using the membership functions given by (2.10), the nonlinear system (2.9) is exactly represented by the TS fuzzy model given by

$$\begin{aligned} \dot{\boldsymbol{x}} &= \sum_{i=1}^{m} w_i(\boldsymbol{z})(A_i\boldsymbol{x} + B_i\boldsymbol{u} + a_i) \\ \boldsymbol{y} &= \sum_{i=1}^{m} w_i(\boldsymbol{z})(C_i\boldsymbol{x} + c_i) \end{aligned} \tag{2.11}$$

Note that (2.9) is not unique and therefore the TS representation of the nonlinear system obtained by the sector nonlinearity approach is not unique.

To illustrate how the sector nonlinearity approach can be employed to construct an exact TS representation of a given nonlinear dynamic system, consider the following example.

Example 2.3. Consider the nonlinear dynamic system with two states, x_1 and x_2, one input u, and one measurement y given as

$$\begin{aligned} \dot{x}_1 &= x_1x_2 - 3x_1^3 + 2x_2 + e^{x_1}u \\ \dot{x}_2 &= x_1 + x_1x_2^2 + u \\ y &= 2x_1^2 + x_2 \end{aligned} \tag{2.12}$$

with the variables defined on the compact set $C = \{\boldsymbol{x},\, y,\, u | u, y \in \mathbb{R},\; |x_i| \leq 1, i = 1,\, 2\}$. This system can be rewritten in the form (2.9), as

$$\begin{aligned} \dot{\boldsymbol{x}} &= \begin{pmatrix} x_2 - 3x_1^2 & 2 \\ 1 & x_1x_2 \end{pmatrix} \boldsymbol{x} + \begin{pmatrix} e^{x_1} \\ 1 \end{pmatrix} u \\ y &= \begin{pmatrix} 2x_1 & 1 \end{pmatrix} \boldsymbol{x} \end{aligned} \tag{2.13}$$

with $\boldsymbol{a} = 0$, and $\boldsymbol{c} = 0$.

The scheduling variables, i.e., the non-constant elements in the matrix functions $\boldsymbol{f}^{\mathrm{m}} = \begin{pmatrix} x_2 - 3x_1^2 & 2 \\ 1 & x_1x_2 \end{pmatrix}$, $\boldsymbol{g}^{\mathrm{m}} = \begin{pmatrix} e^{x_1} \\ 1 \end{pmatrix}$, and $\boldsymbol{h}^{\mathrm{m}} = \begin{pmatrix} 2x_1 & 1 \end{pmatrix}$ are $z_1 = x_2 - 3x_1^2$, $z_2 = x_1x_2$, $z_3 = e^{x_1}$, and $z_4 = 2x_1$. For each of these four terms the two weighting functions and the corresponding matrix elements are computed as follows:

1. $z_1 = x_2 - 3x_1^2 \in [-4,\ 1]$. The first weighting function is

$$\eta_0^1 = \frac{1 - (x_2 - 3x_1^4)}{1 - (-4)} = \frac{1 - x_2 + 3x_1^2}{5}$$

and $\underline{\mathrm{nl}}_1 = -4$. The second weighting function is

$$\eta_1^1 = 1 - \eta_0^1(x_1, x_2) = \frac{4 + x_2 - 3x_1^2}{5}$$

and $\overline{\mathrm{nl}}_1 = 1$. Then, the scheduling variable z_1 is represented as the weighted sum

$$z_1 = -4\eta_0^1(z_1) + 1\eta_1^1(z_1)$$

2. $z_2 = x_1x_2 \in [-1,\ 1]$. The first weighting function is

$$\eta_0^2 = \frac{1 - x_1x_2}{1 - (-1)} = \frac{1 - x_1x_2}{2}$$

and $\underline{\mathrm{nl}}_2 = -1$. The second weighting function is

$$\eta_1^2 = 1 - \eta_0^2(x_1, x_2) = \frac{1 + x_1x_2}{2}$$

and $\overline{\mathrm{nl}}_2 = 1$. The scheduling variable z_2 is represented as

$$z_2 = -1\eta_0^2(z_2) + 1\eta_1^2(z_2)$$

3. $z_3 = e^{x_1} \in [e^{-1},\ e]$. The first weighting function is

$$\eta_0^3 = \frac{e - e^{x_1}}{e - e^{-1}}$$

and $\underline{\mathrm{nl}}_3 = e^{-1}$. The second weighting function is

$$\eta_1^3 = 1 - \eta_0^3(x_1) = \frac{e^{x_1} - e^{-1}}{e - e^{-1}}$$

and $\overline{\mathrm{nl}}_3 = e$. The scheduling variable z_3 is represented as

$$z_3 = e^{-1}\eta_0^3(z_3) + e\eta_1^3(z_3)$$

4. $z_4 = 2x_1 \in [-2, 2]$. The first weighting function is

$$\eta_0^4 = \frac{1 - x_1}{2}$$

and $\underline{\text{nl}}_4 = -2$. The second weighting function is

$$\eta_1^4 = 1 - \eta_0^4(x_1) = \frac{1 + x_1}{2}$$

and $\overline{\text{nl}}_4 = 2$. The scheduling variable z_4 is represented as

$$z_4 = -2\eta_0^4(z_4) + 2\eta_1^4(z_4)$$

For each weighting function, denote the corresponding fuzzy set by $\bar{Z}_i^j$, $j = 1, \ldots, 4$, $i = 0, 1$. For instance, the fuzzy set corresponding to η_0^1 is denoted by $\bar{Z}_0^1$, etc. With these fuzzy sets, the following TS fuzzy model having $2^4 = 16$ rules can be written:

Model rule 1:
If z_1 *is* $\bar{Z}_0^1$ *and* z_2 *is* $\bar{Z}_0^2$ *and* z_3 *is* $\bar{Z}_0^3$ *and* z_4 *is* $\bar{Z}_0^4$ *then*

$$\begin{aligned} \dot{\boldsymbol{x}} &= A_1\boldsymbol{x} + B_1\boldsymbol{u} \\ \boldsymbol{y} &= C_1\boldsymbol{x} \end{aligned}$$

with

$$\begin{aligned} A_1 &= \begin{pmatrix} \underline{\text{nl}}_1 & 2 \\ 1 & \underline{\text{nl}}_2 \end{pmatrix} = \begin{pmatrix} -4 & 2 \\ 1 & -1 \end{pmatrix} \\ B_1 &= \begin{pmatrix} \underline{\text{nl}}_3 \\ 1 \end{pmatrix} = \begin{pmatrix} e^{-1} \\ 1 \end{pmatrix} \\ C_1 &= (\underline{\text{nl}}_4 \ 1) = (-2 \ 1) \end{aligned}$$

and the membership function of the rule computed as $w_1(\boldsymbol{z}) = \eta_0^1\eta_0^2\eta_0^3\eta_0^4$.

Model rule 2:
If z_1 *is* $\bar{Z}_0^1$ *and* z_2 *is* $\bar{Z}_0^2$ *and* z_3 *is* $\bar{Z}_0^3$ *and* z_4 *is* $\bar{Z}_1^4$ *then*

$$\begin{aligned} \dot{\boldsymbol{x}} &= A_2\boldsymbol{x} + B_2\boldsymbol{u} \\ \boldsymbol{y} &= C_2\boldsymbol{x} \end{aligned}$$

with

$$\begin{aligned} A_2 &= \begin{pmatrix} \underline{\text{nl}}_1 & 2 \\ 1 & \underline{\text{nl}}_2 \end{pmatrix} = \begin{pmatrix} -4 & 2 \\ 1 & -1 \end{pmatrix} \\ B_2 &= \begin{pmatrix} \underline{\text{nl}}_3 \\ 1 \end{pmatrix} = \begin{pmatrix} e^{-1} \\ 1 \end{pmatrix} \\ C_2 &= (\overline{\text{nl}}_4 \ 1) = (2 \ 1) \end{aligned}$$

and the membership function of the rule computed as $w_2(\boldsymbol{z}) = \eta_0^1\eta_0^2\eta_0^3\eta_1^4$.

Model rule 3:
If z_1 *is* $\bar{Z}_0^1$ *and* z_2 *is* $\bar{Z}_0^2$ *and* z_3 *is* $\bar{Z}_1^3$ *and* z_4 *is* $\bar{Z}_0^4$ *then*

$$\begin{aligned}\dot{\boldsymbol{x}} &= A_3\boldsymbol{x} + B_3\boldsymbol{u}\\ \boldsymbol{y} &= C_3\boldsymbol{x}\end{aligned}$$

with

$$\begin{aligned}A_3 &= \begin{pmatrix}\underline{\mathrm{nl}}_1 & 2\\ 1 & \underline{\mathrm{nl}}_2\end{pmatrix} = \begin{pmatrix}-4 & 2\\ 1 & -1\end{pmatrix}\\ B_3 &= \begin{pmatrix}\overline{\mathrm{nl}}_3\\ 1\end{pmatrix} = \begin{pmatrix}e\\ 1\end{pmatrix}\\ C_3 &= \begin{pmatrix}\underline{\mathrm{nl}}_4 & 1\end{pmatrix} = \begin{pmatrix}-2 & 1\end{pmatrix}\end{aligned}$$

and the membership function of the rule computed as $w_3(\boldsymbol{z}) = \eta_0^1\eta_0^2\eta_1^3\eta_0^4$.

Model rule 4:
If z_1 *is* $\bar{Z}_0^1$ *and* z_2 *is* $\bar{Z}_0^2$ *and* z_3 *is* $\bar{Z}_1^3$ *and* z_4 *is* $\bar{Z}_1^4$ *then*

$$\begin{aligned}\dot{\boldsymbol{x}} &= A_4\boldsymbol{x} + B_4\boldsymbol{u}\\ \boldsymbol{y} &= C_4\boldsymbol{x}\end{aligned}$$

with

$$\begin{aligned}A_4 &= \begin{pmatrix}\underline{\mathrm{nl}}_1 & 2\\ 1 & \underline{\mathrm{nl}}_2\end{pmatrix} = \begin{pmatrix}-4 & 2\\ 1 & -1\end{pmatrix}\\ B_4 &= \begin{pmatrix}\overline{\mathrm{nl}}_3\\ 1\end{pmatrix} = \begin{pmatrix}e\\ 1\end{pmatrix}\\ C_4 &= \begin{pmatrix}\overline{\mathrm{nl}}_4 & 1\end{pmatrix} = \begin{pmatrix}2 & 1\end{pmatrix}\end{aligned}$$

and the membership function of the rule computed as $w_4(\boldsymbol{z}) = \eta_0^1\eta_0^2\eta_1^3\eta_1^4$.

The remaining rules are defined in a similar manner, corresponding to all 16 combinations.

As already stated, the TS model obtained by the sector nonlinearity approach is in general not a unique fuzzy representation of the nonlinear system. For instance, the nonlinear system (2.12) can be written instead of (2.13) as

$$\begin{aligned}\dot{\boldsymbol{x}} &= \begin{pmatrix}-3x_1^2 & 2 + x_1\\ 1 + x_2^2 & 0\end{pmatrix}\boldsymbol{x} + \begin{pmatrix}e^{x_1}\\ 1\end{pmatrix}u\\ y &= \begin{pmatrix}2x_1 & 1\end{pmatrix}\boldsymbol{x}\end{aligned}$$

in which case a TS model with different 16 rules can be constructed. □

Note that if the assumption that the terms in $\boldsymbol{f}^{\mathrm{m}}$, $\boldsymbol{g}^{\mathrm{m}}$, $\boldsymbol{h}^{\mathrm{m}}$, $\boldsymbol{a}$, and $\boldsymbol{c}$ are bounded, is used, it is no longer necessary for the variables to be defined on a compact set to use the sector nonlinearity approach. However, in such a case, instead of the minimum and maximum, the infimum and supremum of the non-constant terms in the matrix and vector functions have to be used. This case is illustrated using the following example.

Example 2.4. Consider the nonlinear system

$$\dot{x} = \frac{1}{1+e^{-x}}x \tag{2.14}$$

with $x \in [0, \infty]$. As can be seen, $f(x) = \frac{1}{1+e^{-x}}$ is bounded, but the variable x is not defined on a compact set.

The scheduling variable z_1 is defined as $z_1 = \frac{1}{1+e^{-x}} \in \left[\frac{1}{2}, 1\right)$, and the membership functions as

$$\eta_0^1 = 2\left(1 - \frac{1}{1+e^{-x}}\right) = \frac{2e^{-x}}{1+e^{-x}}$$

$$\eta_1^1 = 1 - \eta_0^1 = \frac{1-e^{-x}}{1+e^{-x}}$$

It can easily be seen that $\frac{1}{2}\eta_0^1 x + 1\eta_1^1 x = \frac{1}{1+e^{-x}}x$, i.e., the fuzzy model is equivalent to the nonlinear system (2.14). □

The main advantage of the sector nonlinearity approach is that the obtained TS model is an exact representation of the nonlinear system based on which the TS model has been constructed. However, the approach has two important shortcomings. A first shortcoming is that the obtained consequent linear or affine models are not guaranteed to be stable or observable (detectable), even if the nonlinear system is. Most methods to investigate stability of TS systems require that the linear local models are stable (see Chapter 3). Likewise, the methods for observer design require that the local models are observable or detectable (see Chapter 4). Depending on the nonlinear system considered, instability or unobservability of the local models may be avoided by choosing another representation of the nonlinear system. Otherwise, methods that obtain an approximate fuzzy model, whose local models have the same properties as the nonlinear system, such as the one presented in the next section, can be used.

The second shortcoming is that the number of rules, i.e., the number of local models in the TS model obtained is exponential in the number of nonlinearities. In practical applications, a large number of local models may lead to design problems that are intractable due to either the computational costs or due to the limitations of current algorithms. Therefore, unless instability or unobservability of the local models is an issue, a fuzzy representation with a minimum number of rules should be chosen.

2.3.2 Linearization

One method to obtain a TS fuzzy approximation of a given nonlinear model is local linearization (Johansen et al., 2000). This linearization is in fact a Taylor series expansion in different representative points, which may or may not be equilibria.

Consider the dynamic nonlinear system

$$\begin{aligned} \dot{\boldsymbol{x}} &= \boldsymbol{f}(\boldsymbol{x}, \boldsymbol{u}) \\ \boldsymbol{y} &= \boldsymbol{h}(\boldsymbol{x}) \end{aligned} \tag{2.15}$$

where $\boldsymbol{x} \in \mathbb{R}^{n_x}$ is the vector of state variables, $\boldsymbol{y} \in \mathbb{R}^{n_y}$ is the vector of measurements, $\boldsymbol{u} \in \mathbb{R}^{n_u}$ is the input vector, and $\boldsymbol{f}$ and $\boldsymbol{h}$ are smooth nonlinear vector functions.

The goal is to obtain an approximation of the nonlinear system (2.15) as a set of m rules of the form

Model rule i:
If z_1 *is* Z_1^i *and ... and* z_p *is* Z_p^i *then*

$$\begin{aligned} \dot{\boldsymbol{x}} &= A_i \boldsymbol{x} + B_i \boldsymbol{u} + a_i \\ \boldsymbol{y} &= C_i \boldsymbol{x} + c_i \end{aligned}$$

or, equivalently, a TS model of the form

$$\begin{aligned} \dot{\boldsymbol{x}} &= \sum_{i=1}^{m} w_i(\boldsymbol{z})(A_i \boldsymbol{x} + B_i \boldsymbol{u} + a_i) \\ \boldsymbol{y} &= \sum_{i=1}^{m} w_i(\boldsymbol{z})(C_i \boldsymbol{x} + c_i) \end{aligned} \tag{2.16}$$

where A_i, B_i, a_i, C_i, and c_i are the matrices and biases of the local linear models, $\boldsymbol{z}$ is the scheduling vector that determines which of the rules are active at a certain moment, and $w_i(\boldsymbol{z})$, $i = 1, 2, \ldots, m$ are the normalized membership functions.

First, one has to decide which variables describe the nonlinearities, i.e., which variables should be the scheduling variables. This means deciding on $\boldsymbol{z}$ as a selection of inputs, states, and measurements.

Second, a sufficient number m of linearization points $\boldsymbol{z}_{0,i}$, $i = 1, 2, \ldots, m$ have to be chosen, together with a partition of the space where the variables are defined, and the corresponding membership functions $\omega_{ij}(z_j)$, $i = 1, 2, \ldots, m$. By increasing the number of well-chosen approximation points, the approximation accuracy of the fuzzy model increases. However, by increasing the number of the linearization points, the computational costs of the controller or observer design also increase.

Finally, the consequent matrices are obtained as

$$A_i = \left.\frac{\partial \boldsymbol{f}}{\partial \boldsymbol{x}}\right|_{\boldsymbol{z}_{0,i},0} \qquad B_i = \left.\frac{\partial \boldsymbol{f}}{\partial \boldsymbol{u}}\right|_{\boldsymbol{z}_{0,i},0} \qquad C_i = \left.\frac{\partial \boldsymbol{h}}{\partial \boldsymbol{x}}\right|_{\boldsymbol{z}_{0,i},0}$$

where $|_{\boldsymbol{z}_{0,i},0}$ denotes the evaluation of the expression on the left in the value corresponding to $\boldsymbol{z}_{0,i}$ for those state and input variables that are scheduling variables and 0 for those states and inputs that are not in $\boldsymbol{z}$.

Since generally the linearization is not done in equilibria, affine terms must also be added

$$a_i = \boldsymbol{f}(\boldsymbol{x}, \boldsymbol{u})|_{\boldsymbol{z}_{0,i},0} - (A_i\boldsymbol{x})|_{\boldsymbol{z}_{0,i},0} - (B_i\boldsymbol{u})|_{\boldsymbol{z}_{0,i},0}$$
$$c_i = \boldsymbol{h}(\boldsymbol{x})|_{\boldsymbol{z}_{0,i},0} - (C_i\boldsymbol{x})|_{\boldsymbol{z}_{0,i},0}$$

To obtain the TS system of the form (2.16), the membership functions of each rule are computed using (2.1) and normalized using (2.2). With the normalized membership functions, the TS fuzzy model is expressed as (2.16). The method is illustrated on the following example.

Example 2.5. Consider the nonlinear system (2.12) from Example 2.3, repeated here for convenience

$$\dot{\boldsymbol{x}} = \begin{pmatrix} x_1x_2 - 3x_1^3 + 2x_2 + e^{x_1}u \\ x_1 + x_1x_2^2 + u \end{pmatrix}$$
$$y = 2x_1^2 + x_2$$

with the variables defined on $C = \{\boldsymbol{x},\, y,\, u\,|u,\, y \in \mathbb{R},\, |x_i| \le 1,\, i = 1,\, 2\}$.

Since the state transition function $\boldsymbol{f} = \begin{pmatrix} x_1x_2 - 3x_1^3 + 2x_2 + e^{x_1}u \\ x_1 + x_1x_2^2 + u \end{pmatrix}$ is affine in the input u and nonlinear in the state variables x_1 and x_2, the scheduling vector is chosen as $\boldsymbol{z} = (x_1,\, x_2)^T$.

For this example, we choose the linearization points as $\{(x_1,\, x_2)\,|x_1,\, x_2 \in \{-1, 0, 1\}\}$. A common choice of membership functions is normalized triangular or trapezoidal membership functions that attain their maximum in the linearization points, for each scheduling variable. These functions are easily represented, and for each variable at most two are activated.

In this example, the membership functions chosen are shown in Figure 2.1. Here we denote the corresponding fuzzy sets as *around* 0, *around* -1, and *around* 1.

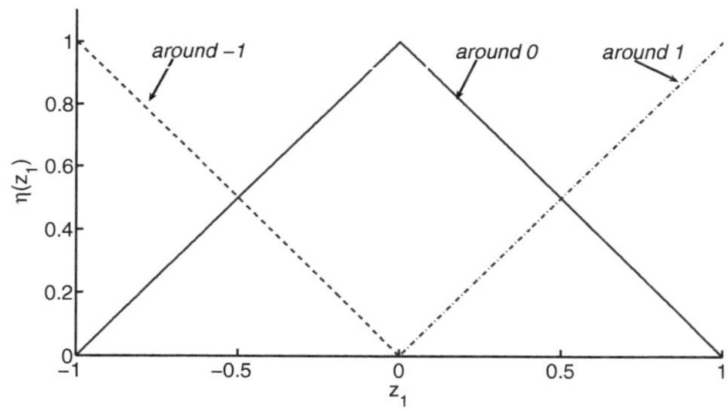

Fig. 2.1 Membership functions for the scheduling variable z_1.

These membership functions for z_1 are defined as

$$\omega_{11}(z_1) = \begin{cases} -z_1, & \text{if } z_1 \leq 0 \\ 0, & \text{otherwise} \end{cases}$$
$$\omega_{21}(z_1) = 1 - |z_1|$$
$$\omega_{31}(z_1) = \begin{cases} z_1, & \text{if } z_1 \geq 0 \\ 0, & \text{otherwise} \end{cases}$$

Similar functions are defined as membership functions for z_2

$$\omega_{12}(z_2) = \begin{cases} -z_2, & \text{if } z_2 \leq 0 \\ 0, & \text{otherwise} \end{cases}$$
$$\omega_{22}(z_2) = 1 - |z_2|$$
$$\omega_{32}(z_2) = \begin{cases} z_2, & \text{if } z_2 \geq 0 \\ 0, & \text{otherwise} \end{cases}$$

The partial derivatives of the state equation with respect to the scheduling vector are

$$\frac{\partial \boldsymbol{f}}{\partial \boldsymbol{x}} = \begin{pmatrix} x_2 - 9x_1^2 + e^{x_1}u & x_1 + 2 \\ 1 + x_2^2 & 2x_1x_2 \end{pmatrix} \qquad \frac{\partial \boldsymbol{f}}{\partial u} = \begin{pmatrix} e^{x_1} \\ 1 \end{pmatrix} \qquad \frac{\partial \boldsymbol{h}}{\partial \boldsymbol{x}} = (4x_1 \ 1)$$

These matrices are evaluated in the selected linearization points to obtain the matrices of the local models. The affine terms are obtained from the evaluation in the linearization points of the expressions

$$\boldsymbol{f}(\boldsymbol{x}, u) - \frac{\partial \boldsymbol{f}}{\partial \boldsymbol{x}}\boldsymbol{x} - \frac{\partial \boldsymbol{f}}{\partial u}u = \begin{pmatrix} 6x_1^3 - x_1x_2 - x_1e^{x_1}u \\ -2x_1x_2^2 \end{pmatrix}$$
$$\boldsymbol{h}(\boldsymbol{x}) - \frac{\partial \boldsymbol{h}}{\partial \boldsymbol{x}}\boldsymbol{x} = -2x_1^2$$

Then, the approximate TS fuzzy model is given by the following $3 \times 3 = 9$ rules:

Model rule 1:
If z_1 *is around* -1 *and* z_2 *is around* -1 *then*

$$\dot{\boldsymbol{x}} = A_1\boldsymbol{x} + B_1\boldsymbol{u} + a_1$$
$$\boldsymbol{y} = C_1\boldsymbol{x} + c_1$$

with

$$
\begin{aligned}
A_1 &= \begin{pmatrix} x_2 - 9x_1^2 + e^{x_1}u & x_1 + 2 \\ 1 + x_2^2 & 2x_1x_2 \end{pmatrix} \Big|_{(\boldsymbol{z}_{0,1},0)} = \begin{pmatrix} -10 & 1 \\ 2 & 2 \end{pmatrix} \\
B_1 &= \begin{pmatrix} e^{x_1} \\ 1 \end{pmatrix} \Big|_{(\boldsymbol{z}_{0,1},0)} = \begin{pmatrix} e^{-1} \\ 1 \end{pmatrix} \\
a_1 &= \begin{pmatrix} 6x_1^3 - x_1x_2 - x_1e^{x_1}u \\ -2x_1x_2^2 \end{pmatrix} \Big|_{(\boldsymbol{z}_{0,1},0)} = \begin{pmatrix} -7 \\ 2 \end{pmatrix} \\
C_1 &= (4x_1 \ 1) \Big|_{(\boldsymbol{z}_{0,1},0)} = (-4 \ 1) \\
c_1 &= -2x_1^2 \Big|_{(\boldsymbol{z}_{0,1},0)} = -2
\end{aligned}
$$

where $\boldsymbol{z}_{0,1} = (-1, -1)$, and the membership function of the rule is computed as $w_1(\boldsymbol{z}) = \omega_{11}\omega_{12}$.

Model rule 2:
If z_1 *is around* -1 *and* z_2 *is around* 0 *then*

$$
\begin{aligned}
\dot{\boldsymbol{x}} &= A_2\boldsymbol{x} + B_2\boldsymbol{u} + a_2 \\
\boldsymbol{y} &= C_2\boldsymbol{x} + c_2
\end{aligned}
$$

with

$$
\begin{aligned}
A_2 &= \begin{pmatrix} x_2 - 9x_1^2 + e^{x_1}u & x_1 + 2 \\ 1 + x_2^2 & 2x_1x_2 \end{pmatrix} \Big|_{(\boldsymbol{z}_{0,2},0)} = \begin{pmatrix} -9 & 1 \\ 1 & 0 \end{pmatrix} \\
B_2 &= \begin{pmatrix} e^{x_1} \\ 1 \end{pmatrix} \Big|_{(\boldsymbol{z}_{0,2},0)} = \begin{pmatrix} e^{-1} \\ 1 \end{pmatrix} \\
a_2 &= \begin{pmatrix} 6x_1^3 - x_1x_2 - x_1e^{x_1}u \\ -2x_1x_2^2 \end{pmatrix} \Big|_{(\boldsymbol{z}_{0,2},0)} = \begin{pmatrix} -6 \\ 0 \end{pmatrix} \\
C_2 &= (4x_1 \ 1) \Big|_{(\boldsymbol{z}_{0,2},0)} = (-4 \ 1) \\
c_2 &= -2x_1^2 \Big|_{(\boldsymbol{z}_{0,2},0)} = -2
\end{aligned}
$$

where $\boldsymbol{z}_{0,2} = (-1, 0)$, and the membership function of the rule is computed as $w_2(\boldsymbol{z}) = \omega_{11}\omega_{22}$.

Model rule 3:
If z_1 *is around* -1 *and* z_2 *is around* 1 *then*

$$
\begin{aligned}
\dot{\boldsymbol{x}} &= A_3\boldsymbol{x} + B_3\boldsymbol{u} + a_3 \\
\boldsymbol{y} &= C_3\boldsymbol{x} + c_3
\end{aligned}
$$

with

$$A_3 = \begin{pmatrix} x_2 - 9x_1^2 + e^{x_1}u & x_1 + 2 \\ 1 + x_2^2 & 2x_1x_2 \end{pmatrix} \Big|_{(z_{0,3},0)} = \begin{pmatrix} -8 & 1 \\ 2 & -2 \end{pmatrix}$$

$$B_3 = \begin{pmatrix} e^{x_1} \\ 1 \end{pmatrix} \Big|_{(z_{0,3},0)} = \begin{pmatrix} e^{-1} \\ 1 \end{pmatrix}$$

$$a_3 = \begin{pmatrix} 6x_1^3 - x_1x_2 - x_1e^{x_1}u \\ -2x_1x_2^2 \end{pmatrix} \Big|_{(z_{0,3},0)} = \begin{pmatrix} -5 \\ 2 \end{pmatrix}$$

$$C_3 = (4x_1 \ 1) \Big|_{(z_{0,3},0)} = (-4 \ 1)$$

$$c_3 = -2x_1^2 \Big|_{(z_{0,3},0)} = -2$$

where $z_{0,3} = (-1,\ 1)$, and the membership function of the rule is computed as $w_3(z) = \omega_{11}\omega_{32}$.

Model rule 4:
If z_1 *is around* 0 *and* z_2 *is around* -1 *then*

$$\begin{aligned} \dot{x} &= A_4x + B_4u + a_4 \\ y &= C_4x + c_4 \end{aligned}$$

with

$$A_4 = \begin{pmatrix} x_2 - 9x_1^2 + e^{x_1}u & x_1 + 2 \\ 1 + x_2^2 & 2x_1x_2 \end{pmatrix} \Big|_{(z_{0,4},0)} = \begin{pmatrix} -1 & 2 \\ 2 & 0 \end{pmatrix}$$

$$B_4 = \begin{pmatrix} e^{x_1} \\ 1 \end{pmatrix} \Big|_{(z_{0,4},0)} = \begin{pmatrix} 1 \\ 1 \end{pmatrix}$$

$$a_4 = \begin{pmatrix} 6x_1^3 - x_1x_2 - x_1e^{x_1}u \\ -2x_1x_2^2 \end{pmatrix} \Big|_{(z_{0,4},0)} = \begin{pmatrix} 0 \\ 0 \end{pmatrix}$$

$$C_4 = (4x_1 \ 1) \Big|_{(z_{0,4},0)} (0 \ 1)$$

$$c_4 = -2x_1^2 \Big|_{(z_{0,4},0)} = 0$$

where $z_{0,4} = (0,\ -1)$ and the membership function of the rule is computed as $w_4(z) = \omega_{21}\omega_{12}$.

The remaining rules are constructed in a similar manner.

Note that since the individual membership functions for z_1 and z_2, respectively, are normal, the obtained, combined membership functions w_i are also normal. □

Using this method, a good approximation can be obtained for nonlinear functions that are analytic in the neighborhood of the chosen linearization points, i.e., functions for which the Taylor series expansion converges to the value of the function. For analytic functions, if the membership functions are chosen such that their value is 1 in the corresponding linearization point, the values of both the nonlinear system and its derivative are equal to that of the fuzzy model and its derivative in the linearization points. To see this, consider an analytic vector function $\boldsymbol{f}(\boldsymbol{x})$, defined on a compact set $C_{\boldsymbol{x}}$, and a set of points $\boldsymbol{x}_{0,i}$, $i = 1, 2, \ldots, m$, in $C_{\boldsymbol{x}}$. Then, for any point $\boldsymbol{x}$ in the neighborhood of one of the points $\boldsymbol{x}_{0,i}$, one can use the Taylor series expansion

$$\boldsymbol{f}(\boldsymbol{x}) \simeq \boldsymbol{f}(\boldsymbol{x}_{0,i}) + \left.\frac{\partial \boldsymbol{f}}{\partial \boldsymbol{x}}\right|_{\boldsymbol{x}_{0,i}} (\boldsymbol{x} - \boldsymbol{x}_{0,i})$$

i.e.,

$$\begin{aligned}
\boldsymbol{f}(\boldsymbol{x}) &\simeq \left.\frac{\partial \boldsymbol{f}}{\partial \boldsymbol{x}}\right|_{\boldsymbol{x}_{0,i}} \boldsymbol{x} + \boldsymbol{f}(\boldsymbol{x}_{0,i}) - \left.\frac{\partial \boldsymbol{f}}{\partial \boldsymbol{x}}\right|_{\boldsymbol{x}_{0,i}} \boldsymbol{x}_{0,i} \\
\boldsymbol{f}(\boldsymbol{x}) &\simeq A_i \boldsymbol{x} + a_i
\end{aligned}$$

where $A_i = \left.\frac{\partial \boldsymbol{f}}{\partial \boldsymbol{x}}\right|_{\boldsymbol{x}_{0,i}}$, and $a_i = \boldsymbol{f}(\boldsymbol{x}_{0,i}) - \left.\frac{\partial \boldsymbol{f}}{\partial \boldsymbol{x}}\right|_{\boldsymbol{x}_{0,i}} \boldsymbol{x}_i$.

Now, consider normalized membership functions $w_i(\boldsymbol{x})$, $i = 1, 2, \ldots, m$, such that $w_i(\boldsymbol{x}_{0,i}) = 1$, $w_i(\boldsymbol{x}_{0,j}) = 0$, $\forall i \neq j$. Then, $\boldsymbol{f}(\boldsymbol{x})$ can be written as

$$\begin{aligned}
\boldsymbol{f}(\boldsymbol{x}) &= \sum_{i=1}^{m} w_i(\boldsymbol{x}) \boldsymbol{f}(\boldsymbol{x}) \\
&\simeq \sum_{i=1}^{m} w_i(\boldsymbol{x}) \Big(\left.\frac{\partial \boldsymbol{f}}{\partial \boldsymbol{x}}\right|_{\boldsymbol{x}_{0,i}} \boldsymbol{x} + \boldsymbol{f}(\boldsymbol{x}_{0,i}) - \left.\frac{\partial \boldsymbol{f}}{\partial \boldsymbol{x}}\right|_{\boldsymbol{x}_{0,i}} \boldsymbol{x}_{0,i} \Big) \\
&\simeq \sum_{i=1}^{m} w_i(\boldsymbol{x}) (A_i \boldsymbol{x} + a_i)
\end{aligned}$$

that is, the nonlinear function $\boldsymbol{f}$ can be approximated by the fuzzy model.

The advantage of constructing TS models using linearization is that although the fuzzy system is only an approximation of the original nonlinear system, the consequents retain important properties of the nonlinear system in the linearization points. For instance, if the nonlinear system is locally observable in a neighborhood of the linearization point, then the corresponding local model is observable or detectable. A disadvantage of the method is that there are no general guidelines on how to chose the linearization points, or how many linearization points should be chosen. Depending on the nonlinearity, a large number of points may be necessary for an accurate approximation, which implies large computational costs.

Since the linearization in general will not be performed in the equilibrium points, with this method, affine TS models will be obtained. This means that stability analysis and controller design become harder to perform. However, as already mentioned, affine models do not present a problem in observer design.

2.4 Summary

This chapter has introduced Takagi-Sugeno (TS) fuzzy models that are used in the sequel. Dynamic TS fuzzy models have been presented, and two methods to construct dynamic TS models given a nonlinear dynamic system have been discussed. The first method, the sector nonlinearity approach can be used to construct exact fuzzy representations of the nonlinear system. Using the second method, linearization, a TS model that approximates the nonlinear system is obtained. This method has the advantage that the local properties of the nonlinear system are retained in the TS model.

Chapter 3
Stability Analysis of TS Fuzzy Systems

3.1 Introduction

The purpose of this chapter is to review various results concerning the stability analysis and control design of Takagi-Sugeno (TS) models. To analyze the stability and to design observers and controllers for TS systems, in general Linear Matrix Inequality (LMI) constraints are used. Therefore, the first part of the chapter presents a brief overview of LMIs and their useful properties. Moreover, many problems encountered can be turned in a multiple-sum co-positivity problem. This is a well-known problem and some results are given for a double-sum co-positivity problem with several possible relaxations.

Generally speaking, Lyapunov's direct method is used to derive stability and stabilization results of TS models. In the literature, for sake of simplicity and in the view of writing the problems in an LMI form, mainly a quadratic Lyapunov function is considered, thus reducing the notion of stability to the notion of quadratic stability. Nevertheless, we also present results that leave the quadratic framework.

Among the numerous possible choices of results (state feedback, output feedback, with uncertainties, H_2, H_∞ performance, delays, etc.) we present works to give the main ideas to the reader of the different possibilities that TS models offer. In particular for state feedback, results for performance through H_∞ attenuation, and robust control of TS models with uncertainties are given as well as the Input-to-State Stability (ISS) property for exogenous signals. The results presented are illustrated on several examples.

3.2 Preliminaries

3.2.1 *Notation*

Let $F = F^T \in \mathbb{R}^{n \times n}$ be a symmetric matrix. In the sequel, $F > 0$ (resp. $F < 0$) stands for positive (resp. negative)-definiteness, i.e., every eigenvalue of F is strictly positive (resp. negative). The notation $F \geq 0$ (resp. $F \leq 0$) stands for

Zs. Lendek et al.: Stability and Observer Design Using TS Models, STUDFUZZ 262, pp. 25–48.
springerlink.com

semi-positive (resp. negative), i.e., the eigenvalues can be positive (resp. negative) or zero. Moreover, whenever an expression is written as $F > 0$, it is assumed that the expression is symmetric, i.e., $F = F^T > 0$, even if the explicit notation is omitted.

With $A,\ B \in \mathbb{R}^{n \times n}$ being two symmetric matrices $A > B$ is equivalent to $A - B > 0$.

A star $(*)$ in a matrix indicates a transposed quantity in the symmetric position. For instance, $\begin{pmatrix} P & (*) \\ A & \tilde{P} \end{pmatrix} < 0$ is equivalent to $\begin{pmatrix} P & A^T \\ A & \tilde{P} \end{pmatrix} < 0$.

The notation *co* stands for the convex hull, i.e., the convex envelope of some vertices: $C = \mathrm{co}\,\{a_1, \ldots, a_n\}$.

3.2.2 Linear Matrix Inequalities

In the sequel, stability and design conditions are presented mainly in the form of LMIs. This section is therefore a very brief introduction to the LMI framework. More details can be found in (Boyd et al., 1994; Scherer and Weiland, 2005).

Overview

In a broad sense an LMI is a set of expressions whose variables are linearly-related matrices. A formal definition of an LMI is (Boyd et al., 1994)

$$F(\boldsymbol{x}) = F_0 + \sum_{i=1}^{m} x_i F_i > 0 \tag{3.1}$$

where $\boldsymbol{x} \in \mathbb{R}^m$ is the vector of decision variables and $F_i = F_i^T \in \mathbb{R}^{n \times n}$, $i = 0, \ldots, m$, are given constant symmetric matrices. As any matrix variable can be decomposed into a base of symmetric matrices, the definition (3.1) involving scalars is easily extended to matrices.

The set of solutions of the LMI (3.1), or the so-called feasibility set, denoted by $S = \{\boldsymbol{x} | \boldsymbol{x} \in \mathbb{R}^m, F(\boldsymbol{x}) > 0\}$ is a convex subset of $\mathbb{R}^m$. Finding a solution to (3.1) is a convex optimization problem avoiding local minima and guaranteeing finite feasibility tests. When no solution exists, the problem is said to be infeasible. The following well-known convex or quasi-convex optimization problems are relevant for the analysis and the synthesis of control systems (Boyd et al., 1994; Scherer and Weiland, 2005).

1. Finding a solution $\boldsymbol{x} \in \mathbb{R}^m$ to the LMI system (3.1) or determining that there is no solution is called the feasibility problem (FP). This problem is equivalent to minimizing the convex function $f : \boldsymbol{x} \rightarrow \lambda_{\min}(F(\boldsymbol{x}))$, where $\lambda_{\min}$ denotes the smallest eigenvalue, and then deciding whether the solution is positive (strictly feasible solution), zero (feasible solution), or negative (unfeasible case).

2. Minimizing a linear combination of the decision variables $\boldsymbol{b}^T\boldsymbol{x}$ subject to (3.1) is called the eigenvalue problem (EVP), also known as an LMI optimization problem.
3. Minimizing the eigenvalues of a pair of matrices that depend affinely on a variable, subject to a set of LMI constraints or determining that the problem is infeasible, i.e., solving the problem: *minimize* λ *subject to*

$$\begin{aligned} &\lambda B(\boldsymbol{x}) - A(\boldsymbol{x}) > 0 \\ &B(\boldsymbol{x}) > 0 \\ &C(\boldsymbol{x}) > 0 \end{aligned}$$

where $A(\boldsymbol{x})$, $B(\boldsymbol{x})$ and $C(\boldsymbol{x})$ are symmetric and affine with respect to $\boldsymbol{x}$, is called a generalized eigenvalue problem (GEVP).

Numerous solvers handle these various optimization problems, such as LMILAB, SeDuMi, SDPT3, VSDP, or LMIRank. In this book, unless otherwise stated, to solve LMI problems, the *SeDuMi* solver within the *Yalmip* toolbox (Löfberg, 2004) is used.

Properties

LMI constraints do not appear "naturally" in control problems. Thanks to the available optimization solutions, one of the goals when encountering control problems is to recast them as LMI expressions. This is done by making use of the properties of LMIs. Some of these properties are enumerated below.

Property 3.1. (Congruence) Given a matrix $P = P^T$ and a full column rank matrix Q it holds that

$$P > 0 \Rightarrow QPQ^T > 0$$

Property 3.2. (Schur complement) Consider a matrix $M = M^T = \begin{pmatrix} M_{11} & M_{12} \\ M_{12}^T & M_{22} \end{pmatrix}$, with M_{11} and M_{22} being square matrices. Then

$$M < 0 \Leftrightarrow \begin{cases} M_{11} < 0 \\ M_{22} - M_{12}^T M_{11}^{-1} M_{12} < 0 \end{cases} \Leftrightarrow \begin{cases} M_{22} < 0 \\ M_{11} - M_{12} M_{22}^{-1} M_{12}^T < 0 \end{cases}$$

Property 3.3. (S-procedure) Consider matrices $F_i = F_i^T \in \mathbb{R}^{n \times n}$, $\boldsymbol{x} \in \mathbb{R}^n$, such that $\boldsymbol{x}^T F_i \boldsymbol{x} \geq 0$, $i = 1, \ldots, p$, and the quadratic inequality condition

$$\boldsymbol{x}^T F_0 \boldsymbol{x} > 0 \tag{3.2}$$

$\boldsymbol{x} \neq 0$. A sufficient condition for (3.2) to hold is: there exist $\tau_i \geq 0$, $i = 1, \ldots, p$, such that $F_0 - \sum_{i=1}^{p} \tau_i F_i > 0$.

Property 3.4. (Completion of squares) Given two matrices X and Y of proper size and $Q = Q^T > 0$, the following inequality holds

$$X^TY + Y^TX \leq X^TQX + Y^TQ^{-1}Y$$

Several of these properties can be useful to recast an expression into LMI constraints; the following example is given as illustration.

Example 3.1. Consider the following problem with $P(\boldsymbol{x})$ and $S(\boldsymbol{x})$ affine functions of $\boldsymbol{x} \in \mathbb{R}^n$ *find* $\boldsymbol{x}$ *such that*

$$\begin{aligned} &P(\boldsymbol{x}) > 0 \\ &\text{trace}\left(S(\boldsymbol{x})^T P^{-1}(\boldsymbol{x}) S(\boldsymbol{x})\right) < 1 \end{aligned}$$

A first step is to replace the condition $\text{trace}\left(S(\boldsymbol{x})^T P^{-1}(\boldsymbol{x}) S(\boldsymbol{x})\right) < 1$ using an auxiliary variable Q with

$$\begin{aligned} &\text{trace}\,(Q) < 1 \\ &S(\boldsymbol{x})^T P^{-1}(\boldsymbol{x}) S(\boldsymbol{x}) < Q \end{aligned}$$

Then, using Property 3.2 (Schur complement) gives:
Find $\boldsymbol{x}$ *and* Q *such that*

$$\begin{aligned} &\text{trace}\,(Q) < 1 \\ &\begin{pmatrix} Q & S(\boldsymbol{x})^T \\ S(\boldsymbol{x}) & P(\boldsymbol{x}) \end{pmatrix} > 0 \end{aligned}$$

Note that $P(\boldsymbol{x}) > 0$ can be omitted as the second condition above holds only if $P(\boldsymbol{x})$ is positive definite. □

Relaxations

Many control and estimation problems can be summarized as a double sum negativity problem

$$\sum_{i=1}^{m}\sum_{j=1}^{m} w_i(\boldsymbol{z}) w_j(\boldsymbol{z}) \Gamma_{ij}(\boldsymbol{x}) < 0, \tag{3.3}$$

with the symmetric matrices $\Gamma_{ij}(\boldsymbol{x})$ being affinely dependent on the unknown variables $\boldsymbol{x} \in \mathbb{R}^{n_x}$ and the functions $w_i(\boldsymbol{z})$ being nonlinear functions that observe the convex sum property, i.e., $\sum_{i=1}^{m} w_i(\boldsymbol{z}) = 1$ and $w_i(\boldsymbol{z}) \geq 0$. The goal is to find the least conservative conditions on Γ_{ij} such that (3.3) holds, using only the convex sum property for the nonlinear functions $w_i(\boldsymbol{z})$. The trivial LMI solution of the problem (3.3) is: $\Gamma_{ij}(\boldsymbol{x}) < 0$, $i, j = 1, \ldots, m$. These conditions can be relaxed by considering that with $w_i(\boldsymbol{z}) \geq 0$ and $w_i(\boldsymbol{z})w_j(\boldsymbol{z}) = w_j(\boldsymbol{z})w_i(\boldsymbol{z})$, a basic sufficient solution is (Wang et al., 1996)

$$\begin{aligned} &\Gamma_{ii} < 0 \\ &\Gamma_{ij} + \Gamma_{ji} < 0 \end{aligned} \tag{3.4}$$

for $i = 1, 2, \ldots, m$, $j = i+1, i+2, \ldots, m$.

The variable $\boldsymbol{x} \in \mathbb{R}^{n_x}$ is omitted in (3.4) for convenience. A refinement of the conditions (3.4) that does not require auxiliary variables has been proposed and it is recalled below.

Lemma 3.1. *(Tuan et al., 2001) Equation* (3.3) *is satisfied provided that the following conditions hold*

$$\begin{aligned} &\Gamma_{ii} < 0 \\ &\frac{2}{m-1}\Gamma_{ii} + \Gamma_{ij} + \Gamma_{ji} < 0 \end{aligned} \tag{3.5}$$

for $i = 1, 2, \ldots, m$, $j = 1, 2, \ldots, m$, $i \neq j$.

Adding auxiliary variables, such as in Example 3.1, can also be useful in order to reduce the conservatism of the conditions. Among all the possible relaxations, the next one can be viewed as a good compromise between the number of additional slack variables and the quality of the solutions.

Lemma 3.2. *(Liu and Zhang, 2003) Condition* (3.3) *is satisfied provided that the following conditions hold: there exist matrices* $Q_{ii} > 0$, $i = 1, 2, \ldots, m$, *and* $Q_{ij} = Q_{ji}^T$, $i = 1, 2, \ldots, m$, $j = i+1, i+2, \ldots, m$ *such that*

$$\begin{aligned} &\Gamma_{ii} + Q_{ii} < 0 \\ &\Gamma_{ij} + \Gamma_{ji} + Q_{ij} + Q_{ji} < 0 \\ &\begin{pmatrix} Q_{11} & Q_{12} & \cdots & Q_{1m} \\ Q_{21} & Q_{22} & \cdots & Q_{2m} \\ \vdots & & \ddots & \vdots \\ Q_{m1} & Q_{m2} & \cdots & Q_{mm} \end{pmatrix} > 0 \end{aligned} \tag{3.6}$$

for $i = 1, 2, \ldots, m$, $j = i+1, i+2, \ldots, m$.

Note that the conditions of Lemmas 3.1 and 3.2 are only sufficient. Nevertheless, some relaxations exist that also become necessary when the number of terms in the summations tends to infinity; Sala and Ariño (2007) proposed conditions that are based on Polya's theorems, whereas Kruszewski et al. (2009) proposed conditions that are based on triangulation. Other works use more properties of the nonlinear functions $w_i(\boldsymbol{z})$, $i = 1, 2, \ldots, m$, such as bounds (Sala and Ariño, 2007) or membership function dependent approaches (Bernal et al., 2009). The main drawback of these results is that the complexity of the LMI problems increases, and they quickly become intractable for the actual LMI solvers.

The following two sections present stability analysis and stabilization of Takagi-Sugeno models using LMI constraints. The TS model used is

$$\begin{aligned} \dot{\boldsymbol{x}} &= \sum_{i=1}^{m} w_i(\boldsymbol{z}) \left(A_i \boldsymbol{x} + B_i \boldsymbol{u} \right) \\ \boldsymbol{y} &= \sum_{i=1}^{m} w_i(\boldsymbol{z}) C_i \boldsymbol{x} \end{aligned} \tag{3.7}$$

3.3 Stability Analysis of TS Systems

This section reviews methods for the stability analysis of TS fuzzy systems.

3.3.1 *Quadratic Stability*

The stability of TS models is investigated using the direct Lyapunov method. The Lyapunov function[1] commonly used is the quadratic one,

$$V(\boldsymbol{x}) = \boldsymbol{x}^T P \boldsymbol{x} \tag{3.8}$$

with $P = P^T > 0$.

When using this Lyapunov function, we speak of "quadratic stability". Note that when a system is quadratically stable it *implies* that it is stable. However, the reverse is *not necessarily* true. Therefore, conditions obtained using the Lyapunov function (3.8) are only sufficient, i.e., if the LMI conditions fail, nothing can be directly said about stability or instability of the TS model.

The unforced ($\boldsymbol{u} = 0$) TS model (3.7) is quadratically stable if the Lyapunov function (3.8) decreases and tends to zero when $t \to \infty$ for all trajectories $\boldsymbol{x}(t)$. The derivative of (3.8) along the trajectories of the unforced model (3.7) is

$$\begin{aligned} \dot{V} &= \left(\sum_{i=1}^{m} w_i(\boldsymbol{z}) A_i \boldsymbol{x} \right)^T P \boldsymbol{x} + \boldsymbol{x}^T P \left(\sum_{i=1}^{m} w_i(\boldsymbol{z}) A_i \boldsymbol{x} \right) \\ &= \sum_{i=1}^{m} w_i(\boldsymbol{z}) \boldsymbol{x}^T \left(A_i^T P + P A_i \right) \boldsymbol{x} \end{aligned} \tag{3.9}$$

Remembering that $w_i(\boldsymbol{z}) \geq 0$, $i = 1, 2, \ldots, m$ the following theorem is straightforwardly obtained.

Theorem 3.1. *(Wang et al., 1996) The unforced model $\dot{\boldsymbol{x}} = \sum_{i=1}^{m} w_i(\boldsymbol{z}) A_i \boldsymbol{x}$ is globally asymptotically stable if there exist a matrix $P = P^T$ such that the following LMI problem is feasible*

[1] In the sequel, whenever it is evident, the explicit dependence of the Lyapunov function and its derivative on the state variables is omitted.

$$\mathcal{H}(PA_i) < 0 \tag{3.10}$$

for $i = 1, 2, \ldots, m$, *where the symbol* $\mathcal{H}$ *denotes the symmetric part, that is* $\mathcal{H}(X) = X + X^T$.

Remark: Notice that the result is strictly equivalent to stability of linear parameter varying (LPV) models, i.e., $\dot{\boldsymbol{x}} = A(\delta)\,\boldsymbol{x}$, $A(\delta) \in \operatorname{co}\{A_1, \ldots, A_m\}$, $\boldsymbol{x} \in \mathbb{R}^{n_x}$.

Theorem 3.1 expresses that it is not enough to have all the vertices globally asymptotically stable (GAS) to ensure the stability of the TS model. The reason for this is that the domain of Hurwitz matrices – matrices whose every eigenvalue has strictly negative real part – is non-convex. To illustrate this, consider the following example.

Example 3.2. Consider the matrices $A_1 = \begin{pmatrix} -1 & 4 \\ 0 & -1 \end{pmatrix}$, $A_2 = \begin{pmatrix} -1 & 0 \\ 4 & -1 \end{pmatrix}$. These matrices are Hurwitz, as they have all eigenvalues at -1. Consider now the convex combination: $A = 0.5 \times A_1 + 0.5 \times A_2 = \begin{pmatrix} -1 & 2 \\ 2 & -1 \end{pmatrix}$, whose eigenvalues are 1 and -3, thus being non-Hurwitz. □

A test of existence of a common matrix $P = P^T$ is described by Johansson et al. (1999) and recalled in what follows.

Property 3.5. If there exist positive definite matrices $R_i = R_i^T > 0, i = 1, 2, \ldots, m,$ such that

$$\sum_{i=1}^{m} \left(A_i^T R_i + R_i A_i\right) > 0 \tag{3.11}$$

then there is no matrix $P = P^T > 0$ such that conditions (3.10) hold.

Example 3.3. For example with the two matrices A_1 and A_2 defined in Example 3.2, a result of the LMI problem (3.11) is $R_1 = \begin{pmatrix} 9 & 4 \\ 4 & 2 \end{pmatrix}$, $R_2 = \begin{pmatrix} 2 & 4 \\ 4 & 9 \end{pmatrix}$, and $\sum_{i=1}^{2}(A_i^T R_i + R_i A_i) = \begin{pmatrix} 10 & 0 \\ 0 & 10 \end{pmatrix} > 0$. Therefore, there is no matrix $P = P^T > 0$ such that conditions (3.10) hold. □

3.3.2 D-Stability

Quadratic stability also ensures an exponential decay rate. Effectively, consider (3.9). If $\dot{V} < 0$ then there always exists an $\varepsilon > 0$ such that

$$\sum_{i=1}^{m} w_i(\boldsymbol{z}) \left(A_i^T P + P A_i\right) + \varepsilon P < 0 \tag{3.12}$$

Thus for all $t \in \mathbb{R}$ and $w_i(\boldsymbol{z}) \geq 0$, $i = 1, 2, \ldots, m$,

$$\dot{V} + \varepsilon V = \boldsymbol{x}^T \left[\sum_{i=1}^{m} w_i(\boldsymbol{z}) \left(A_i^T P + P A_i \right) + \varepsilon P \right] \boldsymbol{x} < 0 \tag{3.13}$$

and integrating (3.13) over $[t_0,\ t]$ yields that $V(\boldsymbol{x})$ has an exponential decay

$$V\left(\boldsymbol{x}(t)\right) \leq V\left(\boldsymbol{x}\left(t_0\right)\right) e^{-\varepsilon(t-t_0)}$$

Since for a positive definite matrix P it holds that $\lambda_{\min}(P) \|\boldsymbol{x}\|^2 \leq \boldsymbol{x}^T P \boldsymbol{x} \leq \lambda_{\max}(P) \|\boldsymbol{x}\|^2$, we have

$$\lambda_{\min}(P) \|\boldsymbol{x}\|^2 \leq V\left(\boldsymbol{x}(t)\right) \leq V\left(\boldsymbol{x}\left(t_0\right)\right) e^{-\varepsilon(t-t_0)} \leq \lambda_{\max}(P) \|\boldsymbol{x}_0\|^2 e^{-\varepsilon(t-t_0)}$$

which means that $\|\boldsymbol{x}(t)\|$ has an exponential decay

$$\|\boldsymbol{x}(t)\|^2 \leq \frac{\lambda_{\max}(P)}{\lambda_{\min}(P)} \|\boldsymbol{x}_0\|^2 e^{-\varepsilon(t-t_0)}$$

Thus, $\lim_{t\to\infty} \boldsymbol{x}(t) = 0$ with exponential decay rate $\varepsilon/2$ irrespective of the initial condition $x_0(t_0)$ for all $w_i(\boldsymbol{z}) \geq 0$, $i = 1, 2, \ldots, m$.

A way to introduce some performance together with quadratic stability is therefore to ensure a certain exponential decay rate, i.e., ask for $\alpha > 0$ such that $\|\boldsymbol{x}(t)\| \leq M \|\boldsymbol{x}_0\| e^{\alpha(t-t_0)}$. In this way, stability can be influenced by defining a region D in the complex plane such that $\mathrm{D}(s) < \alpha$. More generally, LMI regions can be defined (Gahinet et al., 1995).

Definition 3.1. A subset D of the complex plane is called an LMI region if there exists a symmetric matrix $X \in \mathbb{R}^{m\times m}$ and a matrix $Y \in \mathbb{R}^{m\times m}$ such that

$$\mathrm{D} = \{z | z \in \mathcal{C}, f_{\mathrm{D}}(z) < 0\}$$

where

$$f_{\mathrm{D}}(z) = X + zY + \bar{z}Y^T$$

is called the characteristic function of the LMI region.

One can easily see that, because of the form of the function $f_{\mathrm{D}}(z)$, LMI regions are convex and symmetric with respect to the real axis. Useful LMI regions include a vertical strip $[d_{\mathrm{l}}, d_{\mathrm{u}}]$ and a conic sector centered in the origin with inner angle θ (Figure 3.1). If all the eigenvalues of a matrix A are located in a region D, then the matrix A is called D-stable.

A theorem to ensure D-stability of a matrix A is given by Chilali and Gahinet (1996):

Theorem 3.2. *The matrix A is* D*-stable if and only if there exists $P = P^T > 0$ so that*

$$X \otimes P + Y \otimes AP + Y^T \otimes (AP)^T < 0$$

where $\otimes$ is the Kronecker product (Kailath et al., 2000).

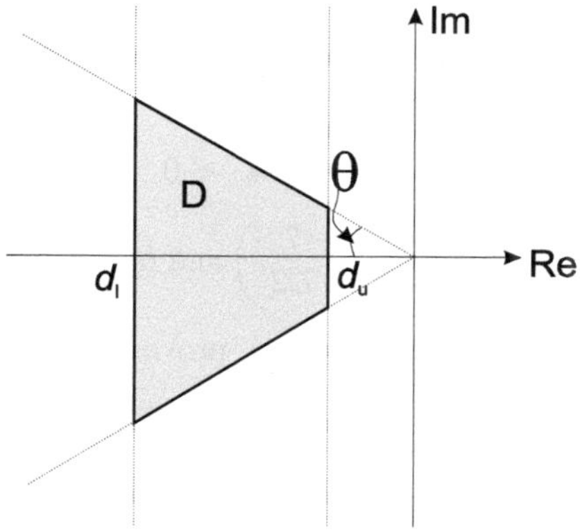

Fig. 3.1 LMI regions.

In the context of stability, using LMI regions to ensure the specific D-stability of the system effectively means adding constraints to the LMI problems, more specifically

$$(X_{j,k}P + Y_{j,k}PA_i + Y_{k,j}A_i^T P) < 0$$
$$j, k = 1, 2, \ldots, m$$

Here, $X_{j,k}$ and $Y_{j,k}$ denote the (j,k)th element of the corresponding matrices.

Remark: Note that the upper limit of the vertical strip, d_{u}, corresponds to the decay rate.

3.3.3 *Leaving the Quadratic Stability Framework*

As previously mentioned, the conditions presented in the previous section are only sufficient conditions. Sources of conservatism are:

1. using only the knowledge of the convex sum property for $w_i(\boldsymbol{z})$;
2. the conditions that are used to ensure the negativeness of the double sum;
3. reducing the stability issue to quadratic stability.

To illustrate this latter case, consider the very simple example inspired by Johansson and Rantzer (1998).

Example 3.4. Consider the model

$$\dot{\boldsymbol{x}} = \begin{cases} A_1\boldsymbol{x}, & \text{if } x_1 < 0 \\ A_2\boldsymbol{x}, & \text{if } x_1 \geq 0 \end{cases} \tag{3.14}$$

with $A_1 = \begin{pmatrix} -5 & -4 \\ -1 & -2 \end{pmatrix}$, $A_2 = \begin{pmatrix} -2 & -4 \\ 20 & -2 \end{pmatrix}$. Quadratic stability reduces to *find* $P > 0$ *such that*

$$\begin{aligned} A_1^T P + PA_1 &< 0 \\ A_2^T P + PA_2 &< 0 \end{aligned}$$

It is easy to show that with $R_1 = \begin{pmatrix} 9.3 & -7.5 \\ -7.5 & 7.2 \end{pmatrix}$ and $R_2 = \begin{pmatrix} 6.5 & 2.7 \\ 2.7 & 1.4 \end{pmatrix}$, $\sum_{i=1}^{2} (A_i^T R_i + R_i A_i) = \begin{pmatrix} 4 & -0.7 \\ -0.7 & 4 \end{pmatrix} > 0$, which proves, according to Property 3.5, equation (3.11), that no such P exists. Therefore quadratic stability fails to demonstrate the stability of (3.14).

Now consider the piecewise Lyapunov function:

$$V(\boldsymbol{x}) = \begin{cases} \boldsymbol{x}^T P \boldsymbol{x}, & \text{if } x_1 < 0 \\ \boldsymbol{x}^T P \boldsymbol{x} + \eta x_1^2, & \text{if } x_1 \geq 0 \end{cases} \tag{3.15}$$

With $C = \begin{pmatrix} 1 & 0 \end{pmatrix}$, it is straightforward to show that the LMI problem for stability is *find* $P = P^T > 0$ *and* η *such that*

$$\begin{aligned} & P + \eta C^T C > 0 \\ & \mathcal{H}(PA_1) < 0 \\ & \mathcal{H}((P + \eta C^T C) A_2) < 0 \end{aligned}$$

where $\mathcal{H}(X) = X + X^T$. With $P = \begin{pmatrix} 1 & 0 \\ 0 & 3 \end{pmatrix}$ and $\eta = 9$ a solution is obtained. Effectively,

$$\begin{aligned} \mathcal{H}(PA_1) &= \begin{pmatrix} -10 & -7 \\ -7 & -12 \end{pmatrix} \\ \mathcal{H}((P + \eta C^T C) A_2) &= \begin{pmatrix} -40 & 20 \\ 20 & -12 \end{pmatrix} \end{aligned}$$

with both matrices being negative definite. Thus, the system (3.14) is asymptotically stable. □

Whereas quadratic stability cannot be proven for this simple example, it shows that introducing some "knowledge" in the Lyapunov function can eliminate some drawbacks. Therefore, pursuing this idea, several ideas can be used. In a sense, a TS model induces a state space partition according to the scheduling variables. The main ideas rely on introducing this partition into the Lyapunov function. This can be done in two ways.

The first way relies on using piecewise quadratic Lyapunov functions (Johansson and Rantzer, 1998; Johansson et al., 1999; Feng, 2003, 2006). The state space is partitioned according to the activation of the linear models, allowing the Lyapunov function to change from one region to another, for instance

$$V(\boldsymbol{x},\boldsymbol{z}) = \boldsymbol{x}^T P(\boldsymbol{z})\boldsymbol{x} \quad P(\boldsymbol{z}) = P_i > 0 \quad \text{if } \boldsymbol{z} \in S_i$$

where S_i are given sets such that $\bigcup_i S_i$ covers the state space. For example, "natural" regions for TS models are: $S_i = \{\boldsymbol{z}|w_i(\boldsymbol{z}) \geq w_j(\boldsymbol{z}),\ j = 1, 2, \ldots, m,\ j \neq i\}, i = 1, 2, \ldots, m$.

The above partition is natural for those TS models that do not have all their linear models activated at once. Unfortunately, this assumption does not hold for TS models built by using the sector nonlinearity approach.

The second way is to use the Lyapunov function (Blanco et al., 2001; Tanaka et al., 2003)

$$V(\boldsymbol{x},\boldsymbol{z}) = \boldsymbol{x}^T \sum_{i=1}^{m} w_i(\boldsymbol{z}) P_i \boldsymbol{x} \tag{3.16}$$

with $P_i > 0, i = 1, 2, \ldots, m$, thus introducing the nonlinear functions $w_i(\boldsymbol{z})$. Nevertheless, the results concerning this approach face the following problem. Taking the derivative of (3.16) gives

$$\dot{V}(\boldsymbol{x},\boldsymbol{z}) = \sum_{i=1}^{m}\sum_{j=1}^{m} w_i(\boldsymbol{z}) w_j(\boldsymbol{z})\, \boldsymbol{x}^T (A_j^T P_i + P_i A_j)\boldsymbol{x} + \sum_{i=1}^{m} \frac{dw_i(\boldsymbol{z})}{dt} \boldsymbol{x}^T P_i \boldsymbol{x}$$

Therefore it involves the derivative of the nonlinear functions $w_i(\boldsymbol{z})$ and

$$\frac{dw_i(\boldsymbol{z})}{dt} = \frac{\partial w_i(\boldsymbol{z})}{\partial \boldsymbol{z}} \dot{\boldsymbol{z}}(t) = \frac{\partial w_i(\boldsymbol{z})}{\partial \boldsymbol{z}} q(\dot{\boldsymbol{x}})$$

where $\dot{\boldsymbol{z}} = q(\dot{\boldsymbol{x}})$ represents the linear or nonlinear mapping between the scheduling and the state vector. Although the quantities $\frac{\partial w_i(\boldsymbol{z})}{\partial \boldsymbol{z}}$ can easily be derived and thus bounded, $q(\dot{\boldsymbol{x}})$ is a priori unknown. Some works (Tanaka et al., 2003; Mozelli et al., 2009) propose to use some bounds as $\left|\frac{dw_i(\boldsymbol{z})}{dt}\right| < \phi_i$. Nevertheless, the main drawback is they need to verify a posteriori that the future trajectory does not escape from the boundaries.

Nowadays, two results seem of real interest that use the Lyapunov function (3.16) or a slightly different one:

$$V(\boldsymbol{x},\boldsymbol{z}) = \boldsymbol{x}^T \left(\sum_{i=1}^{m} w_i(\boldsymbol{z}) P_i\right)^{-1} \boldsymbol{x} \tag{3.17}$$

with $P_i > 0$, $i = 1, 2, \ldots, m$. The first one reduces to a restrictive class of TS models, invoking path independency (Rhee and Won, 2006). The second reduces the problem of global stability to a problem of local stability (Bernal and Guerra, 2010).

Remark: Surprisingly, for the discrete case, very interesting results (Guerra and Vermeiren, 2004; Kruszewski et al., 2008) can be found for Lyapunov functions in the form of (3.16) or (3.17) that seem not to have their counterpart for continuous-time models.

3.4 State Feedback Stabilization

To stabilize a TS system using state feedback, several control laws can be used, among which the linear feedback $\boldsymbol{u} = -L\boldsymbol{x}$. A more general solution is a Parallel Distributed Compensation (PDC) scheme (Wang et al., 1996). The PDC is composed of linear state feedbacks blended together using the nonlinear functions $w_i(\cdot)$ of the model

$$\boldsymbol{u} = -\sum_{i=1}^{m} w_i(\boldsymbol{z}) L_i \boldsymbol{x} \tag{3.18}$$

Therefore, introducing (3.18) in the TS model (3.7) gives the closed loop

$$\begin{aligned}\dot{\boldsymbol{x}} &= \sum_{i=1}^{m} w_i(\boldsymbol{z}) \left(A_i - B_i \sum_{j=1}^{m} w_j(\boldsymbol{z}) L_j \right) \boldsymbol{x} \\ &= \sum_{i=1}^{m} w_i(\boldsymbol{z}) \left(\underbrace{\sum_{j=1}^{m} w_j(\boldsymbol{z})}_{=1} A_i - B_i \sum_{j=1}^{m} w_j(\boldsymbol{z}) L_j \right) \boldsymbol{x}\end{aligned}$$

and finally, the closed loop is composed of m^2 linear models

$$\dot{\boldsymbol{x}} = \sum_{i=1}^{m} \sum_{j=1}^{m} w_i(\boldsymbol{z}) w_j(\boldsymbol{z}) \left(A_i - B_i L_j \right) \boldsymbol{x} \tag{3.19}$$

Going on with quadratic stability (3.8) gives for the derivative of the Lyapunov function along the trajectories of (3.19)

$$\dot{V} = \sum_{i=1}^{m} \sum_{j=1}^{m} w_i(\boldsymbol{z}) w_j(\boldsymbol{z}) \boldsymbol{x}^T \left(\left(A_i - B_i L_j \right)^T P + P \left(A_i - B_i L_j \right) \right) \boldsymbol{x}$$

Therefore $\dot{V} < 0$ is ensured if the double sum negativity problem (3.3) is satisfied, which in this case can be written as

$$\sum_{i=1}^{m} \sum_{j=1}^{m} w_i(\boldsymbol{z}) w_j(\boldsymbol{z}) \left(A_i^T P + P A_i - P B_i L_j - L_j^T B_i^T P \right) < 0 \tag{3.20}$$

Note also that due to the quantity PB_iL_j, this expression is not an LMI. To express it with LMI conditions, the following change of variables can be performed: $X = P^{-1}$, $M_i = L_iX$, $i = 1, 2, \ldots, m$ and with the property of congruence with full rank matrix X (3.20) is equivalent to

$$\sum_{i=1}^{m}\sum_{j=1}^{m} w_i(\boldsymbol{z})w_j(\boldsymbol{z})\left(XA_i^T + A_iX - B_iM_j - M_j^TB_i^T\right) < 0$$

The result is summarized in the following theorem.

Theorem 3.3. *The continuous TS model* (3.7) *with the PDC control law* (3.18) *is GAS if there exist matrices* $X > 0$, *and* M_i, $i = 1, 2, \ldots, m$, *such that with* $\Gamma_{ij} \triangleq XA_i^T + A_iX - M_j^TB_i^T - B_iM_j$ *conditions* (3.5) *or* (3.6) *hold. Moreover, if the conditions are satisfied the PDC gains are:* $L_i = M_iX^{-1}$, $i = 1, 2, \ldots, m$.

Remark: With conditions (3.4), this result can be found in (Tanaka et al., 1998), with (3.5) it corresponds to (Tuan et al., 2001), and with (3.6) to (Liu and Zhang, 2005).

The application of Theorem 3.3 is illustrated on the following example.

Example 3.5. Consider the continuous TS model (3.7) composed of 2 linear models, with matrices $A_1 = \begin{pmatrix} -1 & -10 \\ 0 & 1 \end{pmatrix}$, $A_2 = \begin{pmatrix} -1 & -10 \\ 0 & -1 \end{pmatrix}$, $B_1 = \begin{pmatrix} 2 \\ 1 \end{pmatrix}$, $B_2 = \begin{pmatrix} 2 \\ 2 \end{pmatrix}$, $C_1 = \begin{pmatrix} -1 & 0 \end{pmatrix}$ and $C_2 = \begin{pmatrix} -1 & 1 \end{pmatrix}$. Note that A_1 has an unstable pole located at 1. For simulation purpose the membership functions are chosen as $w_1 = \frac{1}{1-(x_1+x_2)^2}$ and $w_2 = 1 - w_1$.

Using Theorem 3.3 with conditions (3.5) gives the solution[2]: $P = \begin{pmatrix} 2.60 & -0.37 \\ -0.37 & 117.33 \end{pmatrix}$, $\begin{pmatrix} L_1 \\ L_2 \end{pmatrix} = \begin{pmatrix} -0.43 & 7.08 \\ -0.38 & 7.15 \end{pmatrix}$. Figure 3.2(a) presents a result[3] with initial conditions $\boldsymbol{x}(0) = \begin{pmatrix} 1 & -2 \end{pmatrix}^T$. Notice that without additional constraints the performance is rather poor, especially for the state variable x_1. In a second round a decay rate $\alpha = 4$ is added to the constraints. The maximum possible decay rate is $\alpha < 11$. The result is now $P = \begin{pmatrix} 11.16 & -35.03 \\ -35.03 & 125.85 \end{pmatrix}$, $\begin{pmatrix} L_1 \\ L_2 \end{pmatrix} = \begin{pmatrix} -12.62 & 46.04 \\ -3.49 & 13.23 \end{pmatrix}$ and the results are presented in Figure 3.2(b). Obviously, better performance is obtained. □

The classical nonlinear TS model (3.7) represents the simplest case of TS representation. More general forms can be found: uncertain systems (parametric and/or stochastic) (Xu et al., 2007; Tanaka and Wang, 2001) with or without H_2, H_∞ specifications (Tanaka and Wang, 2001; Liu and Zhang, 2003; Delmotte et al., 2007; Wu, 2007), delayed systems (Yoneyama, 2007; Liu et al., 2010), systems with bounded external disturbances (Tanaka and Wang, 2001; Zhou and Feng, 2006),

[2] Throughout the chapter, all variables are rounded to two decimal places.

[3] For numerical integration, the *ode45* Matlab function is used throughout the chapter.

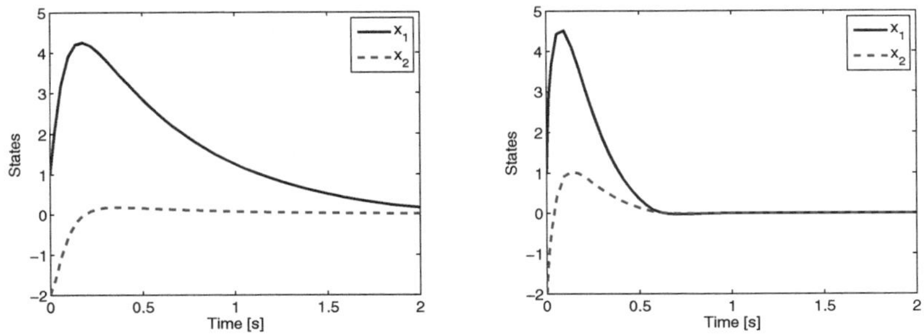

(a) Stabilization without constraints. (b) Stabilization with decay rate $\alpha = 4$.

Fig. 3.2 Simulation results for Example 3.5.

systems in a descriptor form (Taniguchi et al., 2001; Xu et al., 2007; Guelton et al., 2008a,b), periodic systems (Kruszewski and Guerra, 2007), and switching systems (Dong and Yang, 2009; Wang and Qu, 2007; Choi and Park, 2004). For overviews, the reader can refer to (Tanaka and Wang, 2001; Feng, 2006; Sala et al., 2005). For example, consider an uncertain TS model with external disturbances

$$\begin{aligned} \dot{\boldsymbol{x}} &= \sum_{i=1}^{m} w_i(\boldsymbol{z}) \left((A_i + \Delta A_i)\, \boldsymbol{x} + (B_i + \Delta B_i)\, \boldsymbol{u} + B_{di}\boldsymbol{d} \right) \\ \boldsymbol{y} &= \sum_{i=1}^{m} w_i(\boldsymbol{z}) \left((C_i + \Delta C_i)\, \boldsymbol{x} + D_{di}\boldsymbol{d} \right) \end{aligned} \tag{3.21}$$

where $\boldsymbol{d}(t) \in \mathbb{R}^{n_d}$ is the disturbance input and A_i, B_i, C_i, B_{di}, and D_{di}, $i = 1, 2, \ldots, m$, are matrices with appropriate dimensions. The parameter uncertainties are usually written as (Tanaka and Wang, 2001): $\Delta A_i = H_a \Delta a(t) E_{ai}$, $\Delta B_i = H_b \Delta b(t) E_{bi}$, $\Delta C_i = H_c \Delta c(t) E_{ci}$ where the matrices H_a, H_b, H_c, E_{ai}, E_{bi}, and E_{ci} are constants, and $\Delta a(t)$, $\Delta b(t)$, $\Delta c(t)$ satisfy the conditions

$$\Delta a^T(t)\Delta a(t) \leq I \quad \Delta b^T(t)\Delta b(t) \leq I \quad \Delta c^T(t)\Delta c(t) \leq I \tag{3.22}$$

Therefore, according to what parts of the model (3.22) are considered and what the desired goals are (robustness, performances, etc.) numerous results exist. Among them, in order to show the main principles of proofs for these results we consider two particular examples: a first one for H_∞ attenuation of input signals, the second one for robust stabilization conditions according to uncertainties.

3.4.1 H_∞ Attenuation

As previously said, performance-related criteria can be added by using extra LMI constraints. In a more general way, it is possible to cope with H_2 or H_∞ criteria. In this latter case consider model (3.21) without uncertainties, $D_{di} = 0$ and with initial conditions $\boldsymbol{x}(0) = 0$, i.e.,

$$\dot{\boldsymbol{x}} = \sum_{i=1}^{m} w_i(\boldsymbol{z})(A_i\boldsymbol{x} + B_i\boldsymbol{u} + B_{di}\boldsymbol{d})$$
$$\boldsymbol{y} = \sum_{i=1}^{m} w_i(\boldsymbol{z})C_i\boldsymbol{x} \tag{3.23}$$

The goal is to find the best $L_2 \to L_2$ gain, i.e., to minimize the worst case

$$\sup_{\boldsymbol{d}(t)\neq 0} \frac{\|\boldsymbol{y}(t)\|_2}{\|\boldsymbol{d}(t)\|_2} \leq \gamma \tag{3.24}$$

Theorem 3.4. *The continuous TS model* (3.23) *with the PDC control law* (3.18) *is GAS and the attenuation of the disturbance* $\mathbf{d}$ *is at least* γ, *if there exist matrices* $X > 0$, *and* M_i, $i = 1, 2, \ldots, m$, *such that with*

$$\Gamma_{ij} \triangleq \begin{pmatrix} XA_i^T + A_iX - M_j^TB_i^T - B_iM_j & B_{di} & XC_i^T \\ B_{di}^T & -\gamma^2 I & 0 \\ C_iX & 0 & -I \end{pmatrix}$$

conditions (3.5) *or* (3.6) *hold. Moreover, if the conditions are satisfied, then the PDC gains are* $L_i = M_iX^{-1}$, $i = 1, 2, \ldots, m$.

Proof: Consider a quadratic Lyapunov function V such that

$$\dot{V} + \boldsymbol{y}^T\boldsymbol{y} - \gamma^2\boldsymbol{d}^T\boldsymbol{d} \leq 0$$

Thus, integrating this expression leads to $V(\boldsymbol{x}(\infty)) - V(\boldsymbol{x}(0)) \leq \int_0^\infty \left(\gamma^2\boldsymbol{d}^T\boldsymbol{d} - \boldsymbol{y}^T\boldsymbol{y}\right) dt$. Since the TS is assumed to be GAS, $\boldsymbol{x}(\infty) = 0$ and with initial conditions such that $\boldsymbol{x}(0) = 0$, we obtain

$$0 < \int_0^\infty \left(\gamma^2\boldsymbol{d}^T\boldsymbol{d} - \boldsymbol{y}^T\boldsymbol{y}\right) dt$$

which is equivalent to

$$\int_0^\infty \boldsymbol{y}^T\boldsymbol{y}dt < \gamma^2 \int_0^\infty \boldsymbol{d}^T\boldsymbol{d}dt \Leftrightarrow \|\boldsymbol{y}\|_2 < \gamma \|\boldsymbol{d}\|_2$$

and γ satisfies (3.24). Now, the stability conditions are obtained as

$$\begin{aligned}
&\dot{V} + \boldsymbol{y}^T\boldsymbol{y} - \gamma^2\boldsymbol{d}^T\boldsymbol{d} = \\
&2\boldsymbol{x}^T\sum_{i=1}^{m}\sum_{j=1}^{m} w_i(\boldsymbol{z})w_j(\boldsymbol{z})\left(P\left(\left(A_i - B_iL_j\right)\boldsymbol{x} + B_{di}\boldsymbol{d}\right)\right) \\
&+\boldsymbol{x}^T\sum_{i=1}^{m}\sum_{j=1}^{m} w_i(\boldsymbol{z})w_j(\boldsymbol{z})C_i^TC_j\boldsymbol{x} - \gamma^2\boldsymbol{d}^T\boldsymbol{d} \\
&= \begin{pmatrix}\boldsymbol{x}\\ \boldsymbol{d}\end{pmatrix}^T \begin{pmatrix}\sum_{i=1}^{m}\sum_{j=1}^{m} w_i(\boldsymbol{z})w_j(\boldsymbol{z})\left(Y + C_i^TC_j\right) & (*)\\ \sum_i^m w_i(\boldsymbol{z})B_{di}^TP & -\gamma^2 I\end{pmatrix}\begin{pmatrix}\boldsymbol{x}\\ \boldsymbol{d}\end{pmatrix}
\end{aligned}$$

with $Y = \mathcal{H}(P\left(A_i - B_iL_j\right))$, where $\mathcal{H}(X) = X + X^T$. Using the Schur complement, $\dot{V} + \boldsymbol{y}^T\boldsymbol{y} - \gamma^2\boldsymbol{d}^T\boldsymbol{d} < 0$ is equivalent to

$$\begin{pmatrix}\sum_{i=1}^{m}\sum_{j=1}^{m} w_i(\boldsymbol{z})w_j(\boldsymbol{z})\mathcal{H}\left(P\left(A_i - B_iL_j\right)\right) & (*) & (*)\\ \sum_i^m w_i(\boldsymbol{z})B_{di}^TP & -\gamma^2 I & 0\\ \sum_i^m w_i(\boldsymbol{z})C_i & 0 & -I\end{pmatrix} < 0$$

Using the property of congruence with full rank matrix $\begin{pmatrix}X & 0 & 0\\ 0 & I & 0\\ 0 & 0 & I\end{pmatrix}$ gives with $X = P^{-1}$ and $M_i = L_iX$, $i = 1, 2, \ldots, m$,

$$\sum_{i=1}^{m}\sum_{j=1}^{m} w_i(\boldsymbol{z})w_j(\boldsymbol{z})\begin{pmatrix}XA_i^T + A_iX - M_j^TB_i^T - B_iM_j & (*) & (*)\\ B_{di}^T & -\gamma^2 I & 0\\ C_iX & 0 & -I\end{pmatrix} < 0$$

thereby recovering the expressions Γ_{ij} of the theorem and concluding the proof. □

Remark: Several results based on this approach exist (Liu and Zhang, 2005; Delmotte et al., 2007). For example, replacing $\boldsymbol{y} = \sum_{i=1}^{m} w_i\left(\boldsymbol{z}\right)C_i\boldsymbol{x}$ of (3.23) with

$$\boldsymbol{y} = \sum_{i=1}^{m} w_i(\boldsymbol{z})\left(C_i\boldsymbol{x} + D_{di}\boldsymbol{d}\right)$$

leads to a similar result by replacing Γ_{ij} with:

$$\Gamma_{ij} \triangleq \begin{pmatrix}XA_i^T + A_iX - M_j^TB_i^T - B_iM_j & B_{di} & XC_i^T\\ B_{di}^T & -\gamma^2 I & D_{di}^T\\ C_iX & D_{di} & -I\end{pmatrix}$$

The design of a controller with disturbance attenuation is illustrated by using the following example.

Example 3.6. Consider the TS fuzzy system of the form (3.23), with two local models and the matrices given as in Example 3.5, i.e., $A_1 = \begin{pmatrix}-1 & -10\\ 0 & 1\end{pmatrix}$,

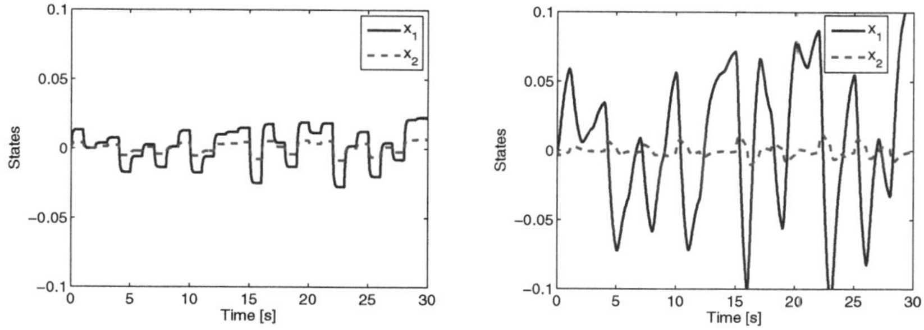

(a) Stabilization with attenuation. (b) Stabilization without attenuation.

Fig. 3.3 Simulation results for Example 3.6.

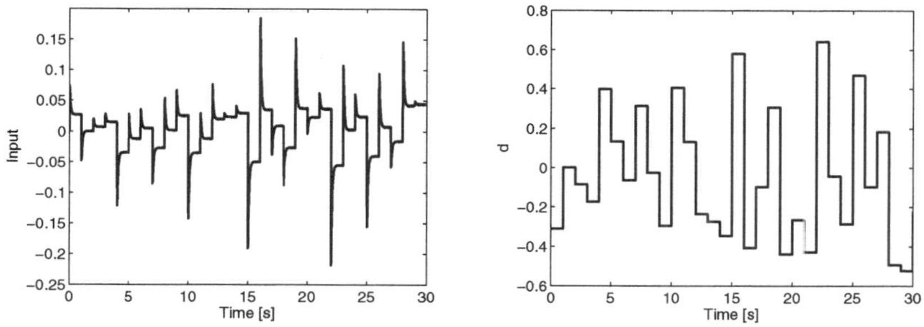

(a) Control signal u. (b) Disturbance signal d.

Fig. 3.4 Control and disturbance signals for Example 3.6.

$A_2 = \begin{pmatrix} -1 & -10 \\ 0 & -1 \end{pmatrix}$, $B_1 = \begin{pmatrix} 2 \\ 1 \end{pmatrix}$, $B_2 = \begin{pmatrix} 2 \\ 2 \end{pmatrix}$, $C_1 = \begin{pmatrix} -1 & 0 \end{pmatrix}$, $C_2 = \begin{pmatrix} -1 & 1 \end{pmatrix}$, $B_{d1} = \begin{pmatrix} 0 \\ 0.1 \end{pmatrix}$ and $B_{d2} = \begin{pmatrix} -0.1 \\ 0.1 \end{pmatrix}$, and membership functions $w_1 = \frac{1}{1+(x_1+x_2)^2}$ and $w_2 = 1 - w_1$.

The highest H_∞ attenuation, i.e., the smallest γ, is $\gamma_{\min} = 0.0342$. For example, with a fixed $\gamma = 0.05$ the results are presented in Figure 3.3(a), while Figures 3.4(a) and 3.4(b) present the control signal and the generated disturbance signal d (band-limited white noise with power 0.1 and sample time 1). For the sake of comparison, Figure 3.3(b) presents the results without considering H_∞ attenuation, with the same gains as those computed in Example 3.5 without the decay rate. □

3.4.2 *Robust Control*

Several results concerning robust control can be found in the literature. In order to show the main principle we will consider the very basic norm-bounded uncertain

TS model (Tanaka and Wang, 2001), (3.21) without external signals, i.e., $B_{di} = 0$ and $D_{di} = 0$:

$$\begin{aligned} \dot{\boldsymbol{x}} &= \sum_{i=1}^{m} w_i(\boldsymbol{z}) \left((A_i + \Delta A_i) \boldsymbol{x} + (B_i + \Delta B_i) \boldsymbol{u} \right) \\ \boldsymbol{y} &= \sum_{i=1}^{m} w_i(\boldsymbol{z}) (C_i + \Delta C_i) \boldsymbol{x} \end{aligned} \tag{3.25}$$

and the uncertainties described as in (3.22). The goal is to derive stability conditions for the closed-loop system for every variation of the uncertainties in their domain of variation. Considering again the PDC law (3.18) the state feedback closed-loop is

$$\dot{\boldsymbol{x}} = \sum_{i=1}^{m} \sum_{j=1}^{m} w_i(\boldsymbol{z}) w_j(\boldsymbol{z}) (A_i - B_i L_j + \Delta A_i - \Delta B_i L_j) \boldsymbol{x}$$

Theorem 3.5. *The continuous TS model* (3.25) *with the PDC control law* (3.18) *is robustly GAS, i.e., is GAS whenever the uncertainties satisfy the boundary conditions* (3.22)*, if there exist matrices* $X > 0$*, and* M_i*,* $i = 1, 2, \ldots, m$*, and scalars* $\tau_a > 0$ *and* $\tau_b > 0$*, such that with*

$$\Gamma_{ij} \triangleq \begin{pmatrix} \mathcal{H}(A_i X - B_i M_j) + \tau_a H_a H_a^T + \tau_b H_b H_b^T & X E_{ai}^T & -M_j^T E_{bi}^T \\ E_{ai} X & -\tau_a I & 0 \\ -E_{bi} M_j & 0 & -\tau_b I \end{pmatrix}$$

the conditions (3.5) *or* (3.6) *hold. Moreover, if the conditions are satisfied, then the PDC gains are* $L_i = M_i X^{-1}$*,* $i = 1, 2, \ldots, m$*.*

Proof: Using the quadratic Lyapunov function (3.8) gives directly

$$\dot{V} = 2\boldsymbol{x}^T \sum_{i=1}^{m} \sum_{j=1}^{m} w_i(\boldsymbol{z}) w_j(\boldsymbol{z}) P (A_i - B_i L_j + \Delta A_i - \Delta B_i L_j) \boldsymbol{x}$$

Therefore $\dot{V} < 0$ if and only if

$$\sum_{i=1}^{m} \sum_{j=1}^{m} w_i(\boldsymbol{z}) w_j(\boldsymbol{z}) \mathcal{H}(P A_i - P B_i L_j + P \Delta A_i - P \Delta B_i L_j) < 0$$

Using the property of congruence with full a rank matrix $X = P^{-1}$ and using the change of variables $M_j = L_j X$, $j = 1, 2, \ldots, m$, gives:

$$\sum_{i=1}^{m} \sum_{j=1}^{m} w_i(\boldsymbol{z}) w_j(\boldsymbol{z}) \mathcal{H}(A_i X - B_i M_j + \Delta A_i X - \Delta B_i M_j) < 0 \tag{3.26}$$

Consider

$$\Theta_j = \left(\sum_{i=1}^{m} w_i(\boldsymbol{z})\Delta A_i\right) X + X\left(\sum_{i=1}^{m} w_i(\boldsymbol{z})\Delta A_i\right)^T + \left(\sum_{i=1}^{m} w_i(\boldsymbol{z})\Delta B_i\right) M_j + M_j^T\left(\sum_{i=1}^{m} w_i(\boldsymbol{z})\Delta B_i\right)^T$$

with the uncertainties described as

$$\Theta_j = (H_a\ H_b)\begin{pmatrix}\Delta_a & 0\\ 0 & \Delta_b\end{pmatrix}\sum_{i=1}^{m} w_i(\boldsymbol{z})\begin{pmatrix}E_{ai}X\\ -E_{bi}M_j\end{pmatrix} + \sum_{i=1}^{m} w_i(\boldsymbol{z})\left(XE_{ai}^T\ -M_j^TE_{bi}^T\right)\begin{pmatrix}\Delta_a^T & 0\\ 0 & \Delta_b^T\end{pmatrix}\begin{pmatrix}H_a^T\\ H_b^T\end{pmatrix}$$

Using Property 3.4 (completion of squares) with $Q = \begin{pmatrix}\tau_a I & 0\\ 0 & \tau_b I\end{pmatrix}$ gives the bound

$$\sum_{j=1}^{m} w_j(\boldsymbol{z})\Theta_j \le (H_a\ H_b)\begin{pmatrix}\Delta_a & 0\\ 0 & \Delta_b\end{pmatrix}\begin{pmatrix}\tau_a I & 0\\ 0 & \tau_b I\end{pmatrix}\begin{pmatrix}\Delta_a^T & 0\\ 0 & \Delta_b^T\end{pmatrix}\begin{pmatrix}H_a^T\\ H_b^T\end{pmatrix} + \left(\sum_{i=1}^{m}\sum_{j=1}^{m} w_i(\boldsymbol{z})w_j(\boldsymbol{z})\left(XE_{ai}^T\ -M_j^TE_{bi}^T\right)\right) \cdot \begin{pmatrix}\tau_a^{-1}I & 0\\ 0 & \tau_b^{-1}I\end{pmatrix}\left(\sum_{i=1}^{m}\sum_{j=1}^{m} w_i(\boldsymbol{z})w_j(\boldsymbol{z})\begin{pmatrix}E_{ai}X\\ -E_{bi}M_j\end{pmatrix}\right)$$

Together with conditions (3.22), we have

$$\sum_{j=1}^{m} w_j(\boldsymbol{z})\Theta_j \le (H_a\ H_b)\begin{pmatrix}\tau_a I & 0\\ 0 & \tau_b I\end{pmatrix}\begin{pmatrix}H_a^T\\ H_b^T\end{pmatrix} + \left(\sum_{i=1}^{m}\sum_{j=1}^{m} w_i(\boldsymbol{z})w_j(\boldsymbol{z})\left(XE_{ai}^T\ -M_j^TE_{bi}^T\right)\right) \cdot \begin{pmatrix}\tau_a^{-1}I & 0\\ 0 & \tau_b^{-1}I\end{pmatrix}\left(\sum_{i=1}^{m}\sum_{j=1}^{m} w_i(\boldsymbol{z})w_j(\boldsymbol{z})\begin{pmatrix}E_{ai}X\\ -E_{bi}M_j\end{pmatrix}\right)$$

and (3.26) is satisfied if

$$\sum_{i=1}^{m}\sum_{j=1}^{m} w_i(\boldsymbol{z})w_j(\boldsymbol{z})\mathcal{H}(A_iX - B_iM_j) + \begin{pmatrix} H_a & H_b \end{pmatrix}\begin{pmatrix} \tau_a I & 0 \\ 0 & \tau_b I \end{pmatrix}\begin{pmatrix} H_a^T \\ H_b^T \end{pmatrix}$$
$$+ \left(\sum_{i=1}^{m}\sum_{j=1}^{m} w_i(\boldsymbol{z})w_j(\boldsymbol{z})\begin{pmatrix} XE_{ai}^T & -M_j^T E_{bi}^T \end{pmatrix}\right)$$
$$\cdot \begin{pmatrix} \tau_a^{-1} I & 0 \\ 0 & \tau_b^{-1} I \end{pmatrix}\left(\sum_{i=1}^{m}\sum_{j=1}^{m} w_i(\boldsymbol{z})w_j(\boldsymbol{z})\begin{pmatrix} E_{ai}X \\ -E_{bi}M_j \end{pmatrix}\right) < 0$$

Using the Schur complement, we obtain

$$\sum_{i=1}^{m}\sum_{j=1}^{m} w_i(\boldsymbol{z})w_j(\boldsymbol{z}) \cdot \begin{pmatrix} \mathcal{H}(A_iX - B_iM_j) + \tau_a H_a H_a^T + \tau_b H_b H_b^T & XE_{ai}^T & -M_j^T E_{bi}^T \\ E_{ai}X & -\tau_a I & 0 \\ -E_{bi}M_j & 0 & -\tau_b I \end{pmatrix} < 0 \tag{3.27}$$

which completes the proof. □

Remark: The conservativeness of the result above can be reduced by using additional slack variables such as $Q = \sum_{i=1}^{m}\sum_{j=1}^{m} w_i(\boldsymbol{z})\, w_j(\boldsymbol{z}) \begin{pmatrix} \tau_{aij} I & 0 \\ 0 & \tau_{bij} I \end{pmatrix}$, which introduces extra degrees of freedom in (3.27).

Example 3.7. Consider the TS fuzzy system of the form (3.25), with two local models and the matrices given as $A_1 = \begin{pmatrix} -1 & -10 \\ 0 & 1 \end{pmatrix}$, $A_2 = \begin{pmatrix} -1 & -10 \\ 0 & -1 \end{pmatrix}$, $B_1 = \begin{pmatrix} 2 \\ 1 \end{pmatrix}$, $B_2 = \begin{pmatrix} 2 \\ 2 \end{pmatrix}$, $C_1 = \begin{pmatrix} -1 & 0 \end{pmatrix}$, $C_2 = \begin{pmatrix} -1 & 1 \end{pmatrix}$, and the uncertainty matrices $H_a = \begin{pmatrix} 0 \\ 1 \end{pmatrix}$, $H_b = 1$, $E_{a1} = \begin{pmatrix} 0 & 0.5 \end{pmatrix}$, $E_{a2} = \begin{pmatrix} -0.5 & 0.5 \end{pmatrix}$ and $E_{b1} = E_{b2} = \begin{pmatrix} 0 & 0.5 \end{pmatrix}^T$, and membership functions $w_1 = \frac{1}{1+(x_1+x_2)^2}$ and $w_2 = 1 - w_1$.

Using Theorem 3.5, the solution obtained is $P = \begin{pmatrix} 0.017 & -0.14 \\ -0.14 & 1.67 \end{pmatrix}$, $\begin{pmatrix} L_1 \\ L_2 \end{pmatrix} = \begin{pmatrix} -10.28 & 112.4 \\ -8.23 & 91.97 \end{pmatrix}$.

Consider also the uncertainty signals $\Delta a(t)$ and $\Delta b(t)$, with $\Delta a(t) = \Delta b(t)$, as a band-limited white noise of power 0.1 with sample time $1s$, given in Figure 3.5(a). The simulation results are presented in Figure 3.5(b). The initial states were $(0\ 0)^T$. In order to the simulation to exhibit a good behavior (i.e., not all signals to converge to zero), the control law includes an external constant signal. In fact, we replace the PDC law (3.18) with $\boldsymbol{u} = H\boldsymbol{y}_c - \sum_{i=1}^{m} w_i(\boldsymbol{z})L_i\boldsymbol{x}$, where $H = 10$, $\boldsymbol{y}_c = -1$. Such input signals are discussed in Section 3.6. □

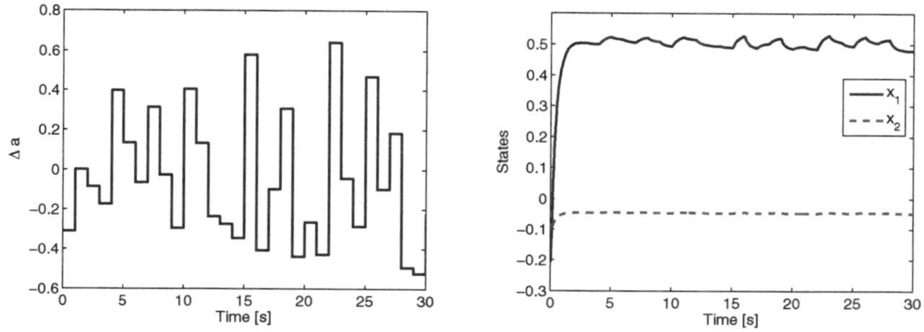

(a) Uncertainty signals ($\Delta a = \Delta b$). (b) Stabilization results.

Fig. 3.5 Simulation results for Example 3.7.

3.5 Output Feedback Stabilization

The state vector is rarely entirely measured and extra work is needed to go from state feedback to output feedback. Two ways are possible: 1) by adding an observer or 2) by considering a static or a dynamic output feedback directly. The former will be described in Chapter 4, the latter is briefly outlined in what follows.

A static output feedback can be written as (Kau et al., 2007; Lo and Lin, 2003)

$$\boldsymbol{u} = -\sum_{i=1}^{m} w_i(\boldsymbol{z}) L_i \boldsymbol{y} \tag{3.28}$$

with $L_i \in \mathbb{R}^{n_u \times n_y}$, $i = 1, 2, \ldots, m$ the control gains. Introducing (3.28) into the simplest model (3.7) leads to the closed-loop:

$$\dot{\boldsymbol{x}} = \sum_{i=1}^{m}\sum_{j=1}^{m}\sum_{k=1}^{m} w_i(\boldsymbol{z}) w_j(\boldsymbol{z}) w_k(\boldsymbol{z}) (A_i - B_i L_j C_k)\boldsymbol{x}$$

Therefore, a triple sum occurs. Several authors relax the problem considering a common output matrix (Lo and Lin, 2003), i.e. $C_i = C$, $i = 1, 2, \ldots, m$.

A dynamic output feedback controller can be written as (Ding, 2009)

$$\begin{aligned} \dot{\boldsymbol{x}}_{\mathrm{c}} &= \sum_{i=1}^{m}\sum_{j=1}^{m} w_i(\boldsymbol{z}) w_j(z) A_{\mathrm{c}ij} \boldsymbol{x}_{\mathrm{c}} + \sum_{i=1}^{m} w_i(\boldsymbol{z}) B_{\mathrm{c}i} \boldsymbol{y} \\ \boldsymbol{u} &= \sum_{i=1}^{m} w_i(\boldsymbol{z}) C_{\mathrm{c}i} \boldsymbol{x}_{\mathrm{c}} + D_{\mathrm{c}} \boldsymbol{y} \end{aligned} \tag{3.29}$$

with $\boldsymbol{x}_{\mathrm{c}} \in \mathbb{R}^{n_{\mathrm{c}}}$, $A_{\mathrm{c}ij}$, $B_{\mathrm{c}i}$, $C_{\mathrm{c}i}$, $i, j = 1, 2, \ldots, m$, and D_{c} matrices of appropriate dimension to be designed.

Then, for example substituting (3.29) in the general model (3.21) without uncertainties leads to the closed-loop

$$\begin{pmatrix} \dot{\boldsymbol{x}} \\ \dot{\boldsymbol{x}}_{\mathrm{c}} \end{pmatrix} = \sum_{i=1}^{m} \sum_{j=1}^{m} w_i(\boldsymbol{z}) w_j(\boldsymbol{z}) \cdot \left(\begin{pmatrix} A_i + B_i D_{\mathrm{c}} C_j & B_i C_{\mathrm{c}j} \\ B_{\mathrm{c}j} C_i & A_{\mathrm{c}ij} \end{pmatrix} \begin{pmatrix} \boldsymbol{x} \\ \boldsymbol{x}_{\mathrm{c}} \end{pmatrix} + \begin{pmatrix} B_i D_{\mathrm{c}} D_{dj} + B_{di} \\ B_{\mathrm{c}j} D_{di} \end{pmatrix} \boldsymbol{d} \right)$$

Note that the matrices are affine in the variables $A_{\mathrm{c}ij}$, $B_{\mathrm{c}i}$, $C_{\mathrm{c}i}$, D_{c} and that the control law is written in a way that the closed-loop has only m^2 models, thus avoiding a triple sum. Several other possibilities exist (Guelton et al., 2009).

The main problem encountered when using these approaches is that an equivalent LMI formulation is very hard, if not impossible to derive, except for restricted classes of TS models (without uncertainties and delays). Some results using a descriptor redundancy approach exist (Guelton et al., 2009) but they are generally over conservative. Moreover, the scheduling vector $\boldsymbol{z}$ is assumed to be measurable, otherwise the problem would be harder.

3.6 Input-to-State Stability

Generally speaking, Lyapunov stability gives a result regarding the response to initial conditions. Of course, when dealing with exogenous inputs, input-to-state stability (ISS) should be considered (Sontag and Wang, 1995). Fortunately, when dealing with TS models, the vector field $\sum_{i=1}^{m} w_i(\boldsymbol{z}) B_i$ is bounded and the ISS property holds for systems that are GAS in the sense of Lyapunov. We recall the ISS property, which also requires the following definition.

Definition 3.2. A continuous function $\alpha : \mathbb{R}^+ \to \mathbb{R}^+$ belongs to *class* $\mathcal{K}$ if and only if it is strictly increasing and $\alpha(0) = 0$. If, in addition, $\alpha(s) \to \infty$ when $s \to \infty$ then α is said to be of *class* $\mathcal{K}^\infty$. □

Property 3.6. (Input-to-state stability) (Sontag, 1995) Consider a nonlinear model $\dot{\boldsymbol{x}} = \boldsymbol{f}(\boldsymbol{x}) + \boldsymbol{g}^{\mathrm{m}}(\boldsymbol{x})\boldsymbol{u}$, and a Lyapunov function candidate V, i.e., $V : \mathbb{R}^{n_x} \to \mathbb{R}^+$, $V(0) = 0$, and $V(\boldsymbol{x}) > 0$, $\forall \boldsymbol{x} \neq 0$). If there exist two $\mathcal{K}^\infty$ functions, $\alpha(\cdot)$ and $\theta(\cdot)$ such that $\dot{V}(\boldsymbol{x}) \leq \theta(\|\boldsymbol{u}\|) - \alpha(\|\boldsymbol{x}\|)$ for all $\boldsymbol{x} \in \mathbb{R}^{n_x}$ and $\boldsymbol{u} \in \mathbb{R}^{n_u}$, then the model is input-to-state stable.

Consider again the uncertain model (3.25) and a control law including an exogenous input $\boldsymbol{y}_{\mathrm{c}}$

$$\boldsymbol{u} = \sum_{i=1}^{m} w_i(\boldsymbol{z}) \left(H_i \boldsymbol{y}_{\mathrm{c}} - L_i \boldsymbol{x} \right) \tag{3.30}$$

and assume that there exists a quadratic Lyapunov function V showing that the model without the exogenous input is GAS, i.e., for $\boldsymbol{y}_c = 0$ we have

$$\dot{V} < 0 \Leftrightarrow \exists \lambda > 0,\ \dot{V} < -\lambda \|\boldsymbol{x}\|^2$$

Then for $\boldsymbol{y}_{\mathrm{c}}(t) \neq 0$ we have

$$\begin{aligned}
\dot{V}(\boldsymbol{x}) &\leq -\lambda \|\boldsymbol{x}\|^2 + 2\boldsymbol{x}^T \sum_{i=1}^{m}\sum_{j=1}^{m} w_i(\boldsymbol{z}) w_j(\boldsymbol{z}) P \left(B_i + \Delta B_i\right) H_j \boldsymbol{y}_{\mathrm{c}} \\
&\leq -\lambda \|\boldsymbol{x}\|^2 + 2 \left\| \sum_{i=1}^{m}\sum_{j=1}^{m} w_i(\boldsymbol{z}) w_j(\boldsymbol{z}) P \left(B_i + \Delta B_i\right) H_j \right\| \|\boldsymbol{y}_{\mathrm{c}}\| \|\boldsymbol{x}\| \\
&= -\lambda \|\boldsymbol{x}\|^2 + 2\delta \|\boldsymbol{y}_{\mathrm{c}}\| \|\boldsymbol{x}\|
\end{aligned} \tag{3.31}$$

The constant δ is bounded by the definition of TS models, and therefore (3.31) can be written as

$$\begin{aligned}
\dot{V}(\boldsymbol{x}) &\leq -\frac{\lambda}{2} \|\boldsymbol{x}\|^2 - \left(\sqrt{\frac{\lambda}{2}} \|\boldsymbol{x}\| - \delta \sqrt{\frac{2}{\lambda}} \|\boldsymbol{y}_{\mathrm{c}}\| \right)^2 + \frac{2\delta^2}{\lambda} \|\boldsymbol{y}_{\mathrm{c}}\|^2 \\
&\leq -\frac{\lambda}{2} \|\boldsymbol{x}\|^2 + \frac{2\delta^2}{\lambda} \|\boldsymbol{y}_{\mathrm{c}}\|^2
\end{aligned}$$

Finally, with $\alpha\left(\|\boldsymbol{x}\|\right) = \frac{\lambda}{2} \|\boldsymbol{x}\|^2$ and $\theta\left(\|\boldsymbol{y}_{\mathrm{c}}\|\right) = \frac{2\delta^2}{\lambda} \|\boldsymbol{y}_{\mathrm{c}}\|^2$, the model has the ISS property.

Remark: The result above implies that finding a solution using a standard LMI problem for TS models will guarantee the ISS stability.

At last, the necessity of adding integrators for steady-state purposes can lead to a more general scheme, such as the one presented in Figure 3.6. Consider the general TS model (3.21). The integral part corresponds to

$$\dot{\boldsymbol{x}}_{\mathrm{I}} = \boldsymbol{y}_{\mathrm{c}} - \sum_{i=1}^{m} w_i(\boldsymbol{z}) \left(\left(C_i + \Delta C_i\right) \boldsymbol{x} + D_{di} \boldsymbol{d} \right)$$

Introducing the extended vector $\bar{\boldsymbol{x}} = \begin{pmatrix} \boldsymbol{x} \\ \boldsymbol{x}_{\mathrm{I}} \end{pmatrix}$ together with control law

$$\boldsymbol{u} = -\sum_{i=1}^{m} w_i(\boldsymbol{z}) \bar{L}_i \bar{\boldsymbol{x}} = -\sum_{i=1}^{m} w_i(\boldsymbol{z}) \left(F_i \ L_i\right) \begin{pmatrix} \boldsymbol{x} \\ \boldsymbol{x}_{\mathrm{I}} \end{pmatrix}$$

it is straightforward to write the extended state representation as

$$\begin{aligned}
\dot{\bar{\boldsymbol{x}}} &= \sum_{i=1}^{m} w_i(\boldsymbol{z}) \left((\bar{A}_i + \Delta \bar{A}_i) \bar{\boldsymbol{x}} + (\bar{B}_i + \Delta \bar{B}_i) \boldsymbol{u} + \bar{B}_{di} \boldsymbol{d} \right) + B_y \boldsymbol{y}_{\mathrm{c}} \\
\boldsymbol{y} &= \sum_{i=1}^{m} w_i(\boldsymbol{z}) \left((\bar{C}_i + \Delta \bar{C}_i) \bar{\boldsymbol{x}} + \bar{D}_{di} \boldsymbol{d} \right)
\end{aligned}$$

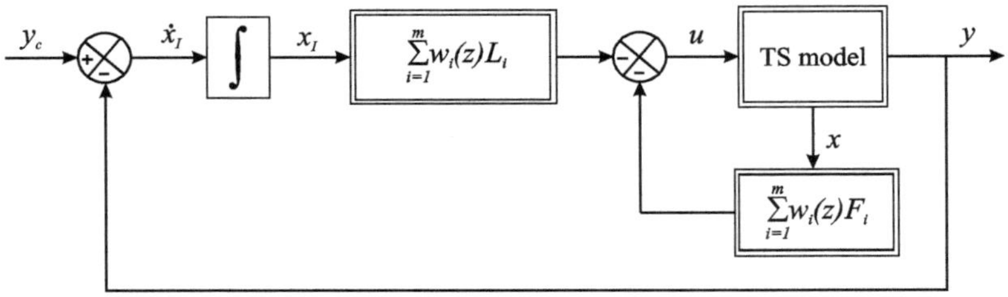

Fig. 3.6 PDC control with an integrator.

with matrices $\bar{A}_i = \begin{pmatrix} A_i & 0 \\ -C_i & 0 \end{pmatrix}$, $\Delta\bar{A}_i = \begin{pmatrix} \Delta A_i & 0 \\ -\Delta C_i & 0 \end{pmatrix}$, $\bar{B}_i = \begin{pmatrix} B_i \\ 0 \end{pmatrix}$, $\Delta\bar{B}_i = \begin{pmatrix} \Delta B_i \\ 0 \end{pmatrix}$, $\bar{B}_{di} = \begin{pmatrix} B_{di} \\ D_{di} \end{pmatrix}$, $B = \begin{pmatrix} 0 \\ I \end{pmatrix}$, $\bar{C}_i = (C_i \ 0)$, and $\Delta\bar{C}_i = (\Delta C_i \ 0)$.

Therefore, all the previous robustness and performance results apply to this extended representation, including the ISS stability according to the exogenous output $\boldsymbol{y}_c$.

3.7 Summary

The aim of this chapter was to show the interest of using Takagi-Sugeno fuzzy models for control. In order to make it short and easily readable, this chapter focused on the stability analysis and state feedback control for particular classes of TS models. Nevertheless, the proofs given along the chapter for H_∞ performance or robust control are generic ones and most of the time they can be easily extended to a more general class of TS models (mixing uncertainties and exogenous signals, introducing delays on the state, the inputs, etc.).

Additional properties such as D-stability, and input-to-state stability were also presented. Natural limitations of the approach, such as quadratic stability and the LMI conditions being independent of the membership functions $w_i(\boldsymbol{z})$ have been discussed. A very important point to achieve the control objectives is therefore to estimate the state vector, thus going from state feedback to output feedback schemes. Therefore, the next chapters focus on the design of observers based on TS models.

Chapter 4
Observers for TS Fuzzy Systems

In practical situations, not all state variables of a given system can be measured. In such cases, an observer has to be designed to estimate the unmeasured states based on the system model and the available input-output data. This chapter introduces the concept of observers used for TS fuzzy systems and reviews methods for designing observers such that the estimation error asymptotically converges to zero. We also briefly describe observer-based stabilization.

4.1 Observer Design for TS Systems

When the whole state information is needed, an observer that is able to estimate the unmeasured variables has to be designed. Once an estimate of the states is available, they can be further used, for instance for control, fault detection, etc.

In this section we review observer design methods for TS fuzzy systems. The observer considered uses the system model and the available input and output measurements. Consider the affine fuzzy system

$$\begin{aligned} \dot{\boldsymbol{x}} &= \sum_{i=1}^{m} w_i(\boldsymbol{z})(A_i\boldsymbol{x} + B_i\boldsymbol{u} + a_i) \\ \boldsymbol{y} &= \sum_{i=1}^{m} w_i(\boldsymbol{z})(C_i\boldsymbol{x} + c_i) \end{aligned} \tag{4.1}$$

where $\boldsymbol{x}$ denotes the state vector, $\boldsymbol{y}$ the measurement vector, and $\boldsymbol{u}$ the input vector, which is known (measured). The observer design problem arises as soon as the measurement vector does not coincide with the state vector, i.e., $\boldsymbol{y} \neq \boldsymbol{x}$. For the model (4.1), several types of observers have been considered, including linear observers, fuzzy Luenberger observers (Palm and Driankov, 1999; Bergsten and Palm, 2000), sliding-mode observers (Palm and Bergsten, 2000; Oudghiri et al., 2007), etc.

Zs. Lendek et al.: Stability and Observer Design Using TS Models, STUDFUZZ 262, pp. 49–71.
springerlink.com

The observer used in this book is of the form

$$\begin{aligned}\dot{\hat{\boldsymbol{x}}} &= \sum_{i=1}^{m} w_i(\hat{\boldsymbol{z}})(A_i\hat{\boldsymbol{x}} + B_i\boldsymbol{u} + a_i + L_i(\boldsymbol{y} - \hat{\boldsymbol{y}})) \\ \hat{\boldsymbol{y}} &= \sum_{i=1}^{m} w_i(\hat{\boldsymbol{z}})(C_i\hat{\boldsymbol{x}} + c_i).\end{aligned} \tag{4.2}$$

where $\hat{\boldsymbol{x}}$ denotes the estimated state vector, $\hat{\boldsymbol{y}}$ denotes the estimated measurement vector, $\hat{\boldsymbol{z}}$ is the vector of the estimated scheduling variables (in the case when the scheduling vector also has to be estimated), and L_i, $i = 1, 2, \ldots, m$, are the observer gains that have to be designed. Note that the observer itself is a TS fuzzy system. When designing an observer, it is generally required that the estimated states converge asymptotically to the true ones, i.e., $\hat{\boldsymbol{x}} \to \boldsymbol{x}$, as $t \to \infty$. This requirement is equivalent to the dynamics of the estimation error $\boldsymbol{e} = \boldsymbol{x} - \hat{\boldsymbol{x}}$ being asymptotically stable, and therefore many design conditions are formulated in terms of the asymptotic stability of the estimation error.

The observability (and similarly controllability) of TS systems is rarely discussed in the literature. TS systems are nonlinear systems, and therefore it does seem straightforward to use the observability criteria for nonlinear systems. However, since the observers are designed such that each rule has a local gain, it is required that the local models are observable or detectable instead of the full nonlinear system. Note that in general this requirement is neither sufficient nor necessary for the nonlinear system to be observable or detectable. However, due to the form of the observer (4.2), it is required and for the design it is implicitly assumed that the local models, i.e., the pairs (A_i, C_i), $i = 1, 2, \ldots, m$, are observable.

The observer (4.2) can be seen as a generalization of the classical Luenberger observer (Luenberger, 1966) to fuzzy systems, and is referred to as a "fuzzy-Luenberger observer" in several publications (Palm and Driankov, 1999; Bergsten and Palm, 2000). In what follows, we refer to it simply as a fuzzy observer.

The dynamics of the estimation error when using observer (4.2) for system (4.1) can be derived as

$$\begin{aligned}\dot{\boldsymbol{e}} &= \dot{\boldsymbol{x}} - \dot{\hat{\boldsymbol{x}}} \\ &= \sum_{i=1}^{m} w_i(\boldsymbol{z})(A_i\boldsymbol{x} + B_i\boldsymbol{u} + a_i) - \sum_{i=1}^{m} w_i(\hat{\boldsymbol{z}})(A_i\hat{\boldsymbol{x}} + B_i\boldsymbol{u} + a_i + L_i(\boldsymbol{y} - \hat{\boldsymbol{y}})) \\ &= \sum_{i=1}^{m} w_i(\hat{\boldsymbol{z}})(A_i\boldsymbol{x} + B_i\boldsymbol{u} + a_i) - \sum_{i=1}^{m} w_i(\hat{\boldsymbol{z}})(A_i\boldsymbol{x} + B_i\boldsymbol{u} + a_i) \\ &\quad + \sum_{i=1}^{m} w_i(\boldsymbol{z})(A_i\boldsymbol{x} + B_i\boldsymbol{u} + a_i) - \sum_{i=1}^{m} w_i(\hat{\boldsymbol{z}})(A_i\hat{\boldsymbol{x}} + B_i\boldsymbol{u} + a_i + L_i(\boldsymbol{y} - \hat{\boldsymbol{y}}))\end{aligned}$$

$$
\begin{aligned}
&= \sum_{i=1}^{m} w_i(\widehat{\boldsymbol{z}})(A_i\boldsymbol{e} - L_i(\boldsymbol{y} - \widehat{\boldsymbol{y}})) + \sum_{i=1}^{m}(w_i(\boldsymbol{z}) - w_i(\widehat{\boldsymbol{z}}))(A_i\boldsymbol{x} + B_i\boldsymbol{u} + a_i)\\
&= \sum_{i=1}^{m} w_i(\widehat{\boldsymbol{z}})\Big(A_i\boldsymbol{e} - L_i\Big(\sum_{j=1}^{m} w_j(\boldsymbol{z})(C_j\boldsymbol{x} + c_j) - \sum_{j=1}^{m} w_j(\widehat{\boldsymbol{z}})(C_j\widehat{\boldsymbol{x}} + c_j)\Big)\Big)\\
&\quad + \sum_{i=1}^{m}(w_i(\boldsymbol{z}) - w_i(\widehat{\boldsymbol{z}}))(A_i\boldsymbol{x} + B_i\boldsymbol{u} + a_i)\\
&= \sum_{i=1}^{m} w_i(\widehat{\boldsymbol{z}})\Big(A_i\boldsymbol{e} - L_i\Big(\sum_{j=1}^{m} w_j(\boldsymbol{z})(C_j\boldsymbol{x} + c_j) - \sum_{j=1}^{m} w_j(\widehat{\boldsymbol{z}})(C_j\boldsymbol{x} + c_j)\\
&\quad + \sum_{j=1}^{m} w_j(\widehat{\boldsymbol{z}})(C_j\boldsymbol{x} + d_j) - \sum_{j=1}^{m} w_j(\widehat{\boldsymbol{z}})(C_j\widehat{\boldsymbol{x}} + c_j)\Big)\Big)\\
&\quad + \sum_{i=1}^{m}(w_i(\boldsymbol{z}) - w_i(\widehat{\boldsymbol{z}}))(A_i\boldsymbol{x} + B_i\boldsymbol{u} + a_i)\\
&= \sum_{i=1}^{m} w_i(\widehat{\boldsymbol{z}})\Big(A_i\boldsymbol{e} - L_i\Big(\sum_{j=1}^{m} w_j(\widehat{\boldsymbol{z}})C_j\boldsymbol{e} + \sum_{j=1}^{m}(w_j(\boldsymbol{z}) - w_j(\widehat{\boldsymbol{z}}))(C_j\boldsymbol{x} + c_j)\Big)\Big)\\
&\quad + \sum_{i=1}^{m}(w_i(\boldsymbol{z}) - w_i(\widehat{\boldsymbol{z}}))(A_i\boldsymbol{x} + B_i\boldsymbol{u} + a_i)
\end{aligned}
$$

since $\sum_{j=1}^{m} w_i(\widehat{\boldsymbol{z}}) = 1$, and ultimately formulated as

$$
\begin{aligned}
\dot{\boldsymbol{e}} = &\sum_{i=1}^{m} w_i(\widehat{\boldsymbol{z}})\sum_{j=1}^{m} w_j(\widehat{\boldsymbol{z}})(A_i - L_iC_j)\boldsymbol{e}\\
&+ \sum_{i=1}^{m} w_i(\widehat{\boldsymbol{z}})L_i\sum_{j=1}^{m}(w_j(\boldsymbol{z}) - w_j(\widehat{\boldsymbol{z}}))(C_j\boldsymbol{x} + c_j)\\
&+ \sum_{i=1}^{m}(w_i(\boldsymbol{z}) - w_i(\widehat{\boldsymbol{z}}))(A_i\boldsymbol{x} + B_i\boldsymbol{u} + a_i)
\end{aligned}
$$

The expression above represents the general (and complex) case, when all the scheduling variables depend on unmeasured state variables, and the measurement is nonlinear. For observer design under such conditions, the interested reader is referred to (Lendek et al., 2010a). In what follows, two cases will be distinguished: 1) the scheduling vector does not depend on unmeasured states, i.e., in the observer the known (measured) scheduling variables can be used; and 2) the scheduling vector depends on states that are not measured. However, for this second case, for the simplicity of the computations, we consider that the measurement matrix is common for all the rules, i.e., the measurements are linear in the states.

Note that the design conditions presented in the sequel are only sufficient conditions. Similarly to the stability analysis, a major advantage of these conditions is that they are cast into an LMI form, and therefore easily solvable. However, this can also be considered a shortcoming of the approaches, since if the LMIs are infeasible, no conclusive result is obtained. Moreover, the dimension of the LMI problem may be exponential in the number of the rules, and therefore computationally involved to solve.

Although the observer (4.2) is most often used for TS fuzzy systems, in particular for output-feedback control, other observers, such as fuzzy sliding mode observers have also been used. For details on sliding mode observers, the interested reader is referred to (Palm and Driankov, 1999; Palm and Bergsten, 2000; Oudghiri et al., 2007).

4.2 Observer Design: Measured Scheduling Vector

Consider first the case when the scheduling vector depends only on measured variables, i.e., it does not depend on states that have to be estimated. In this case, the scheduling vector itself (instead of its estimate) can be used in the observer, and the observer becomes

$$\begin{aligned}\dot{\widehat{\boldsymbol{x}}} &= \sum_{i=1}^{m} w_i(\boldsymbol{z})(A_i\widehat{\boldsymbol{x}} + B_i\boldsymbol{u} + a_i + L_i(\boldsymbol{y} - \widehat{\boldsymbol{y}})) \\ \widehat{\boldsymbol{y}} &= \sum_{i=1}^{m} w_i(\boldsymbol{z})(C_i\widehat{\boldsymbol{x}} + c_i)\end{aligned} \tag{4.3}$$

Using the observer (4.3), the error dynamics can be written as

$$\dot{e} = \sum_{i=1}^{m}\sum_{j=1}^{m} w_i(\boldsymbol{z})w_j(\boldsymbol{z})(A_i - L_iC_j)e \tag{4.4}$$

For this error dynamics several sufficient stability conditions have been formulated, with the large majority derived from the use of the candidate Lyapunov function $V = \boldsymbol{e}^T P\boldsymbol{e}$, with $P = P^T > 0$. In fact, note that the error dynamics (4.4) is itself a fuzzy system. Therefore, the conditions derived for observer design are extensions of the stability conditions presented in Chapter 3. Moreover, this case is the dual of the stabilization problem described in Chapter 3, and therefore the conditions are again formulated as LMIs.

A first result for establishing the stability of (4.4) has been formulated by Wang et al. (1996) as follows:

Theorem 4.1. *(Wang et al., 1996) The estimation error dynamics* (4.4) *are asymptotically stable, if there exist* $P = P^T > 0$*, and* L_i*,* $i = 1, 2, \ldots, m$*, so that*

$$\begin{aligned} &\mathcal{H}(P(A_i - L_iC_i)) < 0 \\ &\mathcal{H}(P(A_i - L_iC_j + A_j - L_jC_i)) \leq 0 \end{aligned} \tag{4.5}$$

for $i = 1, 2, \ldots, m$*,* $j = i+1, i+2, \ldots, m$*, provided that two rules that are simultaneously active*[1]*, i.e.,* $\forall\ i < j \in \{1, 2, \ldots, m\}$ *for which there exists* $\boldsymbol{z} \in \mathcal{C}_{\boldsymbol{z}}$ *such that* $w_i(\boldsymbol{z})w_j(\boldsymbol{z}) \neq 0$*, where* $\mathcal{H}$ *denotes the symmetric part of a matrix, i.e.* $\mathcal{H}(X) = X + X^T$.

Although the conditions (4.5) are conservative, they have the advantage that they are simple and that they can easily be formulated as linear matrix inequalities (LMIs), using the change of variables $M_i = PL_i$, $i = 1, 2, \ldots, m$. Then, the design of the observer (4.3) is reduced to solving the LMI feasibility problem *find* $P = P^T > 0$*, and* M_i*,* $i = 1, 2, \ldots, m$*, such that*

$$\begin{aligned} &\mathcal{H}(PA_i - M_iC_i) < 0 \\ &\mathcal{H}(PA_i + PA_j - M_iC_j - M_jC_i) \leq 0 \end{aligned}$$

for $i = 1, 2, \ldots, m$, $\forall i < j :\ \exists \boldsymbol{z} :\ \ w_i(\boldsymbol{z})w_j(\boldsymbol{z}) \neq 0$.

The design of an observer using the conditions of Theorem 4.1 is illustrated on the following example.

Example 4.1. Consider the nonlinear dynamic system

$$\begin{aligned} \dot{\boldsymbol{x}} &= \begin{pmatrix} -x_1^2 + x_2 \\ x_1^2 x_2 - x_2 \end{pmatrix} + \begin{pmatrix} x_1 x_2 \\ 1 \end{pmatrix} u \\ \boldsymbol{y} &= \begin{pmatrix} x_1 \\ x_1 x_2 \end{pmatrix} \end{aligned}$$

with $x_1,\ x_2 \in [-1, 1]$, $u \in \mathbb{R}$.

This system can be expressed as

$$\begin{aligned} \dot{\boldsymbol{x}} &= \begin{pmatrix} -x_1 & 1 \\ x_1 x_2 & -1 \end{pmatrix} \begin{pmatrix} x_1 \\ x_2 \end{pmatrix} + \begin{pmatrix} x_1 x_2 \\ 1 \end{pmatrix} u \\ \boldsymbol{y} &= \begin{pmatrix} 1 & 0 \\ 0 & x_1 \end{pmatrix} \begin{pmatrix} x_1 \\ x_2 \end{pmatrix} \end{aligned} \tag{4.6}$$

and can be exactly represented (using the sector nonlinearity approach) by a 4-rule fuzzy system with the local matrices

[1] In what follows, for the ease of the notation, this condition will be denoted as $\forall i < j :$ $\exists \boldsymbol{z} :\ \ w_i(\boldsymbol{z})w_j(\boldsymbol{z}) \neq 0$.

$$A_1 = \begin{pmatrix} 1 & 1 \\ -1 & -1 \end{pmatrix} \quad B_1 = \begin{pmatrix} -1 \\ 1 \end{pmatrix} \quad C_1 = \begin{pmatrix} 1 & 0 \\ 0 & -1 \end{pmatrix}$$
$$A_2 = \begin{pmatrix} 1 & 1 \\ 1 & -1 \end{pmatrix} \quad B_2 = \begin{pmatrix} 1 \\ 1 \end{pmatrix} \quad C_2 = \begin{pmatrix} 1 & 0 \\ 0 & -1 \end{pmatrix}$$
$$A_3 = \begin{pmatrix} -1 & 1 \\ -1 & -1 \end{pmatrix} \quad B_3 = \begin{pmatrix} -1 \\ 1 \end{pmatrix} \quad C_3 = \begin{pmatrix} 1 & 0 \\ 0 & 1 \end{pmatrix}$$
$$A_4 = \begin{pmatrix} -1 & 1 \\ 1 & -1 \end{pmatrix} \quad B_4 = \begin{pmatrix} 1 \\ 1 \end{pmatrix} \quad C_4 = \begin{pmatrix} 1 & 0 \\ 0 & 1 \end{pmatrix}$$

and membership functions $w_1 = \eta_0^1 \eta_0^2$, $w_2 = \eta_0^1 \eta_1^2$, $w_3 = \eta_1^1 \eta_0^2$, $w_4 = \eta_1^1 \eta_1^2$, where $\eta_0^1 = \frac{1-x_1}{2}$, $\eta_0^2 = \frac{1-x_1 x_2}{2}$, $\eta_1^1 = 1 - \eta_0^1$, and $\eta_1^2 = 1 - \eta_0^2$. Note that the scheduling variables are $z_1 = x_1$ and $z_2 = x_1 x_2$, which are both measured, and thus can be used in the observer.

To design the observer, the conditions (4.5) are transformed into LMIs using the change of variables $M_i = PL_i$, $i = 1, 2, \ldots, m$. Then, the following LMI feasibility problem is solved[2]: *find* $P = P^T > 0$, M_i, $i = 1, 2, \ldots, m$, *such that*

$$\begin{aligned} &\mathcal{H}(PA_i - M_i C_i) < 0 \\ &\mathcal{H}(PA_i + PA_j - M_i C_j - M_j C_i) \leq 0 \end{aligned}$$

for[3] $i = 1, 2, 3, 4$, $j = i + 1, \ldots, 4$. The observer gains are recovered as $L_i = P^{-1} M_i$, $i = 1, 2, 3, 4$, and are found as[4]

$$L_1 = \begin{pmatrix} 6.56 & 0.38 \\ 4.95 & -0.27 \end{pmatrix} \quad L_2 = \begin{pmatrix} 5.40 & 1.95 \\ 5.18 & 0.84 \end{pmatrix}$$
$$L_3 = \begin{pmatrix} 2.40 & 0.44 \\ 1.68 & 0.60 \end{pmatrix} \quad L_4 = \begin{pmatrix} 1.42 & 1.93 \\ 2.03 & 1.58 \end{pmatrix}$$

A trajectory[5] of the estimation error using the observer gains above is presented in Figure 4.1. For this particular trajectory, the true initial states were $(0.1\ 0.3)^T$, and the estimated initial states were $(-0.2\ 0.1)^T$, and the input $u = 0$. As can be seen, the estimation error converges to zero. □

Similarly to the conditions described in Chapter 3, several results exist, which, by manipulating the convex sum

$$\sum_{i=1}^{m} \sum_{j=1}^{m} w_i(\boldsymbol{z}) w_j(\boldsymbol{z}) \mathcal{H}(P(A_i - L_i C_j))$$

[2] For solving the LMIs in this chapter, the SeDuMi solver within the Yalmip toolbox was used.

[3] When the TS model is obtained using the sector nonlinearity approach, all the rules are simultaneously active.

[4] All numerical results are given rounded to two decimal places.

[5] In this chapter, for numerical integration the *ode45* Matlab function was used.

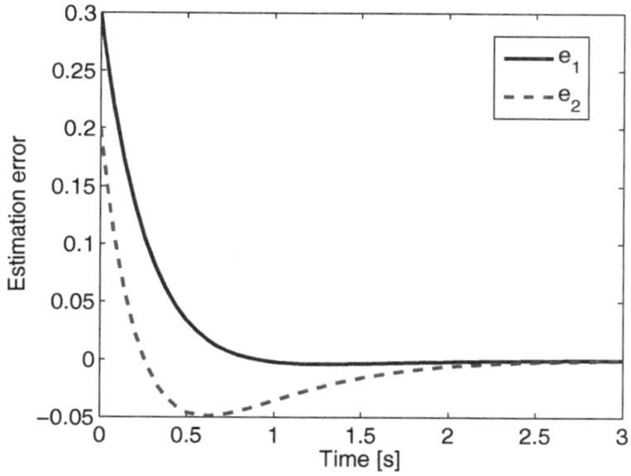

Fig. 4.1 Simulation results for Example 4.1.

aim to reduce the conservativeness of conditions (4.5) for (a class of) systems. For instance, both Kim and Lee (2000) and Bergsten (2001) proposed the following conditions:

Theorem 4.2. *(Bergsten, 2001) The estimation error dynamics* (4.4) *are asymptotically stable, if there exist* $P = P^T > 0$*, and* L_i*,* $i = 1, 2, \ldots, m$*, so that*

$$\begin{pmatrix} H_{11} & H_{12} & \ldots & H_{1m} \\ H_{21} & H_{22} & \ldots & H_{2m} \\ \vdots & \vdots & \ddots & \vdots \\ H_{m1} & H_{m2} & \ldots & H_{mm} \end{pmatrix} < 0$$

where

$$H_{ij} = \begin{cases} \mathcal{H}(P(A_i - L_i C_i)) & \text{if } i = j \\ \mathcal{H}(P(A_i - L_i C_j + A_j - L_j C_i))/2 & \text{otherwise} \end{cases}$$

for $i = 1, 2, \ldots, m$*,* $j = 1, 2, \ldots, m$*.*

Similarly to the conditions of Theorem 4.1, the conditions of Theorem 4.2 can be transformed into LMI conditions with the change of variables $M_i = PL_i$, $i = 1, 2, \ldots, m$.

The observer design using the conditions of Theorem 4.2 are illustrated on the following example.

Example 4.2. Consider the nonlinear system and its fuzzy representation in Example 4.1. Using the conditions of Theorem 4.2, for the observer design one has to solve the LMI: *find* $P = P^T > 0$*,* M_i*,* $i = 1, 2, 3, 4$*, so that*

$$\begin{pmatrix} H_{11} & H_{12} & H_{13} & H_{14} \\ H_{12}^T & H_{22} & H_{23} & H_{24} \\ H_{13}^T & H_{23}^T & H_{33} & H_{34} \\ H_{14}^T & H_{24}^T & H_{34}^T & H_{44} \end{pmatrix} < 0$$

where

$$H_{ij} = \mathcal{H}(PA_i - M_iC_j + PA_j - M_jC_i)/2$$

for $i, j = 1, 2, 3, 4$.

The observer gains are recovered as $L_i = P^{-1}M_i$, $i = 1, 2, 3, 4$, and are found to be

$$L_1 = \begin{pmatrix} 3.01 & 1.42 \\ 1.48 & -0.27 \end{pmatrix} \quad L_2 = \begin{pmatrix} 3.76 & 1.91 \\ 2.00 & -0.73 \end{pmatrix}$$
$$L_3 = \begin{pmatrix} 0.68 & 0.35 \\ 0.33 & 0.28 \end{pmatrix} \quad L_4 = \begin{pmatrix} 0.63 & 1.41 \\ 1.19 & 0.37 \end{pmatrix}$$

A trajectory of the estimation error using the observer gains above is presented in Figure 4.2. This particular trajectory has been obtained with the true initial states being $(0.1\ 0.3)^T$, the estimated initial states being $(-0.2\ 0.1)^T$, and the input $u = 0$. As can be seen, the estimation error converges to zero. □

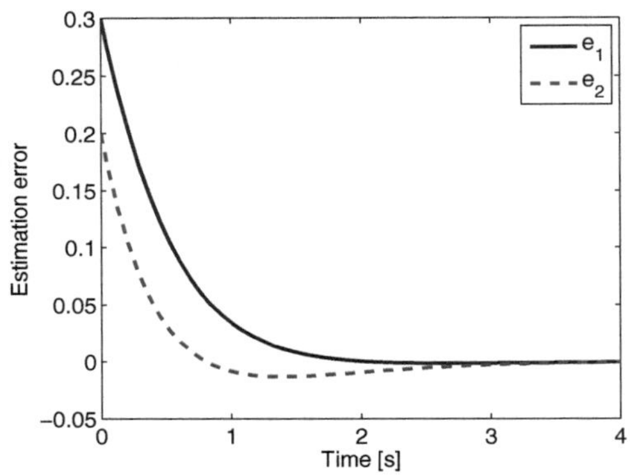

Fig. 4.2 Simulation results for Example 4.2.

A result similar to Theorem 4.2 can be obtained using the relaxation developed by Liu and Zhang (2003). This result requires the introduction of additional decision variables and is formulated as:

Theorem 4.3. *(Liu and Zhang, 2003) The estimation error dynamics* (4.4) *are asymptotically stable, if there exist* $P = P^T > 0$, L_i, Q_{ii}, $i = 1, 2, \ldots, m$, *and* $Q_{ij} = Q_{ji}^T$, $i = 1, 2, \ldots, m$, $j = i+1, i+2, \ldots, m$, *so that*

$$\mathcal{H}(P(A_i - L_iC_i)) + Q_{ii} < 0$$
$$\mathcal{H}(P(A_i - L_iC_j)) + \mathcal{H}(P(A_j - L_jC_i)) + Q_{ij} + Q_{ji} < 0$$

for $i = 1, 2, \ldots, m$, $j = i+1, i+2, \ldots, m$, *and, furthermore*

$$\begin{pmatrix} Q_{11} & Q_{12} & \dots & Q_{1m} \\ Q_{21} & Q_{22} & \dots & Q_{2m} \\ \vdots & \vdots & \ddots & \vdots \\ Q_{m1} & Q_{m2} & \dots & Q_{mm} \end{pmatrix} > 0$$

Similarly to the previously presented results, the conditions of Theorem 4.3 can be formulated as LMIs using the change of variables $M_i = PL_i, i = 1, 2, \dots, m$. The observer design using the conditions of Theorem 4.3 is illustrated on the following example.

Example 4.3. Consider the nonlinear system and its fuzzy representation in Example 4.1. Using the conditions of Theorem 4.3, for the observer design one has to solve the LMIs: *find* $P = P^T > 0$, M_i, Q_{ii}, Q_{ij}, $i = 1, 2, 3, 4$, $j = i+1, \dots, 4$, *so that*

$$\begin{aligned} &\mathcal{H}(PA_i - M_iC_i) + Q_{ii} < 0 \\ &\mathcal{H}(PA_i - M_iC_j + PA_j - M_jC_i) + Q_{ij} + Q_{ij}^T < 0 \\ &\begin{pmatrix} Q_{11} & Q_{12} & Q_{13} & Q_{14} \\ Q_{12}^T & Q_{22} & Q_{23} & Q_{24} \\ Q_{13}^T & Q_{23}^T & Q_{33} & Q_{34} \\ Q_{14}^T & Q_{24}^T & Q_{34}^T & Q_{44} \end{pmatrix} < 0 \end{aligned}$$

for $i = 1, 2, 3, 4$, $j = i+1, \dots, 4$.

The observer gains are recovered as $L_i = P^{-1}M_i$, $i = 1, 2, 3, 4$ and are found as

$$L_1 = \begin{pmatrix} 3.17 & 0.57 \\ 1.53 & -0.48 \end{pmatrix} \quad L_2 = \begin{pmatrix} 3.23 & 1.51 \\ 2.43 & -0.47 \end{pmatrix}$$

$$L_3 = \begin{pmatrix} 0.78 & 0.45 \\ 0.47 & 0.62 \end{pmatrix} \quad L_4 = \begin{pmatrix} 0.49 & 1.48 \\ 1.31 & 0.84 \end{pmatrix}$$

A trajectory of the estimation error using the observer gains above is presented in Figure 4.3. This particular trajectory has been obtained with the true initial states being $(0.1\ 0.3)^T$, the estimated initial states being $(-0.2\ 0.1)^T$, and the input $u = 0$. As can be seen, the estimation error converges to zero. □

Depending on how the fuzzy model (4.1) has been obtained (e.g., using the methods described in Chapter 2), not all the rules may be active at the same time. This property has been used by Tanaka et al. (1998) to reduce the conservativeness of the design as follows.

Theorem 4.4. *(Tanaka et al., 1998) Consider the estimation error dynamics* (4.4), *and let* s, $1 < s \leq m$, *be the maximum number of rules that are simultaneously active. Then, the error dynamics* (4.4) *are asymptotically stable, if there exist* $P = P^T > 0$, L_i, $i = 1, 2, \dots, m$, *and* $Q = Q^T > 0$, *so that:*

$$\begin{aligned} &\mathcal{H}(P(A_i - L_iC_i)) + (s-1)Q < 0 \\ &\mathcal{H}(P(G_{ij} + G_{ji})) - 2Q \leq 0 \end{aligned}$$

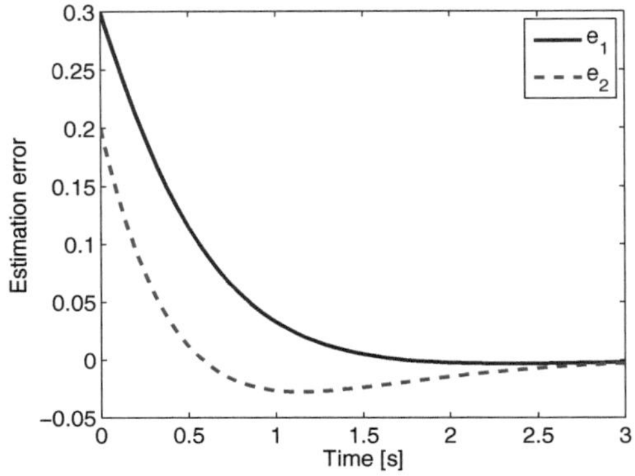

Fig. 4.3 Simulation results for Example 4.3.

for $i = 1, 2, \ldots, m$, $\forall i < j : \ \exists \boldsymbol{z} : \ w_i(\boldsymbol{z})w_j(\boldsymbol{z}) \neq 0$, $j = i+1, i+2, \ldots, m$, *where* $G_{ij} = A_i - L_iC_j$.

This result is useful if triangular or trapezoidal membership functions are used, i.e., when only a relatively small number of rules are active at the same time. Moreover, the conservativeness is reduced in the second condition. The conditions above can also be formulated as LMIs, using the change of variables $M_i = PL_i$, $i = 1, 2, \ldots, m$. The observer design using the conditions of Theorem 4.4 is illustrated on the following example.

Example 4.4. Consider the nonlinear dynamic system

$$\dot{\boldsymbol{x}} = \begin{pmatrix} -x_1 + 2x_2 - 1 \\ x_1x_2 - 3x_2 + 1 \end{pmatrix} \qquad \boldsymbol{y} = \begin{pmatrix} x_1 \\ x_1x_2 \end{pmatrix}$$

with $x_1, x_2 \in [-1, 1]$.

Using x_1 as a (measured) scheduling variable, and choosing the linearization points $\{-1, 0, 1\}$, this system can be approximated by a three-rule fuzzy system using the method from Section 2.3.2. The local matrices are obtained as

$$A_1 = \begin{pmatrix} -1 & 2 \\ 0 & -4 \end{pmatrix} \qquad a_1 = \begin{pmatrix} -1 \\ 1 \end{pmatrix} \qquad C_1 = \begin{pmatrix} 1 & 0 \\ 0 & -1 \end{pmatrix}$$
$$A_2 = \begin{pmatrix} -1 & 2 \\ 0 & -3 \end{pmatrix} \qquad a_2 = \begin{pmatrix} -1 \\ 1 \end{pmatrix} \qquad C_2 = \begin{pmatrix} 1 & 0 \\ 0 & 0 \end{pmatrix}$$
$$A_3 = \begin{pmatrix} -1 & 2 \\ 0 & -2 \end{pmatrix} \qquad a_3 = \begin{pmatrix} -1 \\ 1 \end{pmatrix} \qquad C_3 = \begin{pmatrix} 1 & 0 \\ 0 & 1 \end{pmatrix}$$

$B_i = 0$, $i = 1, 2, 3$, and the triangular membership functions presented in Figure 4.4 are used. Note that using these membership functions, for any value of x_1 at most two rules are activated simultaneously, i.e., only two of the membership functions are non-zero.

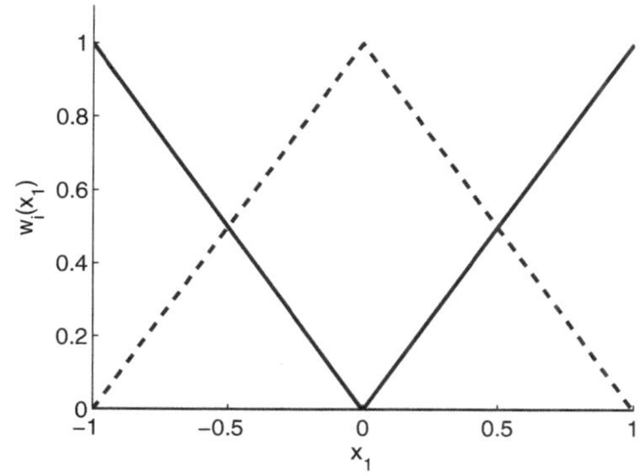

Fig. 4.4 Membership functions used for Example 4.4.

To design the observer, the conditions of Theorem 4.4 are transformed into LMIs using the change of variables $M_i = PL_i$, $i = 1, 2, \ldots, m$. Then, the following LMI feasibility problem is solved: *find* $P = P^T > 0$, $Q = Q^T > 0$, *and* M_i, $i = 1, 2, \ldots, m$, *such that*

$$\begin{aligned}&\mathcal{H}(PA_i - M_iC_i) + Q < 0\\&\mathcal{H}(PA_i + PA_j - M_iC_j - M_jC_i) - 2Q \leq 0\end{aligned}$$

$i = 1, 2, 3, \forall i < j : \quad \exists \boldsymbol{z} : \quad w_i(\boldsymbol{z})w_j(\boldsymbol{z}) \neq 0, j = i+1, i+2, \ldots, m$. The observer gains are recovered as $L_i = P^{-1}M_i$, $i = 1, 2, 3$ and are found as

$$L_1 = \begin{pmatrix} -2.08 & 0.46 \\ 3.51 & -0.05 \end{pmatrix} \quad L_2 = \begin{pmatrix} -2.06 & 0.06 \\ 3.45 & 0.92 \end{pmatrix} \quad L_3 = \begin{pmatrix} -2.02 & -0.33 \\ 3.36 & 1.91 \end{pmatrix}$$

A trajectory of the estimation error using the observer gains above is presented in Figure 4.5. For this particular trajectory, the true initial states were $(0.1\ 0.3)^T$, and the estimated initial states were $(-0.2\ 0.1)^T$. As can be seen, the estimation error converges to zero. □

A similar relaxation for the case when all rules may be simultaneously active was given by Tuan et al. (2001), see Lemma 3.1. Using this relaxation, the observer design problem can be formulated as follows.

Corollary 4.1. *(Tuan et al., 2001) The estimation error dynamics* (4.4), *are asymptotically stable, if there exist* $P = P^T > 0$, *and* L_i, $i = 1, 2, \ldots, m$, *so that*

$$\begin{aligned}&\Gamma_{ii} < 0\\&\frac{1}{m-1}\Gamma_{ii} + \Gamma_{ij} + \Gamma_{ji} < 0\end{aligned} \tag{4.7}$$

for $i = 1, 2, \ldots, m$, $j = 1, 2, \ldots, m$, $i \neq j$, *where* $\Gamma_{ij} = \mathcal{H}(P(A_i - L_iC_j))$.

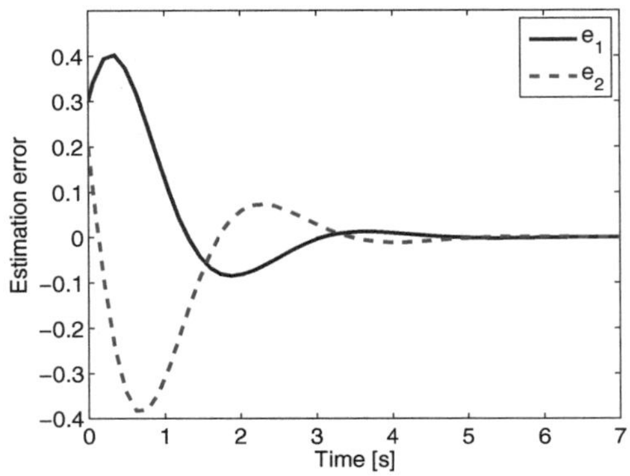

Fig. 4.5 Simulation results for Example 4.4.

The observer design using Corollary 4.1 is illustrated on the following example.

Example 4.5. Consider the nonlinear system and its fuzzy representation in Example 4.1. Using the conditions of Corollary 4.1, to design the observer, the conditions (4.7) are transformed into LMIs, using the change of variables $M_i = PL_i$, $i = 1, 2, 3, 4$, and solved.

The observer gains are recovered as $L_i = P^{-1}M_i$, $i = 1, 2, 3, 4$ and are found to be

$$L_1 = \begin{pmatrix} 2.35 & 0.57 \\ 0.85 & -0.15 \end{pmatrix} \quad L_2 = \begin{pmatrix} 2.64 & 1.92 \\ 2.22 & -0.57 \end{pmatrix}$$

$$L_3 = \begin{pmatrix} 0.28 & 0.12 \\ 0.13 & 0.21 \end{pmatrix} \quad L_4 = \begin{pmatrix} 0.19 & 1.19 \\ 1.12 & 0.40 \end{pmatrix}$$

A trajectory of the estimation error using the observer gains above is presented in Figure 4.6. This particular trajectory has been obtained with the true initial states being $(0.1\ 0.3)^T$, the estimated initial states being $(-0.2\ 0.1)^T$, and the input $u = 0$. As can be seen, the estimation error converges to zero. □

Note that all the conditions presented above require that an observer is designed for each local model, such that the local error dynamics are asymptotically stable (first conditions in Theorems 4.1, 4.4, and 4.1, respectively). Furthermore, the observers have to satisfy a “fuzzy condition”, resulting from the fact that several rules are activated at the same time.

The results presented can be modified such that not only the asymptotic stability is ensured, but also performance measures are satisfied, by using the stability concepts presented in Chapter 3. Of the performance measures, the most well-known concerns the convergence rate of the observer, or, conversely, the decay rate of the estimation error. For instance, the observer design conditions such that a desired decay rate of the estimation error is guaranteed using the conditions of Theorem 4.1 can be formulated as follows:

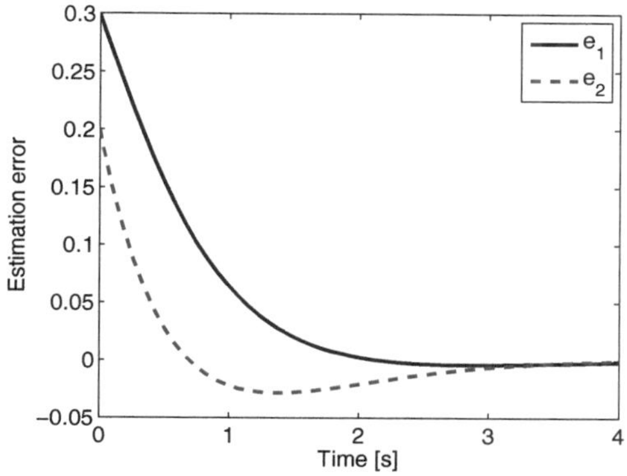

Fig. 4.6 Simulation results for Example 4.5.

Corollary 4.2. *(Wang et al., 1996) The decay rate of the error system* (4.4) *is at least α, if there exist $P = P^T > 0$, and L_i, $i = 1, 2, \ldots, m$, so that*

$$\begin{aligned}&\mathcal{H}(P(A_i - L_iC_i) + 2\alpha P < 0\\&\mathcal{H}(P(A_i - L_iC_j)) + \mathcal{H}(P(A_j - L_jC_i)) + 4\alpha P < 0\end{aligned} \tag{4.8}$$

for $i = 1, 2, \ldots, m$, $j = i+1, i+2, \ldots, m$, $\forall i < j : \; \exists \boldsymbol{z} : \; w_i(\boldsymbol{z})w_j(\boldsymbol{z}) \neq 0$.

The observer design such that a desired convergence rate is obtained is illustrated on the following example.

Example 4.6. Consider the fuzzy system from Example 4.1. The observer may be designed so that the error system has a desired decay rate α by solving the LMIs: *find $P = P^T > 0$, M_i, $i = 1, 2, 3, 4$, so that*

$$\begin{aligned}&\mathcal{H}(PA_i - M_iC_i) + 2\alpha P < 0\\&\mathcal{H}(PA_i + PA_j - M_iC_j - M_jC_i) + 4\alpha P \leq 0\end{aligned}$$

Solving the above LMIs for a desired decay rate $\alpha = 5$, the observer gains are found as

$$L_1 = \begin{pmatrix} 25.30 & 0.12 \\ 191.09 & 0.28 \end{pmatrix} \quad L_2 = \begin{pmatrix} 24.93 & 0.17 \\ 189.38 & 0.71 \end{pmatrix}$$
$$L_3 = \begin{pmatrix} 17.92 & 0.58 \\ 137.34 & 4.96 \end{pmatrix} \quad L_4 = \begin{pmatrix} 17.54 & 0.64 \\ 135.63 & 5.33 \end{pmatrix}$$

A trajectory of the estimation error using the observer gains above is presented in Figure 4.7. For this particular trajectory, the true initial states were $(0.1 \; 0.3)^T$, and the estimated initial states were $(-0.2 \; 0.1)^T$, and the input $u = 0$. As can be seen, the estimation error converges to zero much quicker than in the previous example, however, the overshoot also becomes larger. □

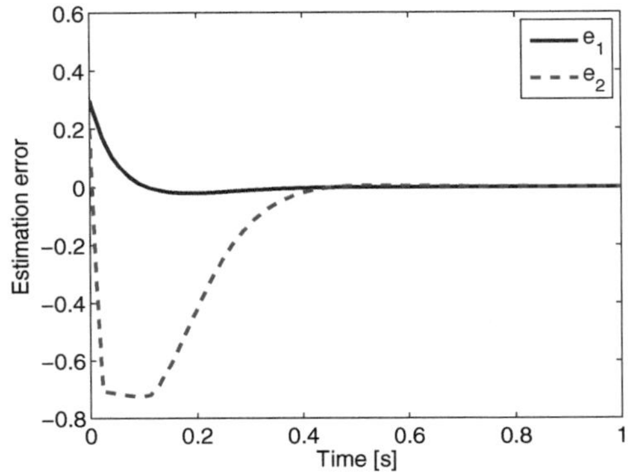

Fig. 4.7 Simulation results for Example 4.6.

Note that if the scheduling vector is known, the (known) input $\boldsymbol{u}$, and the input matrices B_i, $i = 1, 2, \ldots, m$, respectively, do not influence the observer design conditions. Moreover, for this case, non-quadratic stability conditions, such as those presented in Chapter 3, can also be relatively easily extended to observer design. For instance, the results of Bernal et al. (2009) are the dual conditions for control design, while the extension of the results of Guerra and Bernal (2009) to observer design has been reported in (Lendek et al., 2010b).

4.3 Observer Design: Estimated Scheduling Vector

In this section, we consider the observer design problem when the scheduling vector depends on the states to be estimated. Note that in this case the true scheduling variables cannot be used in the observer, and instead their estimated values have to be used. For the simplicity of the notation, only the case with common measurement matrices, i.e., $C_i = C$, $i = 1, 2, \ldots, m$, will be considered. If the measurement matrix is different for each rule, the observer gains may be designed similarly, although the design conditions are more complex. For the complete derivation, the interested reader is referred to (Lendek et al., 2010a).

For common measurement matrices, the observer (4.2) becomes

$$\begin{aligned}\dot{\widehat{\boldsymbol{x}}} &= \sum_{i=1}^{m} w_i(\widehat{\boldsymbol{z}})(A_i\widehat{\boldsymbol{x}} + B_i\boldsymbol{u} + a_i + L_i(\boldsymbol{y} - \widehat{\boldsymbol{y}})) \\ \widehat{\boldsymbol{y}} &= C\widehat{\boldsymbol{x}}\end{aligned} \tag{4.9}$$

and the error dynamics can be expressed as

$$\dot{\boldsymbol{e}} = \sum_{i=1}^{m} w_i(\widehat{\boldsymbol{z}})(A_i - L_iC)\boldsymbol{e} + \sum_{i=1}^{m}(w_i(\boldsymbol{z}) - w_i(\widehat{\boldsymbol{z}}))(A_i\boldsymbol{x} + B_i\boldsymbol{u} + a_i) \tag{4.10}$$

Clearly, there is a time-varying difference between the true and estimated states, $\sum_{i=1}^{m}(w_i(\boldsymbol{z}) - w_i(\widehat{\boldsymbol{z}}))(A_i\boldsymbol{x} + B_i\boldsymbol{u} + a_i)$, which, since the variables are defined in a compact set, goes to zero if $\widehat{\boldsymbol{z}} \to \boldsymbol{z}$. In order for the estimated states to converge to the real ones, the observer has to be robust enough to deal with this difference. For the system (4.10), sufficient stability conditions were given by Bergsten (2001), based on the conditions of Theorem 4.1. Note that similar conditions can be incorporated into any of the theorems presented in Section 4.3. For simplicity, only the simple case of Theorem 4.1 is presented.

Theorem 4.5. *(Bergsten, 2001) Consider the error system* (4.10)*, and assume that*

$$\left\| \sum_{i=1}^{m}(w_i(\boldsymbol{z}) - w_i(\widehat{\boldsymbol{z}}))(A_i\boldsymbol{x} + B_i\boldsymbol{u} + a_i)\right\| \leq \mu\|\boldsymbol{e}\| \tag{4.11}$$

where $\mu > 0$ is a known constant. Then, the error system (4.10) *is exponentially stable, if there exist $P = P^T > 0$, $Q = Q^T > 0$, and L_i, $i = 1, 2, \ldots, m$, so that*

$$\begin{aligned} \mathcal{H}(P(A_i - L_iC)) &\leq -Q \\ \begin{pmatrix} Q - \mu^2 I & P \\ P & I \end{pmatrix} &> 0 \end{aligned} \tag{4.12}$$

for $i = 1, 2, \ldots, m$.

Remark: Note that as long as the membership functions are smooth and the variables are defined on a compact set, there exists $\mu > 0$ so that (4.11) holds. The bounding constant μ in general can be found by solving the optimization problem (Khalil, 2002)

$$\mu = \max_{\boldsymbol{x},\boldsymbol{u},\widehat{\boldsymbol{x}},\widehat{\boldsymbol{z}}} \left\| \frac{\partial(w_i(\boldsymbol{z}) - w_i(\widehat{\boldsymbol{z}}))(A_i\boldsymbol{x} + B_i\boldsymbol{u} + a_i)}{\partial \boldsymbol{e}} \right\|$$

The observer design when the scheduling vector depends on states that have to be estimated is illustrated using the following example.

Example 4.7. Consider the nonlinear dynamic system

$$\dot{\boldsymbol{x}} = \begin{pmatrix} -x_1 + 2x_1^2 x_2 + x_2 \\ x_1^2 x_2 - x_2 \end{pmatrix} \qquad \boldsymbol{y} = x_1$$

with $x_1,\ x_2 \in [-1, 1]$.

This system can be expressed as

$$\begin{aligned} \dot{\boldsymbol{x}} &= \begin{pmatrix} -1 & 2x_1^1 + 1 \\ x_1 x_2 & -1 \end{pmatrix} \begin{pmatrix} x_1 \\ x_2 \end{pmatrix} \\ \boldsymbol{y} &= (1\ 0) \begin{pmatrix} x_1 \\ x_2 \end{pmatrix} \end{aligned}$$

and can be exactly represented (using the sector nonlinearity approach) by a 4-rule fuzzy system with the local matrices

$$A_1 = \begin{pmatrix} -1 & 1 \\ -1 & -1 \end{pmatrix} \qquad A_2 = \begin{pmatrix} -1 & 1 \\ 1 & -1 \end{pmatrix}$$
$$A_3 = \begin{pmatrix} -1 & 3 \\ -1 & -1 \end{pmatrix} \qquad A_4 = \begin{pmatrix} -1 & 3 \\ 1 & -1 \end{pmatrix}$$

and membership functions $w_1 = \eta_0^1\eta_0^2$, $w_2 = \eta_0^1\eta_1^2$, $w_3 = \eta_1^1\eta_0^2$, $w_4 = \eta_1^1\eta_1^2$, where $\eta_0^1 = 1 - x_1^2$, $\eta_0^2 = \frac{1-x_1x_2}{2}$, $\eta_1^1 = 1 - \eta_0^1$, and $\eta_1^2 = 1 - \eta_0^2$. Note that the scheduling variables are $z_1 = x_1$ and $z_2 = x_1x_2$, of which only z_1 is measured, z_2 depending on unmeasured states. Therefore, the estimate od the scheduling variable, $\hat{z}_2$ has to be used in the observer. The equation

$$\left\| \sum_{i=1}^{m} (w_i(\boldsymbol{z}) - w_i(\hat{\boldsymbol{z}}))A_i\boldsymbol{x} \right\| \le \mu\|\boldsymbol{e}\|$$

is satisfied with $\mu = 1$.

To design the observer, the conditions (4.12) are transformed into LMIs using the change of variables $M_i = PL_i$, $i = 1, 2, \ldots, m$. Then, the following LMI feasibility problem is solved: *find* $P = P^T > 0$, *and* M_i, $i = 1, 2, \ldots, m$, *such that*

$$\mathcal{H}(PA_i - M_iC_i) < -Q$$
$$\begin{pmatrix} Q - \mu^2 I & P \\ P & I \end{pmatrix} > 0$$

for $i = 1, 2, 3, 4$. The observer gains are found as

$$L_1 = \begin{pmatrix} 3.42 \\ 1.23 \end{pmatrix} \quad L_2 = \begin{pmatrix} 3.42 \\ 3.23 \end{pmatrix} \quad L_3 = \begin{pmatrix} 4.39 \\ 2.91 \end{pmatrix} \quad L_4 = \begin{pmatrix} 4.39 \\ 4.91 \end{pmatrix}$$

A trajectory of the estimation error using the observer gains above is presented in Figure 4.8. For this particular trajectory, the true initial states were $(0.1\ 0.3)^T$, and

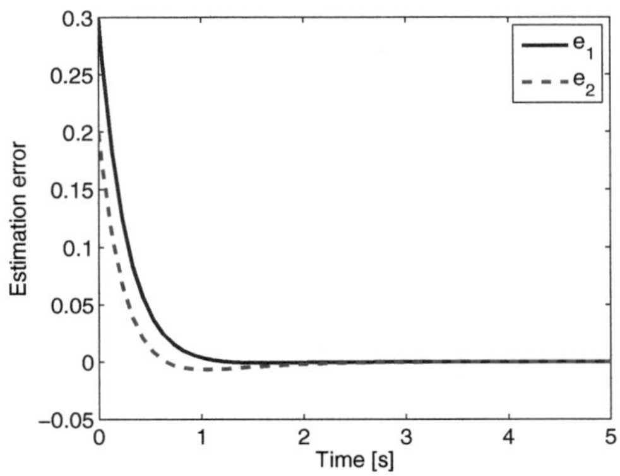

Fig. 4.8 Simulation results for Example 4.7.

the estimated initial states were $(-0.2\ 0.1)^T$. As can be seen, the estimation error converges to zero. □

Since the fuzzy models are in general defined on a compact set and the membership functions are smooth, an upper bound on the Lipschitz constant μ can in general be determined. However, the conditions of Theorem 4.5 are conservative, due to the worst-case assumption of an unstructured, bounded disturbance. In many cases, an observer will work even though the second condition of Theorem 4.5 is not satisfied by the computed bound. Such a case is illustrated on the following example.

Example 4.8. Consider the nonlinear dynamic system

$$\dot{\boldsymbol{x}} = \begin{pmatrix} -x_1 + 2x_1^2 x_2 + x_2 \\ 2x_1^2 x_2 - x_2 \end{pmatrix} \qquad \boldsymbol{y} = x_1$$

with $x_1,\ x_2 \in [-1, 1]$.

Note that this system is not asymptotically stable. Similarly to the system in Example 4.7, this model can be exactly represented (using the sector nonlinearity approach) by a 4-rule fuzzy system with the local matrices

$$A_1 = \begin{pmatrix} -1 & 1 \\ -2 & -1 \end{pmatrix} \qquad A_2 = \begin{pmatrix} -1 & 1 \\ 2 & -1 \end{pmatrix}$$
$$A_3 = \begin{pmatrix} -1 & 3 \\ -2 & -1 \end{pmatrix} \qquad A_4 = \begin{pmatrix} -1 & 3 \\ 2 & -1 \end{pmatrix}$$

and with the same membership functions as those used in Example 4.7. However, the equation

$$\left\| \sum_{i=1}^{m} (w_i(\boldsymbol{z}) - w_i(\hat{\boldsymbol{z}}))A_i \boldsymbol{x} \right\| \leq \mu \|\boldsymbol{e}\|$$

is satisfied with $\mu = 2$.

With this μ, the LMIs

$$\begin{aligned} &\mathcal{H}(PA_i - M_iC_i) < -Q \\ &\begin{pmatrix} Q - \mu^2 I & P \\ P & I \end{pmatrix} > 0 \end{aligned} \tag{4.13}$$

for $i = 1, 2, 3, 4$ are not feasible. The maximum μ for which (4.13) are feasible is $\mu = \sqrt{2}$. Solving (4.13) with $\mu = \sqrt{2}$, the observer gains are found as

$$L_1 = \begin{pmatrix} 8.18 \\ 7.18 \end{pmatrix} \quad L_2 = \begin{pmatrix} 8.18 \\ 11.18 \end{pmatrix} \quad L_3 = \begin{pmatrix} 13.67 \\ 15.82 \end{pmatrix} \quad L_4 = \begin{pmatrix} 13.67 \\ 19.82 \end{pmatrix}$$

Note that the estimate given by the observer found in this way is not guaranteed to converge to the true states. However, simulation results indicate that the estimate does converge, as can be seen in Figure 4.9. For this particular trajectory, the true initial states were $(0.1\ 0.3)^T$, and the estimated initial states were $(-0.2\ 0.1)^T$. □

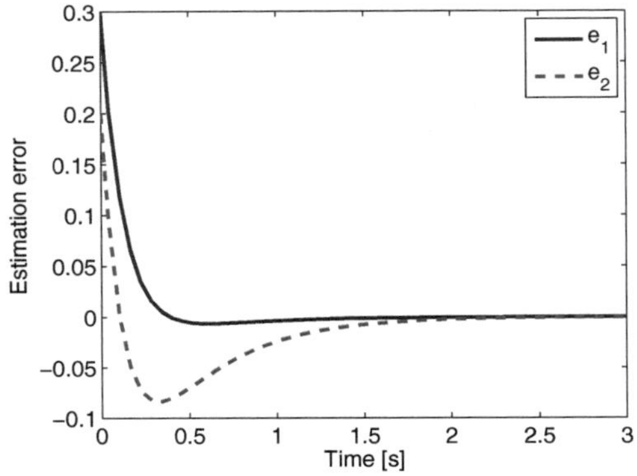

Fig. 4.9 Simulation results for Example 4.8.

In many cases, only some of the scheduling variables depend on states that have to be estimated. In such a case, those scheduling variables that are measured, can be used in the observer. If this means that a structured uncertainty can be constructed as

$$\sum_{i=1}^{m}(w_i(\boldsymbol{z}) - w_i(\widehat{\boldsymbol{z}}))(A_i\boldsymbol{x} + B_i\boldsymbol{u} + a_i) = \mu E_u \bar{\Delta} F_u \boldsymbol{e} \tag{4.14}$$

with E_u and F_u structure matrices of appropriate dimensions, and $\bar{\Delta}^T\bar{\Delta} \leq I$, one can use (4.14) instead of (4.11) to reduce the conservativeness of the approach. Using (4.14), the design conditions can be formulated as follows.

Theorem 4.6. *(Bergsten, 2001) Consider the error system* (4.10), *and assume that*

$$\sum_{i=1}^{m}(w_i(\boldsymbol{z}) - w_i(\widehat{\boldsymbol{z}}))(A_i\boldsymbol{x} + B_i\boldsymbol{u} + a_i) = \mu E_u \bar{\Delta} F_u \boldsymbol{e}$$
$$\bar{\Delta}^T\bar{\Delta} \leq I$$

where $\mu > 0$ *is a known constant, and* E_u *and* F_u *are known structure matrices. Then, the error system* (4.10) *is exponentially stable, if there exist* $P = P^T > 0$, $Q = Q^T > 0$, *and* L_i, $i = 1, 2, \ldots, m$, *so that*

$$\begin{aligned} &\mathcal{H}(P(A_i - L_iC)) \leq -Q \\ &\begin{pmatrix} Q - \mu^2 F_u^T F_u & PE_u \\ E_u^T P & I \end{pmatrix} > 0 \end{aligned} \tag{4.15}$$

for $i = 1, 2, \ldots, m$.

The observer design when a structured uncertainty can be used is illustrated using the following example.

Example 4.9. Consider the nonlinear dynamic system and its fuzzy representation in Example 4.8. Recall that $z_1 = x_1$ is measured (and therefore can be used in the observer), while $z_2 = x_1 x_2$ is not measured, and therefore its estimated value has to be used in the observer. As already stated in Example 4.8, the equation

$$\left\| \sum_{i=1}^{m} (w_i(\boldsymbol{z}) - w_i(\widehat{\boldsymbol{z}})) A_i \boldsymbol{x} \right\| \leq \mu \|\boldsymbol{e}\|$$

is satisfied with $\mu = 2$, a value for which the LMIs used to design the observer are unfeasible. However, this equation can also be written as

$$\begin{aligned}
&\sum_{i=1}^{m} (w_i(\boldsymbol{z}) - w_i(\widehat{\boldsymbol{z}})) A_i \boldsymbol{x} = \\
&\quad = \left(\begin{pmatrix} -1 & 2x_1^2+1 \\ 2x_1x_2 & -1 \end{pmatrix} - \begin{pmatrix} -1 & 2x_1^2+1 \\ 2x_1\hat{x}_2 & -1 \end{pmatrix} \right) \begin{pmatrix} x_1 \\ x_2 \end{pmatrix} \\
&\quad = \begin{pmatrix} 0 & 0 \\ 2x_1(x_2 - \hat{x}_2) & 0 \end{pmatrix} \begin{pmatrix} x_1 \\ x_2 \end{pmatrix} \\
&\quad = \begin{pmatrix} 0 \\ 2x_1^2 e_2 \ 0 \end{pmatrix} \\
&\quad = \begin{pmatrix} 0 & 0 \\ 0 & 2x_1^2 \end{pmatrix} \begin{pmatrix} e_1 \\ e_2 \end{pmatrix} \\
&\quad = 2 \begin{pmatrix} 0 & 0 \\ 0 & 1 \end{pmatrix} \begin{pmatrix} 0 & 0 \\ 0 & x_1^2 \end{pmatrix} \begin{pmatrix} 1 & 0 \\ 0 & 1 \end{pmatrix}
\end{aligned}$$

i.e., $\mu = 2$, $E_u = \begin{pmatrix} 0 & 0 \\ 0 & 1 \end{pmatrix}$, $F_u = I$, and $\bar{\Delta} = \begin{pmatrix} 0 & 0 \\ 0 & x_1^2 \end{pmatrix}$, with $\bar{\Delta}^T \bar{\Delta} \leq I$, since $x_1 \in [-1, 1]$.

Solving the LMI problem: *find* $P = P^T$, $Q = Q^T > 0$, *and* M_i, $i = 1, 2, \ldots, m$, *so that*

$$\begin{aligned}
&\mathcal{H}(PA_i - M_i C) \leq -Q \\
&\begin{pmatrix} Q - \mu^2 F_u^T F_u & PE_u \\ E_u^T P & I \end{pmatrix} > 0
\end{aligned} \tag{4.16}$$

for $i = 1, 2, \ldots, m$, the observer gains are obtained as $L_i = P^{-1} M_i$, $i = 1, 2, 3, 4$

$$L_1 = \begin{pmatrix} 4.24 \\ 16.09 \end{pmatrix} \quad L_2 = \begin{pmatrix} 4.24 \\ 20.09 \end{pmatrix} \quad L_3 = \begin{pmatrix} 11.52 \\ 46.67 \end{pmatrix} \quad L_4 = \begin{pmatrix} 11.52 \\ 50.67 \end{pmatrix}$$

A trajectory of the estimation error using the observer gains above is presented in Figure 4.10. For this particular trajectory, the true initial states were $(0.1 \ 0.3)^T$, and the estimated initial states were $(-0.2 \ 0.1)^T$. As expected, the estimation error converges to zero. □

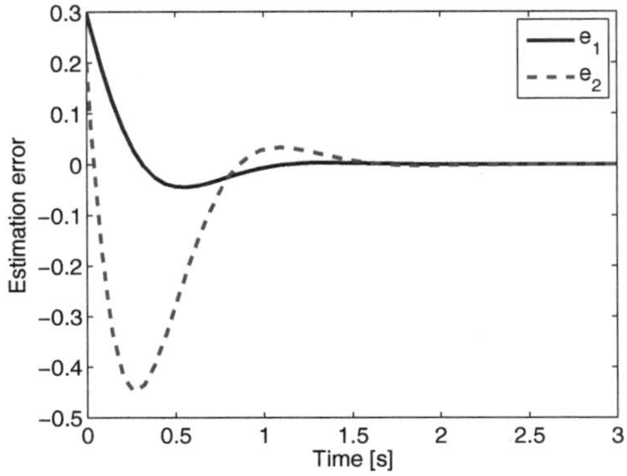

Fig. 4.10 Simulation results for Example 4.9.

Although the remainder of this book is not concerned with controller design, it has to be mentioned that observers for TS fuzzy systems are used extensively in output-feedback controller design. A brief description of observer-based stabilization is given in the following section.

4.4 Observer-Based Stabilization

Several authors have considered the case of joint design of the observer and of the linear state-feedback controller and have developed relaxed stability conditions for the augmented system. The conditions usually lead to (generalized) eigenvalue problems that can be solved using LMIs (Taniguchi et al., 1999b; Tanaka et al., 1998; Taniguchi et al., 1999a), if the scheduling variables are known and the separation principle holds. In the case when the scheduling variables depend on the estimated states, the observer and the controller cannot be designed separately (Tanaka and Sano, 1994; Tanaka and Wang, 2001), and in general a two-step procedure is employed (Uang and Chen, 2000; Tanaka and Wang, 2001; Tseng, 2008). In what follows, we briefly describe observer-based stabilization for the case when the scheduling vector is measured. For the general case when the scheduling vector depends on states to be estimated, the interested reader is referred to (Uang and Chen, 2000; Tanaka and Wang, 2001; Yoneyama et al., 2001; Guerra et al., 2006; Tseng, 2008).

For observer-based stabilization, consider the TS fuzzy model

$$\begin{aligned} \dot{\boldsymbol{x}} &= \sum_{i=1}^{m} w_i(\boldsymbol{z})(A_i\boldsymbol{x} + B_i\boldsymbol{u}) \\ \boldsymbol{y} &= \sum_{i=1}^{m} w_i(\boldsymbol{z})C_i\boldsymbol{x} \end{aligned} \tag{4.17}$$

where $\boldsymbol{x}$ is the state vector, $\boldsymbol{u}$ is the input vector, $\boldsymbol{y}$ is the measurement vector, $\boldsymbol{z}$ is the vector of scheduling variables, which depends only on known (measured) variables. Similarly to stabilization of TS systems (see Section 3.4), the local models are considered linear and the scheduling vector does not depend on the input $\boldsymbol{u}$.

The observer considered is of the form

$$\begin{aligned}\dot{\widehat{\boldsymbol{x}}} &= \sum_{i=1}^{m} w_i(\boldsymbol{z})(A_i\widehat{\boldsymbol{x}} + B_i\boldsymbol{u} + L_i(\boldsymbol{y} - \widehat{\boldsymbol{y}})) \\ \widehat{\boldsymbol{y}} &= \sum_{i=1}^{m} w_i(\boldsymbol{z})C_i\widehat{\boldsymbol{x}}\end{aligned} \tag{4.18}$$

similarly to the observer presented in Section 4.2, and the controller used is

$$\boldsymbol{u} = -\sum_{i=1}^{m} w_i(\boldsymbol{z})K_i\widehat{\boldsymbol{x}} \tag{4.19}$$

Note that since the state vector is not measured, in the controller, the estimated values are used.

The estimation error is obtained as (4.4), repeated here for convenience:

$$\dot{\boldsymbol{e}} = \sum_{i=1}^{m}\sum_{j=1}^{m} w_i(\boldsymbol{z})w_j(\boldsymbol{z})(A_i - L_iC_j)\boldsymbol{e}.$$

The closed-loop dynamics using the estimate-based control law is:

$$\begin{aligned}\dot{\boldsymbol{x}} &= \sum_{i=1}^{m} w_i(\boldsymbol{z})(A_i\boldsymbol{x} - B_i\sum_{j=1}^{m} w_j(\boldsymbol{z})K_j\widehat{\boldsymbol{x}}) \\ &= \sum_{i=1}^{m}\sum_{j=1}^{m} w_i(\boldsymbol{z})w_j(\boldsymbol{z})((A_i - B_iK_j)\boldsymbol{x} - B_iK_j\boldsymbol{e})\end{aligned} \tag{4.20}$$

Combining the dynamics of the estimation error and the state, we obtain

$$\begin{pmatrix}\dot{\boldsymbol{e}} \\ \dot{\boldsymbol{x}}\end{pmatrix} = \sum_{i=1}^{m}\sum_{j=1}^{m}\sum_{k=1}^{m} w_i(\boldsymbol{z})w_j(\boldsymbol{z})w_k(\boldsymbol{z}) \begin{pmatrix}A_i - L_iC_k & 0 \\ -B_jK_k & A_j - B_jK_k\end{pmatrix}\begin{pmatrix}\boldsymbol{e} \\ \boldsymbol{x}\end{pmatrix} \tag{4.21}$$

Then, the combined observer and control design problem consists in finding the gains L_i and K_i, $i = 1, 2, \ldots, m$, such that the system (4.21) is asymptotically stable. For the system (4.21), the separation principle holds (Ma et al., 1998; Yoneyama et al., 2000), which means that it is possible to design separately the observer and the controller, and therefore one can make use of the results and relaxations presented in Sections 3.2.2, 3.4, and 4.2. For instance, using Theorem 3.3 with conditions (3.4) for controller design and Theorem 3.1 for observer design, the following results can be formulated (Tanaka and Wang, 2001):

Theorem 4.7. *The closed-loop dynamics* (4.21) *is asymptotically stable, if there exist* $X = X^T > 0$, Y_i, $P = P^T > 0$, *and* M_i, $i = 1, 2, \ldots, m$, *such that*

$$\begin{aligned} &\mathcal{H}(A_i X - B_i Y_i) < 0 \\ &\mathcal{H}(A_i X - B_i M_i + A_j X - B_j M_j) \leq 0 \\ &\mathcal{H}(P A_i - M_i C_i) < 0 \\ &\mathcal{H}(P A_i + P A_j - M_i C_j - M_j C_i) \leq 0 \end{aligned} \tag{4.22}$$

for $i = 1, 2, \ldots, m$, $j = i + 1, i + 2, \ldots, m$, $\forall i < j : \ \exists \boldsymbol{z} : \ w_i(\boldsymbol{z}) w_j(\boldsymbol{z}) \neq 0$ *hold. Moreover, if the conditions* (4.22) *are satisfied, then the controller gains are* $K_i = Y_i X^{-1}$, *and the observer gains are* $L_i = P^{-1} M_i$, $i = 1, 2, \ldots, m$.

The following example illustrates observer-based control design:

Example 4.10. Consider the nonlinear system

$$\begin{aligned} \dot{\boldsymbol{x}} &= \begin{pmatrix} 1 & x_1^2 + 1 \\ -x_1 - 2 & x_1^2 - 2 \end{pmatrix} \boldsymbol{x} + \begin{pmatrix} 1 \\ 0 \end{pmatrix} \boldsymbol{u} \\ \boldsymbol{y} &= (1 \ ; 0) \, \boldsymbol{x} \end{aligned} \tag{4.23}$$

with $x_1, x_2 \in [-1, 1]$.

This system can be exactly represented (using the sector nonlinearity approach) by a 4-rule fuzzy system with the local matrices

$$A_1 = \begin{pmatrix} 1 & 1 \\ -3 & -2 \end{pmatrix} \quad A_2 = \begin{pmatrix} 1 & 2 \\ -3 & -1 \end{pmatrix} \quad A_3 = \begin{pmatrix} 1 & 1 \\ -1 & -2 \end{pmatrix} \quad A_4 = \begin{pmatrix} 1 & 2 \\ -1 & -1 \end{pmatrix}$$
$$B_i = \begin{pmatrix} 1 \\ 0 \end{pmatrix} \qquad C_i = (1 \ 0) \qquad i = 1, 2, 3, 4$$

and membership functions $w_1 = \eta_0^1 \eta_0^2$, $w_2 = \eta_0^1 \eta_1^2$, $w_3 = \eta_1^1 \eta_0^2$, $w_4 = \eta_1^1 \eta_1^2$, where $\eta_0^1 = \frac{1 - x_1}{2}$, $\eta_0^2 = 1 - x_1^2$, $\eta_1^1 = 1 - \eta_0^1$, and $\eta_1^2 = 1 - \eta_0^2$. Note that the scheduling variables are $z_1 = x_1$ and $z_2 = x_1^2$, which depend on x_1 which is measured, and thus can be used in the observer.

To design the observer, the conditions (4.22) are solved. Note that since both the input matrix and the measurement matrix is common for all the rules, the LMI problem is reduced to

find $P = P^T > 0$, $X = X^T > 0$, Y_i, *and* M_i, $i = 1, 2, \ldots, m$, *so that*

$$\begin{aligned} &\mathcal{H}(A_i X - B_i Y_i) < 0 \\ &\mathcal{H}(P A_i - M_i C_i) < 0 \end{aligned}$$

for $i = 1, 2, \ldots, m$.

The observer gains are recovered as $L_i = P^{-1} M_i$, $i = 1, 2, 3, 4$, and are found to be

$$L_1 = \begin{pmatrix} 4.61 \\ 0.25 \end{pmatrix} \quad L_2 = \begin{pmatrix} 5.29 \\ 2.01 \end{pmatrix} \quad L_3 = \begin{pmatrix} 4.61 \\ 2.25 \end{pmatrix} \quad L_4 = \begin{pmatrix} 5.29 \\ 4.01 \end{pmatrix}$$

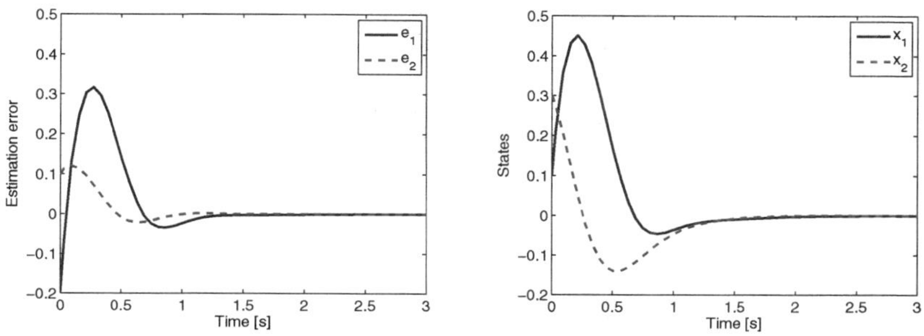

(a) Estimation errors. (b) State variables.

Fig. 4.11 Simulation results for Example 4.10.

and the controller gains as $K_i = Y_i X^{-1}$, $i = 1, 2, 3, 4$, with the values

$$K_1 = \begin{pmatrix}10.65 & -33.56\end{pmatrix} \quad K_2 = \begin{pmatrix}10.14 & -30.17\end{pmatrix}$$
$$K_3 = \begin{pmatrix}5.88 & -11.04\end{pmatrix} \quad K_4 = \begin{pmatrix}5.37 & -7.66\end{pmatrix}$$

A trajectory of the estimation error and of the states of the closed-loop system using the observer and controller gains above is presented in Figures 4.11(a) and 4.11(b). For this particular trajectory, the true initial states were $(0.1\ \ 0.3)^T$, and the estimated initial states were $(-0.2\ 0.1)^T$. As can be seen, all the variables converge to zero. □

4.5 Summary

For TS systems, several types of observers have been developed in the literature. In this chapter, the Luenberger type fuzzy observer has been discussed, together with the design conditions that are used in the following chapters. Regarding the observer design, two cases can be distinguished, depending on whether or not the scheduling vector is a function of the states to be estimated. When the scheduling vector depends on the states to be estimated, an observer that can handle the mismatch between the true and estimated value of the membership functions, has to be designed. We have also briefly described observer-based stabilization of TS systems for the case when the scheduling vector depends only on measured variables.

The presented design conditions are only sufficient conditions. A major advantage of these conditions is that they are cast into an LMI form, which is easily solvable. However, this can also be considered a shortcoming of the approaches, since if the LMIs are infeasible, no conclusive result is obtained.

Chapter 5
Cascaded TS Systems and Observers

Many physical systems are distributed, i.e., they are composed of lower dimensional, interacting subsystems. In this chapter, a special class of distributed systems is considered: cascaded systems. Systems can either be naturally cascaded, or can be transformed into a cascade of submodels by a suitable reordering of the state variables. It is assumed that both the whole system, and also the subsystems are represented by TS fuzzy models, i.e., the systems are cascaded TS fuzzy systems.

This chapter consists of three parts. First, an algorithm to partition a general nonlinear system into the cascade of two subsystems is presented, together with stability conditions for cascaded nonlinear systems. In the second part, we consider the cascaded stability analysis of cascaded TS systems. Finally, cascaded observer design for TS systems is studied.

5.1 Introduction

Many physical systems, such as power systems, communication networks, economic systems, and traffic networks are interconnections of lower-dimensional subsystems. An important class of these systems, such as material processing systems or chemical processes, can be represented as cascaded subsystems (Seibert and Suarez, 1990; Jankovic et al., 1996; Arcak et al., 2002). It has since long been investigated whether based solely on the analysis of the subsystems conclusions referring to the whole system can be drawn or under which conditions such conclusions can be drawn. For instance, for linear systems, the stability of the subsystems implies the stability of the cascaded system (Loria and Panteley, 2005). For nonlinear, or time-varying systems, however, this property does not hold in general. Even global asymptotic stability of the individual subsystems does not always imply the stability of the cascade.

The stability of several types of cascaded systems has already been studied in the literature. Conditions to ensure the overall stability of general cascades, in which both subsystems are nonlinear, have been derived by Sontag (1989b); Seibert and Suarez (1990); Loria and Panteley (2005). Some of these conditions

Zs. Lendek et al.: Stability and Observer Design Using TS Models, STUDFUZZ 262, pp. 73–102.
springerlink.com

represent the basis for the stability analysis of cascaded TS systems and are presented in the next section. In addition, an algorithm is given to determine whether a nonlinear system is the cascade of observable subsystems. The chapter continues with the presentation of stability conditions for cascaded TS systems. Finally, observer design is considered, i.e., conditions for the cascaded design of observers for cascaded TS systems are presented.

5.2 Stability of Cascaded Dynamic Systems

The stability and design conditions for cascaded TS systems presented in this chapter are based on results obtained for general nonlinear dynamical systems. The first motivation to consider cascaded dynamical systems came from the analysis of the models obtained after input-output linearization (Arcak et al., 2002; Loria and Panteley, 2005). Following this, stability conditions have been derived for different types of subsystems. In this section, stability conditions for cascaded nonlinear systems are presented, together with an algorithm for partitioning a system into cascaded subsystems. For the ease of notation and without loss of generality, only two subsystems are considered.

5.2.1 Cascaded Dynamic Systems

Consider the following general nonlinear system:

$$\begin{aligned} \dot{x}_1 &= f_1(\boldsymbol{x}, \boldsymbol{u}) & y_1 &= h_1(\boldsymbol{x}) \\ \dot{x}_2 &= f_2(\boldsymbol{x}, \boldsymbol{u}) & y_2 &= h_2(\boldsymbol{x}) \\ &\vdots & &\vdots \\ \dot{x}_{n_x} &= f_{n_x}(\boldsymbol{x}, \boldsymbol{u}) & y_{n_y} &= h_{n_y}(\boldsymbol{x}) \end{aligned} \tag{5.1}$$

where $\boldsymbol{x} = (x_1, ..., x_{n_x})^T$ is the state vector, $\boldsymbol{u} = (u_1, ..., u_{n_u})^T$ is the input vector, $\boldsymbol{y} = (y_1, ..., y_{n_y})^T$ is the measurement vector, $\boldsymbol{f} = (f_1, ..., f_{n_x})^T$ is the vector of state equations, and $\boldsymbol{h} = (h_1, ..., h_{n_y})^T$ is the vector of measurement equations. This system is the cascade of two subsystems, if it can be written as

$$\begin{aligned} \dot{\boldsymbol{x}}_1 &= \boldsymbol{f}_1(\boldsymbol{x}_1, \boldsymbol{u}) \\ \boldsymbol{y}_1 &= \boldsymbol{h}_1(\boldsymbol{x}_1) \end{aligned} \tag{5.2}$$

and

$$\begin{aligned} \dot{\boldsymbol{x}}_2 &= \boldsymbol{f}_2(\boldsymbol{x}_1, \boldsymbol{x}_2, \boldsymbol{u}) \\ \boldsymbol{y}_2 &= \boldsymbol{h}_2(\boldsymbol{x}_1, \boldsymbol{x}_2) \end{aligned} \tag{5.3}$$

where $\boldsymbol{x} = R_x(\boldsymbol{x}_1^T\ \boldsymbol{x}_2^T)^T$, $\boldsymbol{y} = R_y(\boldsymbol{y}_1^T\ \boldsymbol{y}_2^T)^T$, with R_x and R_y being suitable permutation matrices. In this setting, $\boldsymbol{x}_1$ can also be considered an input for the second subsystem (5.3). In general, such a partition of the model does not necessarily exist, and if it exists, the partition might not be unique. Note that if it is not possible to determine a partition by reordering the variables, it may still be possible to partition the system by a transformation of the state variables. However, in this book such a transformation is not considered.

When partitioning a nonlinear system, two cases can be distinguished, depending on whether stability analysis or observer design is considered. If one considers partitioning the system in order to facilitate stability analysis, only the state transition functions have to be partitioned. Such a cascaded structure is illustrated in Figure 5.1 for the case of two subsystems.

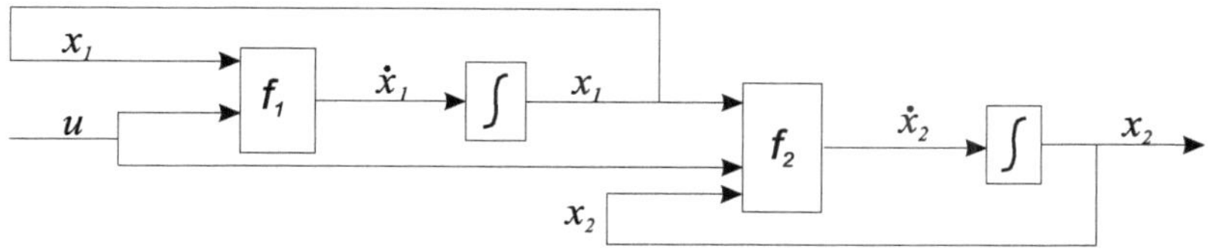

Fig. 5.1 A block diagram of two cascaded subsystems.

On the other hand, if the goal is to design cascaded observers, both the state and measurement equations have to be partitioned, and, furthermore, two conditions have to be satisfied. First of all, the nonlinear system (5.1) has to be observable, and must have at least two measurement equations. Second, the system (5.1) should be partitioned into (5.2) and (5.3), such that (5.2) is observable. In this case, since both (5.1) and (5.2) are observable, the subsystem (5.3) is also observable, given $\boldsymbol{x}_1$. The cascaded structure for two subsystems together with the measurement equations is presented in Figure 5.2.

After a partition has been determined, cascaded observer design can be performed. This means that the observers are designed for the individual subsystems, with some observers using the estimates obtained by other observers. For two subsystems, the cascaded observer structure is depicted in Figure 5.3.

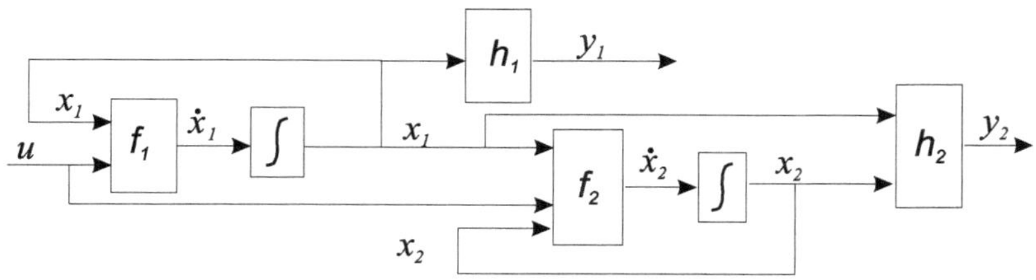

Fig. 5.2 Cascaded subsystems with measurement equations.

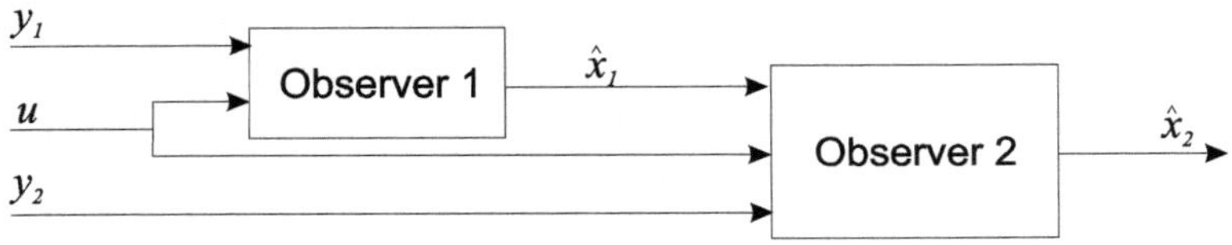

Fig. 5.3 Cascaded observers.

5.2.2 Partitioning a Nonlinear System

Before performing the cascaded analysis or design, one has to determine whether the system considered can be written as the cascade of at least two subsystems. In what follows, partitioning a system for cascaded observer design is considered. Therefore, it is assumed that the nonlinear system (5.1) is observable, and $n_y \geq 2$, i.e., there are at least two different measurement equations. Since the given state variables should be preserved, no coordinate change is allowed.

In what follows, an algorithm that determines whether a system is the cascade of two subsystems is presented. Given the nonlinear system (5.1), for each measurement function, one can determine the variables observable from the respective measurement, thereby constructing sets of observable variables. After these sets are constructed, the problem of determining whether the system is cascaded is reduced to that of partitioning the variable sets. The algorithm can be given as follows:

Algorithm 5.1

1. Construct the variable table presented in Table 5.1, where $v_{1,i}$, $i = 1, 2, \ldots, n_y$ is the set of state variables that appear in the expression of h_i, $v_{2,i}$, $i = 1, 2, \ldots, n_y$ is the set of state variables that appear in the expression of h_i and $L_{\boldsymbol{f}} h_i$, etc., where $L_{\boldsymbol{f}} h_i$ denotes the derivative[1] of h_i with respect to $\boldsymbol{f}$.

Table 5.1 Variable table

	h_1	h_2	$\ldots$	h_{n_y}
h	$v_{1,1}$	$v_{1,2}$	$\ldots$	v_{1,n_y}
$L_{\boldsymbol{f}} h$	$v_{2,1}$	$v_{2,2}$	$\ldots$	v_{2,n_y}
$L_{\boldsymbol{f}}^2 h$	$v_{3,1}$	$v_{3,2}$	$\ldots$	v_{3,n_y}
$\ldots$				

After a maximum of n_x steps, these sets cannot expand anymore, $v_{n_x,i} = v_{n_x+1,i}$. The "worst" case is an n_xth-order integrator, in which case at each step a new variable appears and the expansion stops at exactly the n_xth step.

[1] The derivative of h_i with respect to $\boldsymbol{f}$ is defined as: $L_{\boldsymbol{f}} h_i = \frac{\partial h_i}{\partial \boldsymbol{x}} \boldsymbol{f}$. The second-order derivative is $L_{\boldsymbol{f}}^2 h_i = L_{\boldsymbol{f}}(L_{\boldsymbol{f}} h_i)$, etc.

2. *Denote with ϕ_i the set of state variables corresponding to h_i, $i = 1, 2, \ldots, m$, i.e., $\phi_i = \cup_{j=1}^{n_x} v_{i,j}$. It can be easily seen that, since the system (5.1) is observable, $\cup_{i=1}^{n_y} \phi_i = \Phi$, where Φ corresponds to the set of all state variables, $\Phi = \{x_1, x_2, \ldots, x_{n_x}\}$.*
3. *Group together those measurement equations, which have or include the same set of variables: $h^i = \{h_k | \phi_k \subset \phi_i\}$ and delete the doubles. If only one pair (h^i, ϕ_i) remains, the system cannot be partitioned using this algorithm.*
4. *For each pair (h^i, ϕ_i) for which $\phi_i \neq \Phi$ construct the subsystem*

$$\begin{aligned} \dot{\boldsymbol{x}}^i &= \boldsymbol{f}^i(\boldsymbol{x}^i, \boldsymbol{u}) \\ \boldsymbol{y}^i &= \boldsymbol{h}^i(\boldsymbol{x}^i) \end{aligned} \tag{5.4}$$

where $\boldsymbol{x}^i$ is the vector of the variables in ϕ_i, $\boldsymbol{f}^i$ is the set of the corresponding functions, $\boldsymbol{h}^i$ is obtained at Step 3, and $\boldsymbol{y}^i$ are the measurements given by $\boldsymbol{h}^i$. If any of the systems (5.4) is observable, then it can be considered as one of the subsystems, and the remaining variables and functions form a second subsystem. Otherwise, the system cannot be partitioned using this algorithm.

The procedure is illustrated on the following example.

Example 5.1. Consider the nonlinear system

$$\begin{aligned} \dot{x}_1 &= x_1^2 - x_2 + 3x_3 & y_1 &= x_1 + x_3 \\ \dot{x}_2 &= 2x_2 + x_1 x_2 - 3x_3^2 & y_2 &= x_3 + x_4 \\ \dot{x}_3 &= 3x_3 + 5x_4 & y_3 &= x_4 \\ \dot{x}_4 &= -5x_3 \end{aligned} \tag{5.5}$$

This system is observable. The derivatives of $\boldsymbol{h}$ with respect to $\boldsymbol{f}$ are computed as

$$\begin{aligned} \boldsymbol{h} &= \begin{pmatrix} x_1 + x_3 \\ x_3 + x_4 \\ x_4 \end{pmatrix} \\ L_{\boldsymbol{f}}\boldsymbol{h} &= \begin{pmatrix} x_1^2 - x_2 + 6x_3 + 5x_4 \\ -31x_3 - 10x_4 \\ -5x_3 \end{pmatrix} \\ L_{\boldsymbol{f}}^2\boldsymbol{h} &= \begin{pmatrix} 2x_1^3 - 3x_1x_2 + 6x_1x_3 - 2x_2 + 3x_3^2 - 7x_3 + 30x_4 \\ -16x_3 + 15x_4 \\ -15x_3 - 25x_4 \end{pmatrix} \end{aligned} \tag{5.6}$$

Applying the Algorithm 5.1, the following results are obtained.

1. Based on (5.6), the variable table presented in Table 5.2 can be constructed for system (5.5).

   Note that it is not necessary to compute $L_{\boldsymbol{f}}^3\boldsymbol{h}$, as the sets of variables are no longer changing.

Table 5.2 Variable table for system (5.5)

	h_1	h_2	h_3
$\boldsymbol{h}$	$\{x_1, x_3\}$	$\{x_3, x_3\}$	$\{x_4\}$
$L_f\boldsymbol{h}$	$\{x_1, x_2, x_3, x_4\}$	$\{x_4, x_3\}$	$\{x_4, x_3\}$
$L_f^2\boldsymbol{h}$	$\{x_1, x_2, x_3, x_4\}$	$\{x_4, x_3\}$	$\{x_4, x_3\}$

2. The sets of variables are obtained as $\phi_1 = \{x_1, x_2, x_3, x_4\}$, $\phi_2 = \{x_3, x_4\}$, and $\phi_3 = \{x_3, x_4\}$.
3. The measurement equations are grouped together as: $\boldsymbol{h}^1 = \{h_1, h_2, h_3\}$ with the corresponding ϕ_1 and $\boldsymbol{h}^2 = \{h_2, h_3\}$ with the corresponding ϕ_2.
4. Finally, the system is partitioned as

   First subsystem:

$$\begin{aligned} \dot{x}_3 &= 3x_3 + 5x_4 & \qquad y_2 &= x_3 + x_4 \\ \dot{x}_4 &= -5x_3 & \qquad y_3 &= x_4 \end{aligned} \tag{5.7}$$

   Second subsystem:

$$\begin{aligned} \dot{x}_1 &= x_1^2 - x_2 + 3x_3 & \qquad y_1 &= x_1 + x_3 \\ \dot{x}_2 &= 2x_2 + x_1x_2 - 3x_3^2 \end{aligned}$$

   It can easily be verified that the subsystem (5.7) is observable, and therefore this partition is valid.

Note that this is not the only possible partition. Since the subsystem

$$\begin{aligned} \dot{x}_3 &= 3x_3 + 5x_4 \\ \dot{x}_4 &= -5x_3 \end{aligned}$$

is observable both from the measurement equation $y_2 = x_3 + x_4$ and from $y_3 = x_4$, the partitions

First subsystem:

$$\begin{aligned} \dot{x}_3 &= 3x_3 + 5x_4 & \qquad y_2 &= x_3 + x_4 \\ \dot{x}_4 &= -5x_3 \end{aligned}$$

Second subsystem:

$$\begin{aligned} \dot{x}_1 &= x_1^2 - x_2 + 3x_3 & \qquad y_1 &= x_1 + x_3 \\ \dot{x}_2 &= 2x_2 + x_1x_2 - 3x_3^2 & \qquad y_3 &= x_4 \end{aligned}$$

or

First subsystem:

$$\dot{x}_3 = 3x_3 + 5x_4 \qquad y_2 = x_4$$
$$\dot{x}_4 = -5x_3$$

Second subsystem:

$$\dot{x}_1 = x_1^2 - x_2 + 3x_3 \qquad y_1 = x_1 + x_3$$
$$\dot{x}_2 = 2x_2 + x_1x_2 - 3x_3^2 \qquad y_3 = x_3 + x_4$$

are also possible.

However, the latter two partitions do not use all available information: the measurement equations $y_3 = x_4$ and $y_2 = x_3 + x_4$, when used in the second subsystem do not add new information. In order to use all available information, the measurement equations corresponding to the same set of variables have to be grouped together, as it is done in the algorithm. □

Note that the partitioning of a system, even without loss of information, is in general not unique, as illustrated in the following example.

Example 5.2. Consider the system

$$\dot{x}_1 = x_1 + x_3 \qquad y_1 = x_1$$
$$\dot{x}_2 = x_2 + x_3 \qquad y_2 = x_2$$
$$\dot{x}_3 = u$$

This system is observable, and there are two possible ways to divide it: by using as first subsystem

$$\dot{x}_1 = x_1 + x_3 \qquad y_1 = x_1$$
$$\dot{x}_3 = u$$

or, by using as first subsystem

$$\dot{x}_2 = x_2 + x_3 \qquad y_1 = x_2$$
$$\dot{x}_3 = u$$

both being observable. The corresponding variable sets are $\phi_1 = \{x_1, x_3\}$ and $\phi_2 = \{x_2, x_3\}$. □

Remark: The partitioning of a general nonlinear dynamics system into two observable subsystems does not guarantee that observers can be designed for the subsystems by using a given method. Moreover, the cascade of the observers designed for the individual subsystems is in general not a valid observer for the cascaded system. That is why we study the special case of cascaded TS fuzzy systems.

Remark: Algorithm 5.1 can also be used to partition a system for stability analysis, by using instead of the measurement equations $\boldsymbol{h}$ the state vector $\boldsymbol{x}$. In this case, in Step 4, the observability of the first subsystem does not have to be verified.

5.2.3 Stability of Cascaded Systems

It is well-known that the cascade of stable linear systems is stable (Loria and Panteley, 2005), since the eigenvalues of the joint system are determined only by the eigenvalues of the individual subsystems. Therefore, the stability of the joint, cascaded system is directly implied by the stability of the subsystems. However, the same reasoning does not necessarily hold for nonlinear or time-varying systems. Even global asymptotic stability (GAS) of the individual subsystems does not necessarily imply the stability of the cascade.

General cascades, in which both subsystems are nonlinear, were studied and conditions to ensure overall stability were derived in (Loria and Panteley, 2005). A selection of relevant results is presented below. These results rely on the ISS property (see Property 3.6).

Consider the nonlinear, cascaded, autonomous system

$$\dot{\boldsymbol{x}}_1 = \boldsymbol{f}_1(\boldsymbol{x}_1) \tag{5.8}$$

$$\dot{\boldsymbol{x}}_2 = \boldsymbol{f}_2(\boldsymbol{x}_1, \boldsymbol{x}_2) \tag{5.9}$$

Without loss of generality, $\boldsymbol{x} = 0$ is considered to be the equilibrium point. Conditions for the stability of this system have been derived by Sontag (1989a).

Theorem 5.1. *Consider the nonlinear system* (5.8)–(5.9). *If*

- *the functions $\boldsymbol{f}_1$ and $\boldsymbol{f}_2$ are sufficiently smooth in their arguments,*
- *system* (5.9) *is input-to-state-stable with regard to the input $\boldsymbol{x}_1$, and*
- *systems* (5.8) *and*

$$\dot{\boldsymbol{x}}_2 = \boldsymbol{f}_2(0, \boldsymbol{x}_2) \tag{5.10}$$

  are globally asymptotically stable (GAS),

then the cascade (5.8)-(5.9) *is GAS.*

An equivalent sufficient stability condition is presented by Seibert and Suarez (1990): the cascaded system is GAS, if both subsystems (5.8) and (5.10) are GAS and all trajectories are bounded. The main difficulty with this approach is that in general, boundedness of all the solutions is not easy to determine and the conditions that ensure boundedness may be very conservative.

More relaxed sufficient stability conditions have been derived for systems of the form

$$\begin{aligned} \dot{\boldsymbol{x}}_1 &= \boldsymbol{f}_1(\boldsymbol{x}_1) \\ \dot{\boldsymbol{x}}_2 &= \boldsymbol{f}_2(\boldsymbol{x}_2) + \boldsymbol{g}(\boldsymbol{x}_1, \boldsymbol{x}_2) \end{aligned} \tag{5.11}$$

assuming that the individual subsystems are GAS and, additionally, certain restrictions related to the continuity and/or slope, apply for the interconnection term $\boldsymbol{g}(\boldsymbol{x}_1, \boldsymbol{x}_2)$ (Jankovic et al., 1996; Arcak et al., 2002; Chaillet and Loria, 2006). A theorem for ensuring that the cascaded system (5.11) is uniformly GAS (UGAS) (Loria and Panteley, 2005) is presented below. This result is valid under the following assumptions:

Assumption 5.1. System

$$\dot{\boldsymbol{x}}_2 = \boldsymbol{f}_2(\boldsymbol{x}_2) \tag{5.12}$$

is UGAS.

Assumption 5.2. There exist constants $\gamma_1, \gamma_2, \mu \geq 0$ and a Lyapunov function $V(t, \boldsymbol{x}_2)$ for (5.10) such that V is positive definite, radially unbounded, $\dot{V}(t, \boldsymbol{x}_2) \leq 0$ and

$$\begin{aligned} \left\| \frac{\partial V}{\partial \boldsymbol{x}_2} \right\| \|\boldsymbol{x}_2\| &\leq \gamma_1 V(t, \boldsymbol{x}_2) \qquad && \forall \boldsymbol{x}_2 : \|\boldsymbol{x}_2\| > \mu \\ \left\| \frac{\partial V}{\partial \boldsymbol{x}_2} \right\| &\leq \gamma_2 && \forall \boldsymbol{x}_2 : \|\boldsymbol{x}_2\| \leq \mu \end{aligned} \tag{5.13}$$

Assumption 5.3. There exist two continuous functions $\theta_1, \theta_2 : \mathbb{R}^+ \rightarrow \mathbb{R}^+$ such that $\boldsymbol{g}(\boldsymbol{x}_1, \boldsymbol{x}_2)$ satisfies

$$\|\boldsymbol{g}(\boldsymbol{x}_1, \boldsymbol{x}_2)\| \leq \theta_1(\|\boldsymbol{x}_1\|) + \theta_2(\|\boldsymbol{x}_1\|)\|\boldsymbol{x}_2\| \tag{5.14}$$

Assumption 5.4. There exists a class $\mathcal{K}$ function $\alpha(\cdot)$ so that for all $t_0 \geq 0$, the trajectories of the system

$$\dot{\boldsymbol{x}}_1 = \boldsymbol{f}_1(\boldsymbol{x}_1)$$

satisfy

$$\int_{t_0}^{\infty} \|\boldsymbol{x}_1(t)\| dt \leq \alpha(\|\boldsymbol{x}_1(t_0)\|) \tag{5.15}$$

Using the assumptions above, the following theorem has been formulated (Loria and Panteley, 2005):

Theorem 5.2. *Let Assumption 5.1 hold and suppose that the trajectories of* (5.8) *are uniformly globally bounded. If, in addition, Assumptions 5.2–5.4 are satisfied, then the solutions of system* (5.11) *are uniformly globally bounded. If furthermore, system* (5.8) *is UGAS, then so is the cascaded system* (5.11).

Proposition 5.1. *If in addition to the above assumptions systems* (5.8) *and* (5.12) *are exponentially stable, then the cascaded system* (5.11) *is also exponentially stable.*

Different cases of interconnection terms have been studied in (Loria and Panteley, 2005; Arcak et al., 2002). Stabilizability conditions for cascaded systems have been derived by Bacciotti et al. (1993); Chaillet and Loria (2006); Roebenack and Lynch (2006).

5.3 Cascaded TS Fuzzy Systems

In this section, the stability and convergence rate of cascaded TS systems are analyzed, and sufficient conditions that are based on the stability of the subsystems are presented. The idea behind this type of stability analysis is that many systems are cascaded (e.g., hierarchical large-scale systems, flow processes), while others may be represented as a cascaded system that is less complex than the original system. Since the dimensions of the subsystems of a cascaded system are smaller than that of the whole system, a cascaded analysis involves LMI problems of smaller dimensions, thereby reducing the computational costs.

5.3.1 Stability Analysis of Cascaded TS Systems

First, the stability analysis of cascaded TS systems is considered. The results presented in this section make use of the stability conditions of Section 5.2.3. Consider a general, nonlinear autonomous dynamical system given as

$$\dot{\boldsymbol{x}} = \boldsymbol{f}(\boldsymbol{x})$$

that is the cascade of two subsystems

$$\begin{aligned} \dot{\boldsymbol{x}}_1 &= \boldsymbol{f}_1(\boldsymbol{x}_1) \\ \dot{\boldsymbol{x}}_2 &= \boldsymbol{f}_2(\boldsymbol{x}_1, \boldsymbol{x}_2) \end{aligned} \tag{5.16}$$

Using the methods for constructing TS models presented in Chapter 2, for a cascaded nonlinear system, it is always possible to determine a fuzzy representation or approximation that is also cascaded. This means that it is possible to determine a fuzzy representation or approximation of the form

$$\dot{\boldsymbol{x}} = \sum_{i=1}^{m} w_i(\boldsymbol{z}) A_i \boldsymbol{x} \tag{5.17}$$

where the system matrices for each rule $i = 1, 2, \ldots, m$ can be written as

$$A_i = \begin{pmatrix} A_{1i} & 0 \\ A_{21i} & A_{2i} \end{pmatrix}$$

Hence, the system (5.16) can be expressed or approximated as

$$\begin{aligned} \dot{\boldsymbol{x}}_1 &= \sum_{i=1}^{m_1} w_{1i}(\boldsymbol{z}_1) A_{1i} \boldsymbol{x}_1 \\ \dot{\boldsymbol{x}}_2 &= \sum_{i=1}^{m_2} w_{2i}(\boldsymbol{z}_2)(A_{21i}\boldsymbol{x}_1 + A_{2i}\boldsymbol{x}_2) \end{aligned} \tag{5.18}$$

with normalized membership functions w_{1i} and w_{2i}. Note that system (5.18) can always be written using lower block-triangular local matrices[2]:

$$\begin{aligned}\dot{\boldsymbol{x}} &= \begin{pmatrix} \sum_{i=1}^{m_1} w_{1i}(\boldsymbol{z}_1) A_{1i} \boldsymbol{x}_1 \\ \sum_{i=1}^{m_2} w_{2i}(\boldsymbol{z}_2) A_{21i} \boldsymbol{x}_1 + \sum_{i=1}^{m_2} w_{2i}(\boldsymbol{z}_2) A_{2i} \boldsymbol{x}_2 \end{pmatrix} \\ &= \begin{pmatrix} \sum_{i=1}^{m_1} w_{1i}(\boldsymbol{z}_1) \sum_{j=1}^{m_2} w_{2i}(\boldsymbol{z}_2) A_{1i} & 0 \\ \sum_{i=1}^{m_1} w_{1i}(\boldsymbol{z}_1) \sum_{j=1}^{m_2} w_{2j}(\boldsymbol{z}_2) A_{21j} & \sum_{i=1}^{m_1} w_{1i}(\boldsymbol{z}_1) \sum_{j=1}^{m_2} w_{2j}(\boldsymbol{z}_2) A_{2j} \end{pmatrix} \begin{pmatrix} \boldsymbol{x}_1 \\ \boldsymbol{x}_2 \end{pmatrix} \\ &= \sum_{i=1}^{m_1} w_{1i}(\boldsymbol{z}_1) \sum_{j=1}^{m_2} w_{2j}(\boldsymbol{z}_2) \begin{pmatrix} A_{1i} & 0 \\ A_{21j} & A_{2j} \end{pmatrix} \boldsymbol{x} \end{aligned} \tag{5.19}$$

Since $\boldsymbol{z}_2$ may contain functions of the states of the first subsystem, or even scheduling variables of the first subsystem, the number of rules in (5.19) is larger than the number of rules in a fuzzy system obtained directly from (5.16). The construction of a cascaded TS system is illustrated using the following example.

Example 5.3. Consider the nonlinear system

$$\begin{aligned}\dot{x}_1 &= -2x_2^3 \\ \dot{x}_2 &= x_1 x_2^2 - x_1 \\ \dot{x}_3 &= 2x_1 + x_3 x_2^2 + 2x_4 \\ \dot{x}_4 &= 3x_2 + x_3 - x_4^3 + x_3 x_4\end{aligned} \tag{5.20}$$

with $x_i \in [-1, 1]$. This system can be written as the cascade of the two subsystems:

$$\begin{pmatrix} \dot{x}_1 \\ \dot{x}_2 \end{pmatrix} = \begin{pmatrix} 0 & -2x_2^2 \\ -1 & x_1 \end{pmatrix} \begin{pmatrix} x_1 \\ x_2 \end{pmatrix}$$

and

$$\begin{pmatrix} \dot{x}_3 \\ \dot{x}_4 \end{pmatrix} = \begin{pmatrix} 2 & 0 & x_2^2 & 2 \\ 0 & 3 & 1 & (x_3 - x_4) \end{pmatrix} \begin{pmatrix} x_1 \\ x_2 \\ x_3 \\ x_4 \end{pmatrix}$$

To write the TS fuzzy models for this example, the sector nonlinearity approach is employed. The first subsystem is represented using four rules. The scheduling vector is $\boldsymbol{z}_1 = (x_2^2,\, x_1)^T$, the membership functions are computed as the products of the

[2] Recall that $\sum_{i=1}^{m_1} w_{1i}(\boldsymbol{z}_1) = 1$ and $\sum_{i=1}^{m_2} w_{2i}(\boldsymbol{z}_2) = 1$, $\forall \boldsymbol{z}_1,\, \boldsymbol{z}_2$.

weighting functions $\eta_{10}^1 = 1 - x_2^2$, $\eta_{11}^1 = 1 - \eta_{10}^1$, $\eta_{10}^2 = (1 - x_1)/2$, $\eta_{11}^2 = 1 - \eta_{10}^2$, and the local matrices are

$$A_{11} = \begin{pmatrix} 0 & 0 \\ -1 & -1 \end{pmatrix} \quad A_{12} = \begin{pmatrix} 0 & 0 \\ -1 & 1 \end{pmatrix} \quad A_{13} = \begin{pmatrix} 0 & -2 \\ -1 & -1 \end{pmatrix} \quad A_{14} = \begin{pmatrix} 0 & -2 \\ -1 & 1 \end{pmatrix}$$

The second subsystem is again represented using four rules. The scheduling vector for this subsystem is $\boldsymbol{z}_2 = (x_2^2, x_3 - x_4^T)$, the membership functions are computed as the products of the weighting functions $\eta_{20}^1 = 1 - x_2^2$, $\eta_{21}^1 = 1 - \eta_{20}^1$, $\eta_{20}^2 = (2 - x_3 + x_4)/4$, $\eta_{21}^2 = 1 - \eta_{20}^2$, and the local matrices are

$$A_{21} = \begin{pmatrix} 0 & 2 \\ 1 & -2 \end{pmatrix} \quad A_{22} = \begin{pmatrix} 0 & 2 \\ 1 & 2 \end{pmatrix} \quad A_{23} = \begin{pmatrix} 1 & 2 \\ 1 & -2 \end{pmatrix} \quad A_{24} = \begin{pmatrix} 1 & 2 \\ 1 & 2 \end{pmatrix}$$

The interconnection term is linear, therefore we have $A_{21j} = \begin{pmatrix} 2 & 0 \\ 0 & 3 \end{pmatrix}$, $j = 1, 2, 3, 4$. The term x_2^2 is a scheduling variable both for the first and the second subsystem. Due to this, writing the overall system as (5.19) results in a TS system with $4 \times 4 = 16$ rules, while by using the sector nonlinearity approach directly for the system (5.20) results in a cascaded TS system with only 8 rules. Nevertheless, both systems are exact fuzzy representations of the nonlinear system (5.20). □

As shown in the example above, in general it is computationally more efficient to use a system description of the form (5.18), instead of (5.17) with the matrices having a cascaded form. Therefore, the system (5.18) is considered in the remainder of this section. Using the results presented in Section 5.2.3 and assuming that the subsystems

$$\dot{\boldsymbol{x}}_1 = \sum_{i=1}^{m} w_{1i}(\boldsymbol{z}_1) A_{1i} \boldsymbol{x}_1 \tag{5.21}$$

and

$$\dot{\boldsymbol{x}}_2 = \sum_{i=1}^{m} w_{2i}(\boldsymbol{z}_2) A_{2i} \boldsymbol{x}_2 \tag{5.22}$$

are uniformly globally asymptotically stable (UGAS), the following basic result can be formulated.

Theorem 5.3. *If there exist two Lyapunov functions of the form* $V_1(\boldsymbol{x}_1) = \boldsymbol{x}_1^T P_1 \boldsymbol{x}_1$ *and* $V_2(\boldsymbol{x}_2) = \boldsymbol{x}_2^T P_2 \boldsymbol{x}_2$ *so that the subsystems* (5.21) *and* (5.22) *are UGAS, then the cascaded system* (5.18) *is also UGAS.*

Proof: The Lyapunov functions $V_1(\boldsymbol{x}_1) = \boldsymbol{x}_1^T P_1 \boldsymbol{x}_1$ and $V_2(\boldsymbol{x}_2) = \boldsymbol{x}_2^T P_2 \boldsymbol{x}_2$ for the subsystems (5.21) and (5.22) satisfy Assumption 5.1. They also ensure that Assumption 5.4 is satisfied. Note that although (5.18) is not of the form (5.11), as the fuzzy term corresponding to $\boldsymbol{f}_2(\boldsymbol{x}_2)$ also contains $\boldsymbol{x}_1$, the common quadratic Lyapunov function $V_2(\boldsymbol{x}_2) = \boldsymbol{x}_2^T P_2 \boldsymbol{x}_2$ ensures the global asymptotic stability (5.22), irrespective of the value and the dynamics of $\boldsymbol{x}_1$.

Assumption 5.2 is satisfied as: $\forall \boldsymbol{x}_2 : \|\boldsymbol{x}_2\| > \mu$,

$$\left\| \frac{\partial V_2}{\partial \boldsymbol{x}_2} \right\| \|\boldsymbol{x}_2\| = 2\|\boldsymbol{x}_2^T\| \|P_2\| \|\boldsymbol{x}_2\| \leq 2\lambda_{\max}(P_2)\|\boldsymbol{x}_2\|^2 \leq \gamma_1 V_2(\boldsymbol{x}_2)$$

for any $\gamma_1 \geq \frac{2\lambda_{\max}(P_2)}{\lambda_{\min}(P_2)}$. For the second condition of Assumption 5.2, we have $\forall \boldsymbol{x}_2 : \|\boldsymbol{x}_2\| \leq \mu$,

$$\left\| \frac{\partial V_2}{\partial \boldsymbol{x}_2} \right\| = \|2\boldsymbol{x}_2^T P_2\| \leq 2\|\boldsymbol{x}_2\| \|P_2\| \leq 2\mu\lambda_{\max}(P_2) = \gamma_2$$

The interconnection term $\boldsymbol{g}(\boldsymbol{x}_1, \boldsymbol{x}_2)$ is a nonlinear combination of local linear models, $\boldsymbol{g}(\boldsymbol{x}_1, \boldsymbol{x}_2) = \sum_{i=1}^{m_2} w_{2i}(\boldsymbol{z}_2) A_{12}\boldsymbol{x}_1$. To satisfy Assumption 5.3, consider continuous, positive functions $\theta_1(\|\boldsymbol{x}_1\|) = \max_{\boldsymbol{z}} \|A_{21}(\boldsymbol{z})\| \|\boldsymbol{x}_1\|$ and $\theta_2(\|\boldsymbol{x}_1\|) = 0$, where $A_{21}(\boldsymbol{z}) = \sum_{i=1}^{m_2} w_{2i}(\boldsymbol{z}_2) A_{21i}$. By selecting these functions, it is ensured that $\|\boldsymbol{g}(\boldsymbol{x}_1, \boldsymbol{x}_2)\| = \|\sum_{i=1}^{m} w_i(\boldsymbol{z}_2) A_{21i}\boldsymbol{x}_1\| \leq \theta_1(\|\boldsymbol{x}_1\|) + \theta_2(\|\boldsymbol{x}_1\|)\|\boldsymbol{x}_2\|$ and therefore Assumption 5.3 is satisfied.

Since the conditions of Theorem 5.2 are satisfied, the cascaded system is UGAS. Furthermore, if these Lyapunov functions ensure exponential stability of the subsystems, based on Proposition 5.1, the cascaded system is also exponentially stable. □

Although the stability of the cascaded system is ensured by the above conditions, finding a Lyapunov function valid for the cascaded system is not trivial. In general, the sum of the Lyapunov functions of the individual subsystems is not a valid Lyapunov function for the whole system. A global Lyapunov function of the form

$$V_0(\boldsymbol{x}_1, \boldsymbol{x}_2) = V_1(\boldsymbol{x}_1) + V_2(\boldsymbol{x}_2) + \Psi(\boldsymbol{x}_1, \boldsymbol{x}_2) \tag{5.23}$$

has been proposed by Jankovic et al. (1996), with V_1 and V_2 being Lyapunov functions for the systems (5.21) and (5.22), respectively.

For the case when the first subsystem is linear and time-invariant, it has been proven by Jankovic et al. (1996) that the cross-term $\Psi(\boldsymbol{x}_1, \boldsymbol{x}_2)$ exists and is continuous, and V_0 is positive definite and radially unbounded. Moreover, if (5.21) is globally exponentially stable, the result can be extended to the system (5.18). The cross-term Ψ has been formally proven to be

$$\Psi(\boldsymbol{x}_1, \boldsymbol{x}_2) = \int_0^\infty \frac{\partial V_2}{\partial \boldsymbol{x}_2}(\tilde{\boldsymbol{x}}_2(s)) A_{21}(\boldsymbol{z}_2(s))\tilde{\boldsymbol{x}}_1(s) ds$$

where $\tilde{\boldsymbol{x}}_1$ and $\tilde{\boldsymbol{x}}_2$ are the trajectories of systems (5.21) and (5.22), respectively, and $A_{21}(\boldsymbol{z}_2) = \sum_{i=1}^{m_2} w_{2i}(\boldsymbol{z}_2) A_{21i}$.

The cascaded approach can also be used when the TS system is subject to vanishing disturbances, i.e., disturbances that go to zero as $\boldsymbol{x} \to 0$ (see (Bergsten, 2001)) or the local models are affine (Johansson et al., 1999). The resulting stability conditions are presented below.

To formulate stability conditions in the cascaded setting when the system is subject to vanishing disturbances, consider the cascaded TS system:

$$\begin{aligned}\dot{\boldsymbol{x}}_1 &= \sum_{i=1}^{m_1} w_{1i}(\boldsymbol{z}_1) A_{1i} \boldsymbol{x}_1 + D_1(\boldsymbol{x}_1) \\ \dot{\boldsymbol{x}}_2 &= \sum_{i=1}^{m_2} w_{2i}(\boldsymbol{z}_2)(A_{21i}\boldsymbol{x}_1 + A_{2i}\boldsymbol{x}_2) + D_{21}(\boldsymbol{x}_1) + D_2(\boldsymbol{x}_2)\end{aligned} \tag{5.24}$$

where $D_1(\boldsymbol{x}_1)$, $D_{21}(\boldsymbol{x}_1)$, and $D_2(\boldsymbol{x}_2)$ denote disturbance terms that are bounded as

$$\begin{aligned}\|D_1(\boldsymbol{x}_1)\| &\leq \mu_1 \|\boldsymbol{x}_1\| \\ \|D_{21}(\boldsymbol{x}_1)\| &\leq \mu_{21} \|\boldsymbol{x}_1\| \\ \|D_2(\boldsymbol{x}_2)\| &\leq \mu_2 \|\boldsymbol{x}_2\|\end{aligned}$$

where μ_1, μ_{21}, and μ_2 are known non-negative finite constants.

Then, the stability conditions can be stated as:

Theorem 5.4. *The system* (5.24) *is UGAS, if there exist* $P_1 = P_1^T > 0$, $P_2 = P_2^T > 0$, $Q_1 = Q_1^T > 0$, *and* $Q_2 = Q_2^T > 0$ *so that*

$$\begin{aligned}&\mathcal{H}(P_1 A_{1i}) < -Q_1 \qquad i = 1, 2, \ldots, m_1 \\ &\begin{pmatrix} Q_1 - \mu_1^2 I & P_1 \\ P_1 & I \end{pmatrix} > 0 \\ &\mathcal{H}(P_2 A_{2i}) < -Q_2 \qquad i = 1, 2, \ldots, m_2 \\ &\begin{pmatrix} Q_2 - \mu_2^2 I & P_2 \\ P_2 & I \end{pmatrix} > 0\end{aligned}$$

Proof: The Lyapunov functions $V_1 = \boldsymbol{x}_1^T P_1 \boldsymbol{x}_1$ and $V_2 = \boldsymbol{x}_2^T P_2 \boldsymbol{x}_2$ prove the UGAS of the subsystems with the corresponding disturbance term

$$\dot{\boldsymbol{x}}_1 = \sum_{i=1}^{m_1} w_{1i}(\boldsymbol{z}_1) A_{1i} \boldsymbol{x}_1 + D_1(\boldsymbol{x}_1)$$

and

$$\dot{\boldsymbol{x}}_2 = \sum_{i=1}^{m_2} w_{2i}(\boldsymbol{z}_2) A_{2i} \boldsymbol{x}_2 + D_2(\boldsymbol{x}_2)$$

The rest of the proof is similar to that of Theorem 5.3. □

The cascaded approach can also be used for affine TS systems, by using the approach of Johansson et al. (1999). For this, consider the affine cascaded fuzzy model

$$\begin{aligned}\dot{\boldsymbol{x}}_1 &= \sum_{i=1}^{m_1} w_{1i}(\boldsymbol{z}_1)(A_{1i}\boldsymbol{x}_1 + a_{1i}) \\ \dot{\boldsymbol{x}}_2 &= \sum_{i=1}^{m_2} w_{2i}(\boldsymbol{z}_2)(A_{21i}\boldsymbol{x}_1 + A_{2i}\boldsymbol{x}_2 + a_{2i})\end{aligned} \tag{5.25}$$

Similarly to the result of Johansson et al. (1999), the analysis relies on dividing the state-space of each individual subsystem into polyhedral partitions. Let K_1 and K_2 be the number of operating and interpolation regions (see Johansson et al. (1999)) for the individual subsystems, with K_1^i and K_2^j the index sets corresponding to the local models of the subsystems active in the regions X_1^i and X_2^j. In general, the number of regions generated in such a way is smaller than the number of regions for the global system, i.e., $K_1 + K_2 \leq K$ and therefore, the number of LMIs to be solved is smaller. Then, the conditions can be expressed as:

Proposition 5.2. *The system* (5.18) *is UGAS, if there exist matrices* $P_1^i = (P_1^i)^T > 0$, $P_2^j = (P_2^j)^T > 0$, $H_1 = H_1^T > 0$, $H_2 = H_2^T > 0$, F_1^i, *and* F_2^j, *so that:*

$$\begin{aligned}
&P_1^i = (F_1^i)^T H_1 F_1^i \\
&P_2^j = (F_2^j)^T H_2 F_2^j \\
&F_1^i \boldsymbol{x}_1 = F_1^t \boldsymbol{x}_1 && \forall \boldsymbol{x}_1 \in X_1^i \cap X_1^t \\
&F_2^j \boldsymbol{x}_2 = F_2^l \boldsymbol{x}_2 && \forall \boldsymbol{x}_2 \in X_2^j \cap X_2^l \\
&\mathcal{H}(P_1^i A_{1k}) < 0 && \forall k \in K_1^i \\
&\mathcal{H}(P_2^j A_{2k}) < 0 && \forall k \in K_2^j
\end{aligned} \tag{5.26}$$

for all $i, t = 1, 2, \ldots, K_1$, $j, l = 1, 2, \ldots, K_2$.

The conditions above are still only sufficient conditions for the stability of cascaded fuzzy systems. Nevertheless, by taking advantage of the special form of the system, i.e., by considering the subsystems instead of the overall fuzzy system, the dimension of the associated LMI problem is reduced with respect to a centralized approach, as illustrated by the following example.

Example 5.4. Consider the nonlinear system:

$$\begin{aligned}\dot{x}_1 &= x_1^2 - 3x_1 \\ \dot{x}_2 &= \frac{x_1^2}{2 + x_2} - 2x_2^3 - x_2\end{aligned}$$

where $x_1, x_2 \in [-1, 1]$. This system has a locally asymptotically stable equilibrium point in $\boldsymbol{x} = 0$, provable e.g., by using the Lyapunov function $V = x_1^2/2 + x_2^2/2$.

A fuzzy representation of this system can be obtained by using the sector nonlinearity approach. In total, 8 rules are obtained, with the state matrices being:

$$A_1=\begin{pmatrix}-4 & 0\\ -1 & -3\end{pmatrix} \quad A_2=\begin{pmatrix}-4 & 0\\ 1 & -3\end{pmatrix} \quad A_3=\begin{pmatrix}-4 & 0\\ -1 & -1\end{pmatrix} \quad A_4=\begin{pmatrix}-4 & 0\\ 1 & -1\end{pmatrix}$$

$$A_5=\begin{pmatrix}-2 & 0\\ -1 & -3\end{pmatrix} \quad A_6=\begin{pmatrix}-2 & 0\\ 1 & -3\end{pmatrix} \quad A_7=\begin{pmatrix}-2 & 0\\ -1 & -1\end{pmatrix} \quad A_8=\begin{pmatrix}-2 & 0\\ 1 & -1\end{pmatrix}$$

i.e., there are 8 local linear models. Using Theorem 3.1, this means that 8 LMIs have to be solved. Using the cascaded approach, the problem is reduced to finding $P_1 > 0$, and $P_2 > 0$, such that

$$\begin{aligned} -4P_1 + P_1(-4) &< 0 \\ -2P_1 + P_1(-2) &< 0 \end{aligned}$$

$$\begin{aligned} -3P_2 + P_2(-3) &< 0 \\ -1P_2 + P_2(-1) &< 0 \end{aligned}$$

As can be seen, by analyzing the subsystems instead of the global fuzzy system, both the number of LMIs and their size can be reduced. □

5.3.2 *Convergence Rate of Cascaded Systems*

In this section, the convergence rate of the system (5.18) compared to the convergence rate of the individual subsystems (5.21) and (5.22) is studied.

For this, consider that both subsystems are exponentially stable, i.e., there exist $k_1, k_2, \alpha_1, \alpha_2 > 0$ so that

$$\|\boldsymbol{x}_1\| \leq k_1 \|\boldsymbol{x}_{10}\| e^{-\alpha_1 t} \tag{5.27}$$

$$\|\boldsymbol{x}_2\| \leq k_2 \|\boldsymbol{x}_{20}\| e^{-\alpha_2 t} \tag{5.28}$$

Since a Lyapunov function of the form $V(\boldsymbol{x}) = V_1(\boldsymbol{x}_1) + V_2(\boldsymbol{x}_2)$, where V_1 and V_2 are Lyapunov functions for the individual subsystems is not a Lyapunov function for the cascaded system, other approaches to determine the convergence rate of the cascaded system have to be considered.

The convergence rate of the system (5.18) is at least γ/β if there exists a Lyapunov function $V = \boldsymbol{x}^T P \boldsymbol{x}$, $P = P^T > 0$, and $\gamma > 0$, so that:

$$\begin{aligned} \alpha\|\boldsymbol{x}\|^2 \leq V &\leq \beta\|\boldsymbol{x}\|^2 \\ \dot{V} &\leq -\gamma\|\boldsymbol{x}\|^2 \end{aligned}$$

In terms of the subsystems, the above conditions are satisfied, if there exist two Lyapunov functions $V_1 = \boldsymbol{x}_1^T P_1 \boldsymbol{x}_1$ and $V_2 = \boldsymbol{x}_2^T P_2 \boldsymbol{x}_2$ that ensure the stability of the subsystems, and, furthermore

- $\alpha \leq \min(\lambda_{\min}(P_1), \lambda_{\min}(P_2))$,
- $\beta \geq \max(\lambda_{\max}(P_1), \lambda_{\max}(P_2))$, and
- $\lambda_{\max}(\text{diag}[\mathcal{H}(P_1 A_{1i}),\ \mathcal{H}(P_2 A_{2j})]) \leq -\gamma$

However, the above condition can be relaxed, and the convergence rate of the joint system can be expressed as follows.

Theorem 5.5. *The convergence rate of the system* (5.18) *is at least* $\max\{-\alpha_1, -\alpha_2\} + \epsilon$*, for an arbitrary* $\epsilon > 0$ *if*

1. *system* (5.21) *is exponentially stable, with convergence rate* $-\alpha_1$*,*
2. *system* (5.22) *is exponentially stable, with convergence rate* $-\alpha_2$*, and*
3. *the matrices* A_{21j} *are bounded, i.e., there exists* $M \in \mathbb{R}$*, so that*

$$\|A_{21j}\| \leq M, \qquad j = 1, 2, \ldots, m_2$$

Proof: Condition 1 above can be written as $\|\boldsymbol{x}_1(t)\| \leq k_1 \|\boldsymbol{x}_{10}\| e^{-\alpha_1 t}$, for some $k_1 > 0$, i.e., as (5.27). The solution of the system (5.22) is the homogeneous solution $\boldsymbol{x}_{2\mathrm{h}}(t)$ of the system

$$\dot{\boldsymbol{x}}_2 = \sum_{i=1}^{m_2} w_{2i}(\boldsymbol{z}_2)(A_{21i}\boldsymbol{x}_1 + A_{2i}\boldsymbol{x}_2) \tag{5.29}$$

and therefore it satisfies $\|\boldsymbol{x}_{2\mathrm{h}}(t)\| \leq k_2 \|\boldsymbol{x}_{20}\| e^{-\alpha_2 t}$, for some $k_2 > 0$. The particular solution of equation (5.29) can be expressed as

$$\boldsymbol{x}_{2\mathrm{p}} = \int_{t_0}^{t} \boldsymbol{x}_{2\mathrm{h}}(t-s) A_{21}(\boldsymbol{z}_2(s)) \boldsymbol{x}_1(s) ds$$

with $A_{21}(\boldsymbol{z}_2) = \sum_{i=1}^{m_2} w_{2i}(\boldsymbol{z}_2) A_{21i}$.

Hence,

$$\begin{aligned}
\|\boldsymbol{x}_{2\mathrm{p}}\| &= \left\| \int_{t_0}^{t} \boldsymbol{x}_{2\mathrm{h}}(t-s) A_{21}(\boldsymbol{z}(s)) \boldsymbol{x}_1(s) ds \right\| \\
&\leq \int_{t_0}^{t} \|\boldsymbol{x}_{2\mathrm{h}}(t-s)\| \|A_{21}(\boldsymbol{z}(s))\| \|\boldsymbol{x}_1(s)\| ds \\
&\leq \int_{t_0}^{t} k_2 \|\boldsymbol{x}_{20}\| e^{-\alpha_2(t-s)} M k_1 \|\boldsymbol{x}_{10}\| e^{-\alpha_1 s} ds \\
&\leq k_1 k_2 M \|\boldsymbol{x}_{10}\| \|\boldsymbol{x}_{20}\| e^{-\alpha_2 t} \int_{t_0}^{t} e^{(\alpha_2 - \alpha_1)s} ds
\end{aligned}$$

If $\alpha_2 \neq \alpha_1$, we have

$$\begin{aligned}
\|\boldsymbol{x}_{2\mathrm{p}}\| &\leq k_1 k_2 M \|\boldsymbol{x}_{10}\| \|\boldsymbol{x}_{20}\| |\alpha_2 - \alpha_1|^{-1} \cdot e^{-\alpha_2 t} |e^{(\alpha_2-\alpha_1)t} - e^{(\alpha_2-\alpha_1)t_0}| \\
&\leq k_1 k_2 M \|\boldsymbol{x}_{10}\| \|\boldsymbol{x}_{20}\| |\alpha_2 - \alpha_1|^{-1} |e^{-\alpha_1 t} - \gamma_1 e^{-\alpha_2 t}|
\end{aligned}$$

where $\gamma_1 = e^{(\alpha_2-\alpha_1)t_0}$.

So in this case, a bound on the general solution of (5.29) is

$$\begin{aligned}\|\boldsymbol{x}_2\| &\leq \|\boldsymbol{x}_{2\mathrm{h}}\| + \|\boldsymbol{x}_{2\mathrm{p}}\| \\ &\leq k_2\|\boldsymbol{x}_{20}\|e^{-\alpha_2 t} + k_1 k_2 M\|\boldsymbol{x}_{10}\|\|\boldsymbol{x}_{20}\| \cdot |\alpha_2 - \alpha_1|^{-1}|e^{-\alpha_1 t} - \gamma_1 e^{-\alpha_2 t}| \\ &\leq \gamma_2 e^{\max\{-\alpha_1, -\alpha_2\}t}\end{aligned}$$

where $\gamma_2 = \max\{k_2\|\boldsymbol{x}_{20}\|(1+k_1 M\|\boldsymbol{x}_{10}\||\alpha_2-\alpha_1|^{-1}\gamma_1),\ k_1 k_2\|\boldsymbol{x}_{10}\|\|\boldsymbol{x}_{20}\|M|\alpha_2-\alpha_1|^{-1}\}$, i.e., the convergence rate of (5.29) is at least $\max\{-\alpha_1, -\alpha_2\}$.

For $\alpha_1 = \alpha_2 = \alpha$, we have

$$\begin{aligned}\|\boldsymbol{x}_{2\mathrm{p}}\| &\leq k_1 k_2 M\|\boldsymbol{x}_{10}\|\|\boldsymbol{x}_{20}\|e^{-\alpha t}(t - t_0) \\ \|\boldsymbol{x}_2\| &\leq \|\boldsymbol{x}_{2h}\| + \|\boldsymbol{x}_{2p}\| \\ &\leq k_2\|\boldsymbol{x}_{20}\|e^{-\alpha t} + k_1 k_2 M\|\boldsymbol{x}_{10}\|\|\boldsymbol{x}_{20}\|e^{-\alpha t}(t - t_0) \\ &\leq \gamma_3 e^{-\alpha t} + \gamma_4 t e^{-\alpha t}\end{aligned} \tag{5.30}$$

with $\gamma_3 = k_2\|\boldsymbol{x}_{20}\|$ and $\gamma_4 = k_1 k_2\|\boldsymbol{x}_{10}\|\|\boldsymbol{x}_{20}\|M$. For the bound $\gamma_3 e^{-\alpha t} + \gamma_4 t e^{-\alpha t}$ on (5.30) it has been shown in (Baddou et al., 2006) that the convergence rate is at least $-\alpha + \epsilon$, for an arbitrary $\epsilon > 0$, i.e., the one stated in Theorem 5.5.

This means that the convergence rate of the system (5.29), and, therefore, of the system (5.18) is determined by the convergence rate of the individual subsystems. □

5.4 Cascaded TS Fuzzy Observers

This section presents the cascaded approach applied to observer design for TS fuzzy systems, i.e., for cascaded TS systems observers are designed in a cascaded manner. The benefit of this type of estimation is that separate observers can be designed for the individual subsystems, which makes their tuning easier. Moreover, different types of observers can be combined, depending on the subsystem considered.

For observer design, consider the fuzzy system with normalized membership functions:

$$\begin{aligned}\dot{\boldsymbol{x}} &= \sum_{i=1}^{m} w_i(\boldsymbol{z})(A_i\boldsymbol{x} + B_i\boldsymbol{u} + a_i) \\ \boldsymbol{y} &= \sum_{i=1}^{m} w_i(\boldsymbol{z})(C_i\boldsymbol{x} + c_i)\end{aligned} \tag{5.31}$$

with the system matrices for each rule $i = 1, 2, \ldots, m$ having the lower block-triangular form

$$A_i = \begin{pmatrix} A_{1i} & 0 \\ A_{21i} & A_{2i} \end{pmatrix}$$

$$C_i = \begin{pmatrix} C_{1i} & 0 \\ C_{21i} & C_{2i} \end{pmatrix}$$

or, consider the equivalent cascaded TS system, with the first subsystem being

$$\begin{aligned}\dot{\boldsymbol{x}}_1 &= \sum_{i=1}^{m_1} w_{1i}(\boldsymbol{z}_1)(A_{1i}\boldsymbol{x}_1 + B_{1i}\boldsymbol{u} + a_{1i})\\ \boldsymbol{y}_1 &= \sum_{i=1}^{m_1} w_{1i}(\boldsymbol{z}_1)(C_{1i}\boldsymbol{x}_1 + c_{1i})\end{aligned} \tag{5.32}$$

and the second subsystem

$$\begin{aligned}\dot{\boldsymbol{x}}_2 &= \sum_{i=1}^{m_2} w_{2i}(\boldsymbol{z}_2)(A_{21i}\boldsymbol{x}_1 + A_{2i}\boldsymbol{x}_2 + B_{2i}\boldsymbol{u} + a_{2i})\\ \boldsymbol{y}_2 &= \sum_{i=1}^{m_2} w_{2i}(\boldsymbol{z}_2)(C_{21i}\boldsymbol{x}_1 + C_{2i}\boldsymbol{x}_2 + c_{2i})\end{aligned} \tag{5.33}$$

For system (5.31), a fuzzy observer of the form

$$\begin{aligned}\dot{\widehat{\boldsymbol{x}}} &= \sum_{i=1}^{m} w_i(\widehat{\boldsymbol{z}})(A_i\widehat{\boldsymbol{x}} + B_i\boldsymbol{u} + a_i + L_i(\boldsymbol{y} - \widehat{\boldsymbol{y}}))\\ \widehat{\boldsymbol{y}} &= \sum_{i=1}^{m} w_i(\widehat{\boldsymbol{z}})(C_i\widehat{\boldsymbol{x}} + c_i)\end{aligned} \tag{5.34}$$

is considered, with the overall observer gain having the form $L_i = \begin{pmatrix} L_{1i} & 0 \\ 0 & L_{2i} \end{pmatrix}$, where i denotes the rule number. Such an observer is equivalent to the cascaded observer

$$\begin{aligned}\widehat{\boldsymbol{x}}_1 &= \sum_{i=1}^{m_1} w_{1i}(\widehat{\boldsymbol{z}}_1)(A_{1i}\widehat{\boldsymbol{x}}_1 + B_{1i}\boldsymbol{u} + a_{1i} + L_{1i}(\boldsymbol{y}_1 - \widehat{\boldsymbol{y}}_1))\\ \widehat{\boldsymbol{y}}_1 &= \sum_{i=1}^{m_1} w_{1i}(\widehat{\boldsymbol{z}}_1)(C_{1i}\widehat{\boldsymbol{x}}_1 + c_{1i})\end{aligned} \tag{5.35}$$

for the first subsystem (5.32), and

$$\begin{aligned}\widehat{\boldsymbol{x}}_2 &= \sum_{i=1}^{m_2} w_{2i}(\widehat{\boldsymbol{z}}_2)(A_{21i}\widehat{\boldsymbol{x}}_1 + A_{2i}\widehat{\boldsymbol{x}}_2 + B_{2i}\boldsymbol{u} + a_{2i} + L_{2i}(\boldsymbol{y}_2 - \widehat{\boldsymbol{y}}_2))\\ \widehat{\boldsymbol{y}}_2 &= \sum_{i=1}^{m_2} w_{2i}(\widehat{\boldsymbol{z}}_2)(C_{21i}\widehat{\boldsymbol{x}}_1 + C_{2i}\widehat{\boldsymbol{x}}_2 + c_{2i})\end{aligned} \tag{5.36}$$

for the second subsystem, (5.33).

For the design two cases are distinguished: 1) the scheduling vector depends only on measured variables and 2) the scheduling vector depends on states that have to be estimated.

5.4.1 Measured Scheduling Vector

If the weights do not depend on the states to be estimated, one can use the known scheduling variables in the observer. Then, the error dynamics can be written as

$$\begin{aligned}\dot{e} &= \sum_{i=1}^{m}\sum_{j=1}^{m} w_i(z)w_j(z)(A_i - L_iC_j)e \\ &= \sum_{i=1}^{m}\sum_{j=1}^{m} w_i(z)w_j(z)\begin{pmatrix} A_{1i} - L_{1i}C_{1j} & 0 \\ A_{21i} - L_{2i}C_{21j} & A_{2i} - L_{2i}C_{2j}\end{pmatrix} e\end{aligned} \tag{5.37}$$

or, in the cascaded form

$$\begin{aligned}\dot{e}_1 &= \sum_{i=1}^{m_1} w_{1i}(z_1)\sum_{j=1}^{m_1} w_{1j}(z_1)(A_{1i} - L_{1i}C_{1j})e_1 \\ \dot{e}_2 &= \sum_{i=1}^{m_2} w_{2i}(z_2)\sum_{j=1}^{m_2} w_{2j}(z_2)\Big[(A_{21i} - L_{2i}C_{21j})e_1 + (A_{2i} - L_{2i}C_{2j})e_2\Big]\end{aligned} \tag{5.38}$$

This system is of the form (5.18), for which the stability conditions from Section 5.3 can be used. If the C matrix is common for all the rules, then the theorems of Section 5.3 can be directly applied, and the following theorem can be formulated:

Theorem 5.6. *Consider the system* (5.38), *with* $C_{1i} = C_1$, $i = 1, 2, \ldots, m_1$, $C_{21i} = C_{21}$, *and* $C_{2i} = C_2$, $i = 1, 2, \ldots, m_2$. *If there exist* $P_1 = P_1^T > 0$, L_{1i}, $i = 1, 2, \ldots, m_1$, $P_2 = P_2^T > 0$, *and* L_{2i}, $i = 1, 2, \ldots, m_2$, *so that*

$$\mathcal{H}(P_1(A_{1i} - L_{1i}C_1)) < 0 \tag{5.39}$$

for $i = 1, 2, \ldots, m_1$, *and*

$$\mathcal{H}(P_2(A_{2i} - L_{2i}C_2)) < 0 \tag{5.40}$$

for $i = 1, 2, \ldots, m_1$, *respectively, then the cascaded system*

$$\begin{aligned}\dot{e}_1 &= \sum_{i=1}^{m_1} w_{1i}(z_1)(A_{1i} - L_{1i}C_1)e_1 \\ \dot{e}_2 &= \sum_{i=1}^{m_2} w_{2i}(z_2)[(A_{21i} - L_{2i}C_{21})e_1 + (A_{2i} - L_{2i}C_2)e_2]\end{aligned} \tag{5.41}$$

is UGAS.

If the measurement matrices are different for each rule, relaxed conditions, such as those of Theorem 4.1 or of Corollary 4.1 can be combined with those of Theorem 5.3. Using the conditions of Theorem 4.1 we have

Theorem 5.7. *If there exist* $P_1 = P_1^T > 0$, L_{1i}, $i = 1, 2, \ldots, m_1$, $P_2 = P_2^T > 0$, *and* L_{2i}, $i = 1, 2, \ldots, m_2$, *so that*

$$\begin{aligned} &\mathcal{H}(P_1(A_{1i} - L_{1i}C_{1i})) < 0 \quad i = 1, 2, \ldots, m_1 \\ &\mathcal{H}(P_1(A_{1i} - L_{1i}C_{1j}) + P_1(A_{1j} - L_{1j}C_{1i})) < 0 \\ &i = 1, 2, \ldots, m_1 \quad j = i + 1, \ldots, m_1 \\ &\forall i < j: \ \exists \boldsymbol{z}_1 : \ w_{1i}(\boldsymbol{z}_1) w_{1j}(\boldsymbol{z}_1) \neq 0 \end{aligned} \tag{5.42}$$

and

$$\begin{aligned} &\mathcal{H}(P_2(A_{2i} - L_{2i}C_{2i})) < 0 \quad i = 1, 2, \ldots, m_2 \\ &\mathcal{H}(P_2(A_{2i} - L_{2i}C_{2j}) + P_2(A_{2j} - L_{2j}C_{2i})) < 0 \\ &i = 1, 2, \ldots, m_2 \quad j = i + 1, \ldots, m_2 \\ &\forall i < j: \ \exists \boldsymbol{z}_2 : \ w_{2i}(\boldsymbol{z}_2) w_{2j}(\boldsymbol{z}_2) \neq 0 \end{aligned} \tag{5.43}$$

respectively, then the cascaded error system (5.38) *is UGAS.*

If Corollary 4.1 is used, the following result can be formulated.

Theorem 5.8. *If there exist* $P_1 = P_1^T > 0$, L_{1i}, $i = 1, 2, \ldots, m_1$, $P_2 = P_2^T > 0$, *and* L_{2i}, $i = 1, 2, \ldots, m_2$, *so that*

$$\begin{aligned} &\mathcal{H}(P_1(A_{1i} - L_{1i}C_{1i})) < 0 \quad i = 1, 2, \ldots, m_1 \\ &\mathcal{H}(\frac{2}{m_1 - 1} P_1(A_{1i} - L_{1i}C_{1i}) + P_1(A_{1i} - L_{1i}C_{1j}) + P_1(A_{1j} - L_{1j}C_{1i})) < 0 \\ &i = 1, 2, \ldots, m_1 \quad j = 1, \ldots, m_1 \quad i \neq j \end{aligned} \tag{5.44}$$

and

$$\begin{aligned} &\mathcal{H}(P_2(A_{2i} - L_{2i}C_{2i})) < 0 \quad i = 1, 2, \ldots, m_2 \\ &\mathcal{H}(\frac{2}{m_2 - 1} P_2(A_{2i} - L_{2i}C_{2i}) + P_2(A_{2i} - L_{2i}C_{2j}) + P_2(A_{2j} - L_{2j}C_{2i})) < 0 \\ &i = 1, 2, \ldots, m_2 \quad j = 1, \ldots, m_2 \quad i \neq j \end{aligned} \tag{5.45}$$

respectively, then the cascaded error system (5.38) *is UGAS.*

In the case when the scheduling vector does not depend on the states to be estimated, Theorem 5.5 can also be applied to the design of observers with a guaranteed convergence rate, using the following conditions.

Theorem 5.9. *The decay rate of the error system* (5.37) *is at least* α, *if there exist* $P_1 = P_1^T > 0$, L_{1i}, $i = 1, 2, \ldots, m_1$, $P_2 = P_2^T > 0$, *and* L_{2i}, $i = 1, 2, \ldots, m_2$, *so that*

$$
\begin{aligned}
&\mathcal{H}(P_1(A_{1i} - L_{1i}C_{1i})) + 2\alpha P_1 < 0 \quad i = 1, 2, \ldots, m_1\\
&\mathcal{H}(P_2(A_{2i} - L_{2i}C_{2i})) + 2\alpha P_2 < 0 \quad i = 1, 2, \ldots, m_2\\
&\mathcal{H}(P_1(A_{1i} - L_{1i}C_{1j}) + P_1(A_{1j} - L_{1j}C_{1i})) + 4\alpha P_1 < 0\\
&i = 1, 2, \ldots, m_1 \quad j = i+1, \ldots, m_1\\
&\forall i < j: \ \exists \boldsymbol{z}_1 : \ w_{1i}(\boldsymbol{z}_1)w_{1j}(\boldsymbol{z}_1) \neq 0\\
&\mathcal{H}(P_2(A_{2i} - L_{2i}C_{2j}) + P_2(A_{2j} - L_{2j}C_{2i})) + 4\alpha P_2 < 0\\
&i = 1, 2, \ldots, m_2 \quad j = i+1, \ldots, m_2\\
&\forall i < j: \ \exists \boldsymbol{z}_2 : \ w_{2i}(\boldsymbol{z}_2)w_{2j}(\boldsymbol{z}_2) \neq 0
\end{aligned}
$$

The above conditions explicitly state that in order to design a global observer with a desired convergence rate, it is sufficient to design observers for the subsystems with the same convergence rate.

The cascaded observer design for systems where the scheduling vector does not depend on states to be estimated is illustrated using the following example.

Example 5.5. Consider the following cascaded system

$$
\begin{aligned}
\dot{x}_1 &= -2x_2 & y_1 &= x_2\\
\dot{x}_2 &= x_1 + 3x_2^2 & y_2 &= x_3\\
\dot{x}_3 &= x_2^2 + 5x_2 + 6x_3 - 7x_3x_4 & &\\
\dot{x}_4 &= -2x_2^2 + x_3^3 - 2x_3 - x_4 & &
\end{aligned}
\tag{5.46}
$$

with $x_i \in [-1, 1]$, for which an observer has to be designed.

This system is the cascade of two observable subsystems, with the first subsystem being

$$
\begin{aligned}
\dot{x}_1 &= -2x_2 & y_1 &= x_2\\
\dot{x}_2 &= x_1 + 3x_2^2 & &
\end{aligned}
$$

and the second subsystem

$$
\begin{aligned}
\dot{x}_3 &= x_2^2 + 5x_2 + 6x_3 - 7x_3x_4 & y_2 &= x_3\\
\dot{x}_4 &= -2x_2^2 + x_3^3 - 2x_3 - x_4 & &
\end{aligned}
$$

An exact fuzzy fuzzy representation of the subsystems can be obtained using the sector nonlinearity approach as follows. For the first subsystem, there is one nonlinearity, and therefore one scheduling variable, x_2. Note that the state x_2 is measured, in fact $y_1 = x_2$. Then, the weighting functions, and also the membership functions, are (see Section 2.3.1)

$$
\begin{aligned}
w_{11} &= \eta_0^1 = \frac{1 - x_2}{2} = \frac{1 - y_1}{2}\\
w_{12} &= \eta_1^1 = \frac{1 + y_1}{2}
\end{aligned}
$$

and the corresponding matrices are

$$A_{11} = \begin{pmatrix} 0 & -2 \\ 1 & -3 \end{pmatrix} \quad A_{12} = \begin{pmatrix} 0 & -2 \\ 1 & 3 \end{pmatrix}$$

In the second subsystem there are three nonlinear terms: x_2 (note that $x_2 + 5$ and $-x_2$ lead to the same weighting functions and therefore they are not considered separately), x_3, and x_3^2. However, of these, x_3 is measured (and consequently x_3^2 is also known) and x_2 is part of the interconnection term, not of the individual second subsystem. Therefore, the scheduling vector of the second subsystems is $\boldsymbol{z}_2 = (x_3, x_3^2)^T$, and the individual second subsystem can be represented using the weighting functions

$$\eta_0^1 = \frac{1 - x_3}{2} = \frac{1 - y_2}{2} \qquad \eta_1^1 = \frac{1 + y_2}{2}$$
$$\eta_0^2 = 1 - y_2^2 \qquad \eta_1^2 = y_2^2$$

Then, the membership functions and the corresponding matrices are

$$w_{21} = \eta_0^1 \eta_0^2 \qquad A_{21} = \begin{pmatrix} 6 & -7 \\ -2 & -1 \end{pmatrix}$$

$$w_{22} = \eta_0^1 \eta_1^2 \qquad A_{22} = \begin{pmatrix} 6 & -7 \\ -1 & -1 \end{pmatrix}$$

$$w_{23} = \eta_1^1 \eta_0^2 \qquad A_{23} = \begin{pmatrix} 6 & 7 \\ -2 & -1 \end{pmatrix}$$

$$w_{24} = \eta_1^1 \eta_1^2 \qquad A_{24} = \begin{pmatrix} 6 & 7 \\ -1 & -1 \end{pmatrix}$$

The observer gains are found by solving the conditions of Theorem 5.6. The observer gains[3] of the first subsystem are

$$L_{11} = \begin{pmatrix} 2.63 \\ 6.43 \end{pmatrix} \quad L_{12} = \begin{pmatrix} 2.6 \\ 0.43 \end{pmatrix}$$

and the gains of the second subsystem are

$$L_{21} = \begin{pmatrix} 7.82 \\ 1.25 \end{pmatrix} \quad L_{22} = \begin{pmatrix} 7.82 \\ 2.25 \end{pmatrix} \quad L_{23} = \begin{pmatrix} 7.82 \\ -4.25 \end{pmatrix} \quad L_{24} = \begin{pmatrix} 7.82 \\ 5.25 \end{pmatrix}$$

A trajectory[4] of the estimation error is presented in Figure 5.4. This trajectory has been obtained with the initial states being $[0.25, 0.25, 0.25, 0.25]^T$, and the estimated initial states being zero. As expected, the estimation error converges to zero.

---

[3] All values are given rounded to two decimal places.

[4] Unless otherwise stated, for numerical integration in this chapter the *ode45* Matlab function was used.

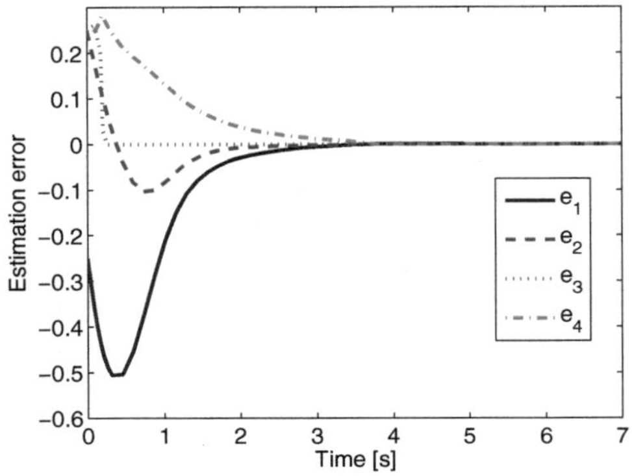

Fig. 5.4 Estimation error trajectory using the cascaded observer in Example 5.5.

Using Theorem 4.1 for a TS model of the centralized system (5.46), the observer gains for the centralized observer are

$$L_1 = 10^4 \cdot \begin{pmatrix} 1.77 & 0.05 \\ 3.22 & -0.12 \\ -0.17 & 0.59 \\ -0.77 & 0.02 \end{pmatrix} \quad L_2 = 10^4 \cdot \begin{pmatrix} 1.76 & -0.07 \\ 3.22 & -0.15 \\ -0.17 & 0.60 \\ -0.77 & 0.03 \end{pmatrix}$$

$$L_3 = 10^4 \cdot \begin{pmatrix} 1.74 & -0.06 \\ 3.21 & -0.14 \\ -0.17 & 0.66 \\ -0.77 & 0.05 \end{pmatrix} \quad L_4 = 10^4 \cdot \begin{pmatrix} 1.73 & -0.05 \\ 3.20 & -0.13 \\ -0.16 & 0.67 \\ -0.77 & 0.05 \end{pmatrix}$$

$$L_5 = 10^4 \cdot \begin{pmatrix} 1.79 & -0.05 \\ 3.30 & -0.11 \\ -0.16 & 0.63 \\ -0.78 & 0.03 \end{pmatrix} \quad L_6 = 10^4 \cdot \begin{pmatrix} 1.79 & -0.09 \\ 3.30 & -0.18 \\ -0.18 & 0.63 \\ -0.79 & 0.04 \end{pmatrix}$$

$$L_7 = 10^4 \cdot \begin{pmatrix} 1.69 & -0.06 \\ 3.13 & -0.15 \\ -0.16 & 0.67 \\ -0.74 & 0.05 \end{pmatrix} \quad L_8 = 10^4 \cdot \begin{pmatrix} 1.71 & -0.07 \\ 3.17 & -0.16 \\ -0.17 & 0.67 \\ -0.75 & 0.06 \end{pmatrix}$$

The trajectory of the estimation error using the centralized observer and the same initial conditions as for the cascaded observer is presented in Figure 5.5. Due to the large gains, for numerical integrations the *ode15s* Matlab function has been used. □

5.4.2 *Estimated Scheduling Vector*

Now, consider the case when the parameters z depend on the states to be estimated, i.e., in the observer, the estimated values of the scheduling variables have to be used. For the simplicity of the computations, only the case with common measurement

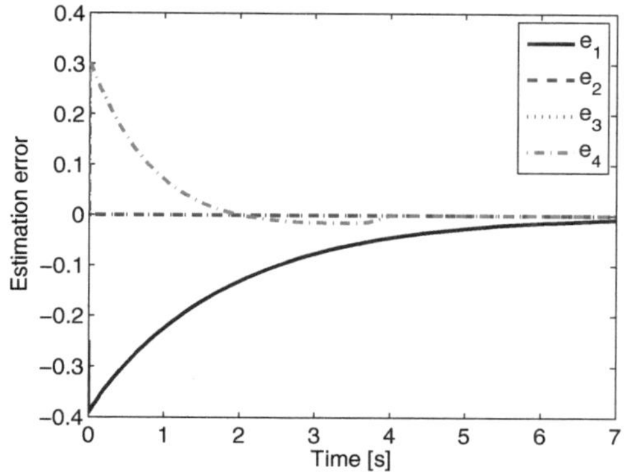

Fig. 5.5 Estimation error trajectory using a centralized observer for Example 5.5.

matrix is considered. For different measurement matrices, similar, although more complex conditions are obtained. Consider the fuzzy system expressed as

$$\begin{aligned}\dot{\boldsymbol{x}} &= \sum_{i=1}^{m} w_i(\boldsymbol{z})(A_i\boldsymbol{x} + B_i\boldsymbol{u} + a_i)\\ \boldsymbol{y} &= C\boldsymbol{x} + c,\end{aligned} \tag{5.47}$$

Using cascaded observers, with gains L_{1i}, $i = 1, 2, \ldots, m_1$, for the first observer and L_{2i}, $i = 1, 2, \ldots, m_2$, for the second observer, the error system can be written as:

$$\begin{aligned}\dot{\boldsymbol{e}} =& \sum_{i=1}^{m} w_i(\widehat{\boldsymbol{z}}) \begin{pmatrix} A_{1i} - L_{1i}C_1 & 0 \\ A_{21i} - L_{2i}C_{21} & A_{2i} - L_{2i}C_2 \end{pmatrix} \boldsymbol{e}\\ &+ \sum_{i=1}^{m}(w_i(\boldsymbol{z}) - w_i(\widehat{\boldsymbol{z}}))(A_i\boldsymbol{x} + B_i\boldsymbol{u} + a_i)\end{aligned} \tag{5.48}$$

or, considering the individual subsystems

$$\begin{aligned}\dot{\boldsymbol{e}}_1 =& \sum_{i=1}^{m_1} w_{1i}(\widehat{\boldsymbol{z}}_1)(A_{1i} - L_{1i}C_1)\boldsymbol{e}_1\\ &+ \sum_{i=1}^{m_1}(w_{1i}(\boldsymbol{z}_1) - w_{1i}(\widehat{\boldsymbol{z}}_1))(A_{1i}\boldsymbol{x}_1 + B_{1i}\boldsymbol{u} + a_{1i})\\ \dot{\boldsymbol{e}}_2 =& \sum_{i=1}^{m_2} w_{2i}(\widehat{\boldsymbol{z}}_2)[(A_{21i} - L_{2i}C_{21i})\boldsymbol{e}_1 + (A_{2i} - L_{2i}C_{2i})\boldsymbol{e}_2]\\ &+ \sum_{i=1}^{m_2}(w_{2i}(\boldsymbol{z}_2) - w_{2i}(\widehat{\boldsymbol{z}}_2))((A_{21i}\boldsymbol{x}_1 + A_{2i}\boldsymbol{x}_2 + B_{2i}\boldsymbol{u} + a_{2i})\end{aligned}$$

Using the cascaded approach for this error system, Theorem 5.3 can be combined with the conditions of Theorem 4.5 and the following result can be stated:

Theorem 5.10. *The cascaded error system* (5.48) *is UGAS, if there exist a Lyapunov function* $V_1(\boldsymbol{x}_1)$, $P_2 = P_2^T > 0$ *and two continuous functions* $\theta_1, \theta_2 : \mathbb{R}^+ \to \mathbb{R}^+$ *such that:*

1. *The Lyapunov function* V_1 *ensures exponential stability of the error system*

$$\begin{aligned}\dot{\boldsymbol{e}}_1 = &\sum_{i=1}^{m} w_{1i}(\widehat{\boldsymbol{z}}_1)(A_{1i} - L_{1i}C_{1i})\boldsymbol{e}_1 + \\ &+ (w_{1i}(\boldsymbol{z}_1) - w_{1i}(\widehat{\boldsymbol{z}}_1))(A_{1i}\boldsymbol{x}_1 + B_{1i}\boldsymbol{u} + a_{1i})\end{aligned} \tag{5.49}$$

2. P_2 *satisfies* $\mathcal{H}(P_2 A_{2i}) < 0$, $i = 1, 2, \ldots, m$, *and*
3. $\|\sum_{i=1}^{m}(w_{2i}(\boldsymbol{z}_2) - w_{2i}(\widehat{\boldsymbol{z}}_2))(A_{21i}\boldsymbol{x}_1 + A_{2i}\boldsymbol{x}_2 + B_{2i}\boldsymbol{u} + a_{2i})\| \leq \theta_1(\|\boldsymbol{e}_1\|) + \theta_2(\|\boldsymbol{e}_1\|)\|\boldsymbol{e}_2\|$.

Proof: The proof follows the same line of thought as that of Theorem 5.3 and makes use of Assumptions 5.1–5.4, as follows.

Since $\mathcal{H}(P_2 A_{2i}) < 0$, $i = 1, 2, \ldots, m$, V_2 is a Lyapunov function for

$$\dot{\boldsymbol{e}}_2 = \sum_{i=1}^{m} w_{2i}(\widehat{\boldsymbol{z}}_2)(A_{2i} - L_{2i}C_{2i})\boldsymbol{e}_2 \tag{5.50}$$

and this system is UGAS (Assumption 5.1). Let $\gamma_1 = 2\frac{\lambda_{\max}(P_2)}{\lambda_{\min}(P_2)}$, $\gamma_2 = 2\eta\lambda_{\max}(P_2)$. With these constants, Assumption 5.2 is satisfied. The Lyapunov function V_1 satisfies Assumption 5.4.

Now, the interconnection term in the second subsystem can be written as

$$\begin{aligned}g(\boldsymbol{e}_1, \boldsymbol{e}_2) = &\sum_{i=1}^{m} w_{2i}(\widehat{\boldsymbol{z}}_2)(A_{21i} - L_{2i}C_{2i})\boldsymbol{e}_1 + \\ &+ \sum_{i=1}^{m}(w_{2i}(\boldsymbol{z}_2) - w_{2i}(\widehat{\boldsymbol{z}}_2))(A_{21i}\boldsymbol{x}_1 + A_{2i}\boldsymbol{x}_2 + B_{2i}\boldsymbol{u} + a_{2i}) \\ \|g(\boldsymbol{e}_1, \boldsymbol{e}_2)\| \leq &\sum_{i=1}^{m} \|w_{2i}(\widehat{\boldsymbol{z}}_2)\|\|A_{21i} - L_{2i}C_{2i}\|\|\boldsymbol{e}_1\| + \theta_1(\|\boldsymbol{e}_1\|) + \theta_2(\|\boldsymbol{e}_1\|)\|\boldsymbol{e}_2\| \\ \|g(\boldsymbol{e}_1, \boldsymbol{e}_2)\| \leq &\tau\|\boldsymbol{e}_1\| + \theta_1(\|\boldsymbol{e}_1\|) + \theta_2(\|\boldsymbol{e}_1\|)\|\boldsymbol{e}_2\| \\ \|g(\boldsymbol{e}_1, \boldsymbol{e}_2)\| \leq &\theta_1'(\|\boldsymbol{e}_1\|) + \theta_2(\|\boldsymbol{e}_1\|)\|\boldsymbol{e}_2\|\end{aligned}$$

where $\theta_1'(\|\boldsymbol{e}_1\|) = \tau\|\boldsymbol{e}_1\| + \theta_1(\|\boldsymbol{e}_1\|)$. With this, Assumption 5.3 (see (5.14)) is satisfied, and based on Theorem 5.2, the cascaded system is UGAS. Moreover, since the first subsystem is exponentially stable, the cascaded system is also exponentially stable (see Proposition 5.1). □

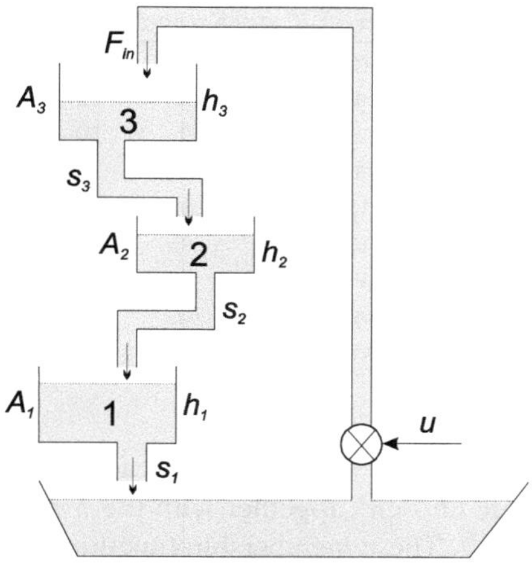

Fig. 5.6 Cascaded tanks system.

For real-world systems, the membership functions, and therefore the scheduling variables will in general be dependent on the states. The observer design for such a case is illustrated in the following example (Waurajitti et al., 2000).

Example 5.6. Consider the three tanks connected in a cascade as shown in Figure 5.6. Water is pumped from a reservoir into the upper tank (3). From this tank, the water flows to the middle tank (2) and to the lower tank (1) and from the lower tank back to the reservoir. The system has one control input u, which is the voltage applied to the motor of the pump and two measured outputs: the water levels h_3 in the upper tank and h_1 in the lowest tank. The flow rate F_{in}, provided by the pump, and the water level h_2 in the middle tank need to be estimated, and therefore, an observer has to be designed. The differential equations describing the dynamics of this system are the following (Waurajitti et al., 2000):

$$\begin{aligned}
\tau\dot{F}_{\text{in}} &= -F_{\text{in}} + Q_s \cdot u \\
\dot{h}_3 &= \frac{F_{\text{in}}}{A_3} - \frac{s_3\sqrt{2gh_3}}{A_3} \\
\dot{h}_2 &= \frac{s_3\sqrt{2gh_3}}{A_2} - \frac{s_2\sqrt{2gh_2}}{A_2} \\
\dot{h}_1 &= \frac{s_2\sqrt{2gh_2}}{A_1} - \frac{s_1\sqrt{2gh_1}}{A_1}
\end{aligned} \tag{5.51}$$

The parameter values are listed in Table 5.3.

It is assumed that the tanks have the same height, $h_{\max} = 2\,\text{m}$, and if a tank is full the overflowing water does not affect the level of the water in the other tanks. Therefore, all levels are bounded, $h_i \in [0, h_{\max}]$. We construct an approximate TS model using linearization.

Table 5.3 Parameter values used.

Parameter	Symbol	Value	Units
Acceleration due to gravity	g	9.81	m/s^2
Cross-sectional area of tank 1	A_1	12	m^2
Cross-sectional area of tank 2	A_2	10	m^2
Cross-sectional area of tank 3	A_3	15	m^2
Outlet area of tank 1	s_1	0.1	m^2
Outlet area of tank 2	s_2	0.5	m^2
Outlet area of tank 3	s_3	0.3	m^2
Input to flow gain	Q_s	0.3	m^3/s/V
Motor time constant	τ	3	s

To obtain a good coverage of the levels, for each level h_i, four points $h_i \in \{0.1, 0.55, 1.05, 1.6\}$ are chosen[5], together with the π-shaped membership functions depicted in Figure 5.7. These membership functions are defined as

$$\omega(h;a,b,c,d) = \begin{cases} 2\left(\frac{h-a}{b-a}\right)^2 & \text{if } a \le h \le \frac{a+b}{2} \\ 1-2\left(\frac{h-b}{b-a}\right)^2 & \text{if } \frac{a+b}{2} \le h \le b \\ 1 & \text{if } b \le h \le c \\ 1-2\left(\frac{h-c}{d-c}\right)^2 & \text{if } c \le h \le \frac{c+d}{2} \\ 2\left(\frac{h-d}{d-c}\right)^2 & \text{if } \frac{c+d}{2} \le h \le d \\ 0 & \text{otherwise} \end{cases}$$

The scheduling vector consists of the levels h_1, h_2, and h_3, which are the states to be estimated.

The system (5.51) is linearized (see Section 2.3.2) for each combination of the chosen points. Since the linearization is not done in equilibria, the consequents are affine. For instance, the rule obtained by linearizing in $h_1 = 0.55$, $h_2 = 0.1$ and $h_3 = 0.55$ is:

If h_1 *is approximately* 0.55 *and* h_2 *is approximately* 0.1 *and* h_3 *is approximately* 0.55, *then* $\dot{\boldsymbol{x}} = A\boldsymbol{x} + Bu + a$, with

$$A = \begin{pmatrix} -0.3333 & 0 & 0 & 0 \\ 0.1111 & -0.0995 & 0 & 0 \\ 0 & 0.1120 & -0.1751 & 0 \\ 0 & 0 & 0.1401 & -0.0747 \end{pmatrix} \quad B = \begin{pmatrix} 0.1120 \\ 0 \\ 0 \\ 0 \end{pmatrix}$$
$$a = (0 \;\; -0.0547 \;\; 0.0441 \;\; -0.0271)^T$$

where $\boldsymbol{x} = (F_{\text{in}} \;\; h_3 \;\; h_2 \;\; h_1)^T$. The membership degree of the scheduling vector is computed as the product of the individual membership degrees of the variables.

[5] The value $h_i = 0.1$ is chosen because the system is not linearizable in $h_i = 0, i = 1, 2, 3$.

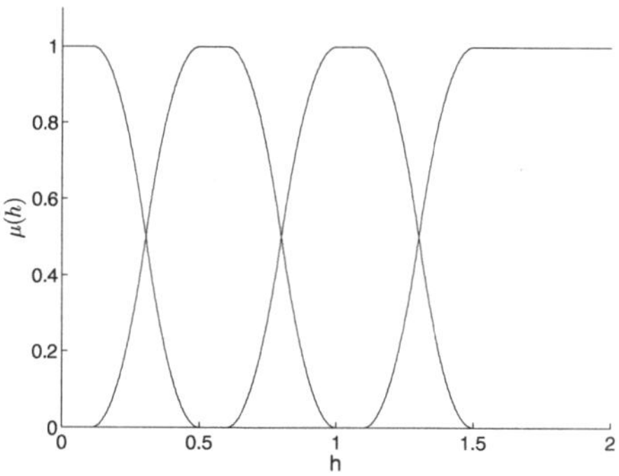

Fig. 5.7 Membership functions for the heights.

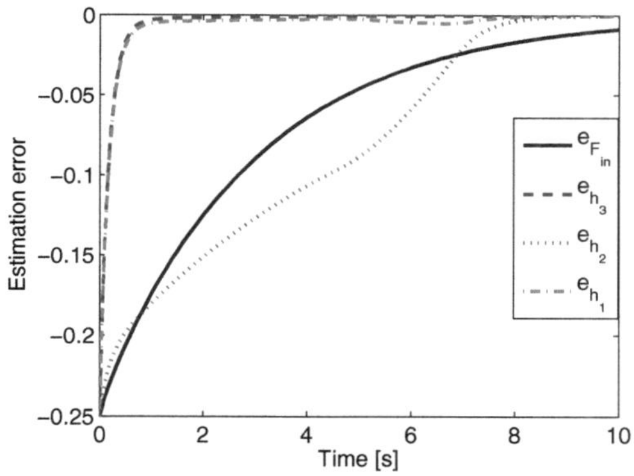

Fig. 5.8 Estimation errors using cascaded observers for Example 5.6.

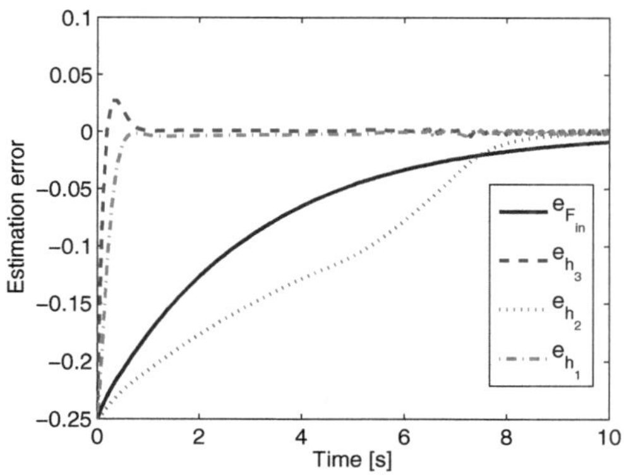

Fig. 5.9 Estimation errors using a centralized observer for Example 5.6.

The system is cascaded, with $\boldsymbol{x}_1 = (F_{\text{in}}\ h_3)^T$ and $\boldsymbol{x}_2 = (h_2\ h_1)^T$. Therefore, observers can be designed separately for the individual subsystems. The observers are designed both for the whole system and for the individual subsystems using the same conditions. Both observers have the form

$$\dot{\widehat{\boldsymbol{x}}} = \sum_{i=1}^{m} w_i(\widehat{\boldsymbol{z}})(A_i\widehat{\boldsymbol{x}} + B_i\boldsymbol{u} + a_i + L_i(\boldsymbol{y} - \widehat{\boldsymbol{y}}))$$
$$\widehat{\boldsymbol{y}} = \sum_{i=1}^{m} w_i(\widehat{\boldsymbol{z}})(C_i\widehat{\boldsymbol{x}} + d_i).$$

When simulating the system, the estimated values given by the fuzzy model and those given by the observer were saturated at 0 and $h_{\max}$. A piecewise constant input has been used. The initial conditions were $(0.25\ 0.25\ 0.25\ 0.25)^T$, while the estimated initial states were $(0.5\ 0.5\ 0.5\ 0.5)^T$. Due to the larger number of LMIs to be solved to obtain the centralized observer (64 4-by-4 LMIs), the CPU time needed to solve the LMIs for the centralized observer was 5 times larger than the time needed to design the cascaded observer.

The estimation errors obtained by the cascaded and centralized observers, are presented in Figures 5.8 and 5.9, respectively. In both cases, the estimation error converges to zero, as expected. □

5.5 Summary

In many real-life applications, a complex process model can be decomposed into simpler, cascaded subsystems. This partitioning of a process leads to increased modularity and a reduced complexity of the problem, while also making the analysis easier.

In this chapter, a cascaded approach for the stability analysis and observer design for TS fuzzy systems has been presented. First, an algorithm for partitioning a nonlinear system and stability conditions for cascaded nonlinear systems have been introduced. Next, cascaded TS fuzzy systems were considered. Based on the stability conditions for general nonlinear systems, stability conditions for cascaded TS systems have been presented. It has been shown that, similarly to linear systems, the exponential stability of the individual subsystems implies the stability of the cascaded system. Moreover, the convergence rate of the cascaded TS system is the maximum of the convergence rates of the individual subsystems. In addition, the cascaded stability analysis reduces the dimension of the LMI problems to be solved.

Observer design has also been performed in the cascaded setting. If the system under consideration can be represented as a cascade of TS fuzzy systems, observers can be designed in a cascaded fashion. This partitioning of a process and observer leads to increased modularity and reduced complexity of the problem, which results in reduced computational costs.

Chapter 6
Distributed TS Systems and Observers

The previous chapter has presented stability analysis and observer design for the special case of distributed systems that can be represented as a cascade of subsystems. In this chapter, a more general case is considered, namely when the subsystems are coupled. We present methods for the distributed stability analysis and observer design for TS fuzzy systems. The results presented in general rely on common quadratic Lyapunov functions, and LMI conditions are derived that are easy to solve. For large-scale or distributed systems, such an approach presents several advantages compared to the centralized approach, among which flexibility and reduced computational costs.

6.1 Introduction

Large-scale or distributed systems are composed of a number of subsystems that influence each other. In addition, in many cases, the structure of the overall system is not fixed, i.e., subsystems may be added online, and therefore a centralized analysis and/or design may be computationally intractable. For such systems, decentralized analysis and control design has received much attention (Akar and Özgüner, 2000; Krishnamurthy and Khorrami, 2003; Wang and Luoh, 2004; Liu and Zhang, 2005; Haijun et al., 2006; Zhang et al., 2006; Liu et al., 2007). In general, conditions developed for distributed stability analysis and controller or observer design are more conservative than those developed for centralized analysis or design (Bavafa-Toosi et al., 2006), but have the benefit of a reduced computational complexity. For control purposes, the decentralized design presents several advantages: flexibility, fault tolerance, simplified design, and easier tuning.

Decentralized control has been successfully employed in economic systems, power systems, large space-structures, traffic control, and process control. Although decentralized control has received much attention (Sandell et al., 1978; Akar and Özgüner, 2000; Jiang, 2000; Krishnamurthy and Khorrami, 2003; Wang and Chai, 2005; Zhang et al., 2006; Bavafa-Toosi et al., 2006) in this context, decentralized state estimation has not been addressed as much. Moreover, decentralized state estimation has rarely been addressed for TS systems, although

Zs. Lendek et al.: Stability and Observer Design Using TS Models, STUDFUZZ 262, pp. 103–147.
springerlink.com

numerous results exist for linear and stochastic systems. In case of linear or stochastic systems, in general sensor fusion is considered, with a network architecture of sensor nodes (López-Orozco et al., 2000; Roumeliotis and Bekey, 2002; Schmitt et al., 2002), such that each node shares information with other nodes and computes a local estimate. Observers used include, but are not limited to linear observers (Sundareshan and Elbanna, 1990; Saif and Guan, 1992; Hou and Müller, 1994), Kalman filter variants (Durrant-Whyte et al., 1990; Benigni et al., 2008), and particle filters (Bolic et al., 2004; Coates, 2004).

In this chapter we consider the distributed stability analysis and observer design for a system composed of interconnected subsystems. Each subsystem is represented by a TS fuzzy model. The coupling between the subsystems is realized through their states, i.e., the states of a subsystem may influence the dynamics of another subsystem. Note that we do not treat the issue of decomposing a given system into subsystems. This is because in many systems, the decomposition is given, i.e., the system considered is naturally composed of interacting subsystems. When this is not the case, decomposition techniques such as the ones presented by Gegov and Frank (1995) or Michel et al. (1978) can be used. Of these techniques, the one developed by Gegov and Frank (1995) specifically treats fuzzy systems, while the method given by Michel et al. (1978) relies on the transformation of the system into lower-block triangular form, effectively a transformation into a cascaded system.

Consider a distributed system consisting of n_s interacting subsystems, with each subsystem l represented by the TS model

$$\dot{\boldsymbol{x}}_l = \sum_{i=1}^{m_l} w_{li}(\boldsymbol{z}_l)\Big(A_{li}\boldsymbol{x}_l + \sum_{j=1, j\neq l}^{n_s} \boldsymbol{f}_{lij}(\boldsymbol{x}_j)\Big) \tag{6.1}$$

for stability analysis and

$$\begin{aligned}
\dot{\boldsymbol{x}}_l &= \sum_{i=1}^{m_l} w_{li}(\boldsymbol{z}_l)\Big(A_{li}\boldsymbol{x}_l + B_{li}\boldsymbol{u}_l + a_{li} + \sum_{j=1, j\neq l}^{n_s} \boldsymbol{f}_{lij}(\boldsymbol{x}_j)\Big)\\
\boldsymbol{y}_l &= \sum_{i=1}^{m_l} w_{li}(\boldsymbol{z}_l)\Big(C_{li}\boldsymbol{x}_l + c_{li} + \sum_{j=1, j\neq l}^{n_s} \boldsymbol{h}_{lij}(\boldsymbol{x}_j)\Big)
\end{aligned} \tag{6.2}$$

for observer design, where $\boldsymbol{x}_l$, $\boldsymbol{u}_l$, and $\boldsymbol{z}_l$, $l = 1, 2, \ldots, n_s$ denote the state, input, and scheduling vectors of the lth subsystem, m_l is the number of rules in the fuzzy representation of the lth subsystem, A_{li}, B_{li}, C_{li}, a_{li}, and c_{li} are the corresponding local matrices and biases, and $\boldsymbol{f}_{lij}$ and $\boldsymbol{h}_{lij}$ denote the interconnection from the jth subsystem to the ith rule of the lth subsystem, and can be nonlinear functions. A general assumption is that these interconnection terms are Lipschitz in the states, i.e., there exist $\mu^f_{lij} \geq 0$ and $\mu^h_{lij} \geq 0$ such that $\|\boldsymbol{f}_{lij}(\boldsymbol{x}_j)\| \leq \mu^f_{lij}\|\boldsymbol{x}_j\|$, and $\|\boldsymbol{h}_{lij}(\boldsymbol{x}_j)\| \leq \mu^h_{lij}\|\boldsymbol{x}_j\|$.

Current approaches to decentralized stability analysis and controller or observer design for distributed TS systems can be classified as perturbation methods (see

Sandell et al. (1978); Bakule (2008), and the references therein), i.e., the interconnection between the subsystems is treated as a perturbation. The stability analysis and the design are in general performed according to the following steps:

1. The distributed system considered consists of interconnected subsystems. Depending on the approach, the description of these subsystems may be available a priori or become available during the analysis or design. In this book, we refer to the case when the description of all the subsystems is available a priori as *parallel* analysis and design. If subsystem may be added to an existing system, we use the term *sequential* analysis or design.
2. It is assumed that each individual subsystem (i.e., without the interconnection terms) is stable. Moreover, some measure of stability, in general, a bound on the derivative of the Lyapunov function, is available.
3. Given that the individual subsystems are stable, a measure on their stability is available, and that the strength of the interconnection terms is known, determine the conditions under which the interconnected system is stable.

Both in stability analysis and in controller or observer design the most common case considered is weak coupling between the subsystems, that is the case when the bound on the interconnection terms is sufficiently small to not influence the stability of the subsystems. In this case, the stability analysis is actually reduced to the robust stability analysis of the subsystems. In state feedback controller design this translates into robust stabilization.

While controller and observer design are generally considered dual problems, note that for decentralized design this is not the case due to the "knowledge" constraints: a well-designed distributed stabilizing controller is able to stabilize the whole system without taking into account the influence between the subsystems, assuming that the influence is small enough, but if the system is not stabilized to zero, without knowing (at least) the estimates of the influencing subsystems, the estimate given by the observer will never converge to the true values. Although in observer design for centralized TS systems approaches that decouple unknown inputs exist – which, for distributed TS systems would translate to decoupling the influence of other subsystems – these approaches have not yet been extended to distributed systems. Moreover, complete decoupling of the unknown input is cumbersome when the measurement matrices differ for each rule. This is one of the reasons why, in the context of distributed TS systems, decentralized state estimation, without the control counterpart, has rarely been addressed. Observers have been used in the context of observer-based control. However, in case of observer-based control, the observer and controller design are coupled: in most cases, the observer cannot be used without the controller (Uang and Chen, 2000; Tseng and Chen, 2001; Tseng, 2008).

This chapter first presents current results for the stability analysis of distributed TS system, and afterwards results for observer design for distributed TS fuzzy systems. These results in general rely on the stability of the independent subsystems and the "sufficient weakness" of the interconnection terms. Results are presented both for parallel and sequential analysis.

Note that although in the literature the term "large-scale fuzzy system" is often used, we refer to the TS systems used here as "distributed TS systems".

6.2 Distributed Stability Analysis of TS Systems

To study the stability of a decentralized system, consider a continuous-time distributed TS system composed of n_s subsystems, with each subsystem given as

$$\dot{\boldsymbol{x}}_l = \sum_{i=1}^{m_l} w_{li}(\boldsymbol{z}_l)\Big(A_{li}\boldsymbol{x}_l + \sum_{j=1,j\neq l}^{n_s} \boldsymbol{f}_{lij}(\boldsymbol{x}_j)\Big) \tag{6.3}$$

where $\boldsymbol{x}_l$ is the state vector and $\boldsymbol{z}_l$ is the vector of scheduling parameters of the lth subsystem, m_l denotes the number of rules, A_{li} are the local matrices of the lth subsystem, and $\boldsymbol{f}_{lij}(\boldsymbol{x}_j)$ are the interconnection terms between the ith rule of the lth subsystem and the jth subsystem.

While in centralized stability analysis of TS fuzzy systems, several types of Lyapunov functions have been employed, stability analysis of distributed TS systems mainly relies on the existence of a common quadratic Lyapunov function for each subsystem. Most results make use of the assumption that the number of subsystems and some bounds on the interconnection terms are known a priori. Although such an assumption restricts the class of systems considered, in the sense that new subsystems cannot be added, it allows for a parallel analysis of the subsystems. Some of these results are presented in what follows.

6.2.1 Parallel Stability Analysis

An early result that relies on the existence of an M-matrix[1] or positive definite matrices has been formulated as follows (Akar and Özgüner, 2000; Wang and Lin, 2005). Let each subsystem of the decentralized system be given as (6.3), and the interconnection terms be bounded as $\|f_{lij}(\boldsymbol{x}_j)\| \leq \mu^f_{lj}\|\boldsymbol{x}_j\|$, with μ^f_{lj}, $l, j = 1, 2, \ldots, n_s$, $i = 1, 2, \ldots, m_l$, known positive constants. Then,

Theorem 6.1. *(Akar and Özgüner, 2000) The distributed system, with each subsystem given by* (6.3) *is asymptotically stable, if there exist* $P_l = P_l^T > 0$, *and* $Q_l = Q_l^T > 0$, $l = 1, 2, \ldots, n_s$, *such that*

$$\mathcal{H}(P_l A_{li}) < -Q_l \qquad i = 1, 2, \ldots, m_l \tag{6.4}$$

where $\mathcal{H}(X) = X + X^T$, *and, furthermore, the matrix* M *defined as*

$$m_{li} = \begin{cases} \lambda_{\min}(Q_l) - 2\mu^f_{ll}\|P_l\|, & \textit{if } l = i \\ -2\mu^f_{li}\|P_l\|, & \textit{if } l \neq i \end{cases} \tag{6.5}$$

[1] A square matrix M is an M-matrix if the off-diagonal elements are all negative and all the eigenvalues of M have non-negative real part.

is an M-matrix, where $\lambda_{\min}$ denotes the eigenvalue with the smallest absolute magnitude.

A similar result has been given by Wang and Lin (2003), where it is required that the matrix M is positive definite instead of being an M-matrix. Requiring that M is positive definite is less restrictive than requiring that it is an M-matrix (Wang and Lin, 2003). For discrete-time TS systems, a similar result has been presented by Hsiao and Hwang (2002).

Note that the conditions of Theorem 6.1 are in fact not distributed. Although a distributed system is considered and (6.4) can be solved in parallel for all the subsystems, the stability of the interconnected system is established only after testing the matrix M. The application of Theorem 6.1 is illustrated using the following example.

Example 6.1. Consider a distributed system composed of two subsystems and the corresponding interconnection terms, as follows:

$$\begin{aligned} \dot{\boldsymbol{x}}_1 &= \boldsymbol{A}_1(\boldsymbol{x}_1)\boldsymbol{x}_1 + \boldsymbol{f}^{\mathrm{m}}_{12}(\boldsymbol{x}_1, \boldsymbol{x}_2)\boldsymbol{x}_2 \\ \dot{\boldsymbol{x}}_2 &= \boldsymbol{A}_2(\boldsymbol{x}_2)\boldsymbol{x}_2 + \boldsymbol{f}^{\mathrm{m}}_{21}(\boldsymbol{x}_1, \boldsymbol{x}_2)\boldsymbol{x}_1 \end{aligned} \tag{6.6}$$

where

$$\begin{aligned} &\boldsymbol{x}_1 = (x_1\ x_2)^T && \boldsymbol{x}_2 = (x_3\ x_4)^T \\ &\boldsymbol{A}_1(\boldsymbol{x}_1) = \begin{pmatrix} -1 & x_2 \\ -1 & -5 + 2x_2^2 \end{pmatrix} && \boldsymbol{A}_2(\boldsymbol{x}_2) = \begin{pmatrix} -4 & 2 + 2x_3^2 \\ -2x_4^2 & -1 \end{pmatrix} \\ &\boldsymbol{f}^{\mathrm{m}}_{12}(\boldsymbol{x}_1, \boldsymbol{x}_2) = \begin{pmatrix} 0 & 1/5 \\ -x_1^2 & 0 \end{pmatrix} && \boldsymbol{f}^{\mathrm{m}}_{21}(\boldsymbol{x}_1, \boldsymbol{x}_2) = \begin{pmatrix} 1/5 & 0 \\ -1/5 & 2x_2^4 \end{pmatrix} \\ &x_i \in [-1, 1] \quad i = 1,\ 2,\ 3,\ 4 \end{aligned}$$

This interconnected system has an asymptotically stable equilibrium point in 0, as it can be proven using a centralized common quadratic Lyapunov function.

To obtain a TS system of the form (6.3), the sector nonlinearity approach can be used on the two individual subsystems. For each subsystem (without the interconnection terms), 4 local models are obtained, with the matrices of the first subsystem:

$$A_{11} = \begin{pmatrix} -1 & -1 \\ -1 & -5 \end{pmatrix} \quad A_{12} = \begin{pmatrix} -1 & -1 \\ -1 & -3 \end{pmatrix} \quad A_{13} = \begin{pmatrix} -1 & 1 \\ -1 & -5 \end{pmatrix} \quad A_{14} = \begin{pmatrix} -1 & 1 \\ -1 & -3 \end{pmatrix}$$

and the matrices of the second subsystems:

$$A_{21} = \begin{pmatrix} -4 & 2 \\ -2 & -1 \end{pmatrix} \quad A_{22} = \begin{pmatrix} -4 & 2 \\ 0 & -1 \end{pmatrix} \quad A_{23} = \begin{pmatrix} -4 & 4 \\ -2 & -1 \end{pmatrix} \quad A_{24} = \begin{pmatrix} -4 & 4 \\ 0 & -1 \end{pmatrix}$$

The interconnection terms are bounded, as $\|\boldsymbol{f}^{\mathrm{m}}_{12}(\boldsymbol{x}_1, \boldsymbol{x}_2)\boldsymbol{x}_2\| \leq \mu^f_{12}\|\boldsymbol{x}_2\|$, and $\|\boldsymbol{f}^{\mathrm{m}}_{21}(\boldsymbol{x}_1, \boldsymbol{x}_2)\boldsymbol{x}_1\| \leq \mu^f_{21}\|\boldsymbol{x}_1\|$, for $x_i \in [-1, 1]$, $i = 1,\ 2,\ 3,\ 4$, with $\mu_{12} = 0.136$ and $\mu_{21} = 0.133$. Moreover $\mu_{11} = 0$ and $\mu_{22} = 0$, as the rules of the subsystems are known without uncertainties (there are no "interconnection" terms from the first subsystem to the first subsystem, and from the second subsystem to the second subsystem, respectively).

Solving[2] the LMIs (6.4), one obtains[3] $Q_1 = Q_2 = I$, $P_1 = \begin{pmatrix} 0.85 & -0.02 \\ -0.02 & 0.80 \end{pmatrix}$, $\|P_1\| = 0.85$, $P_2 = \begin{pmatrix} 0.32 & -0.02 \\ -0.02 & 1.03 \end{pmatrix}$, $\|P_2\| = 1.03$.

Using the above computed values, the elements of the matrix M are $M = \begin{pmatrix} 1.00 & -0.34 \\ -0.95 & 1.00 \end{pmatrix}$, and it is an M-matrix, having the eigenvalues 1.57 and 0.43. Therefore, the stability of the distributed system (6.6) has been established[4]. □

Hybrid linear-fuzzy systems have also been considered in the literature. For instance, the approach by Xu et al. (2006) concerns stability of distributed systems with n_l subsystems of the form

$$\dot{\boldsymbol{x}}_l = A_l \boldsymbol{x}_l + \boldsymbol{f}_l(\boldsymbol{x}_l) + \sum_{i=1, i \neq l}^{n_l + n_t} (D_{li} \boldsymbol{x}_i + \boldsymbol{f}_{li}(\boldsymbol{x}_i)) \tag{6.7}$$

with $l = 1, \ldots, n_l$ and n_t subsystems of the form

$$\dot{\boldsymbol{x}}_l = \sum_{i=1}^{m_l} w_{li}(\boldsymbol{z}_l)(A_{li} \boldsymbol{x}_l + \sum_{j=1, j \neq l}^{n_l + n_t} A_{lij} \boldsymbol{x}_j) \tag{6.8}$$

where $l = n_l + 1, \ldots, n_l + n_t$, A_l, A_{li}, A_{lij}, and D_{li} are matrices with appropriate dimensions, and $\boldsymbol{f}_l$ and $\boldsymbol{f}_{li}$ are nonlinear functions satisfying $\|\boldsymbol{f}_l(\boldsymbol{x}_l)\| \leq \mu_l^f \|\boldsymbol{x}_l\|$ and $\|\boldsymbol{f}_{li}(\boldsymbol{x}_i)\| \leq \mu_{li}^f \|\boldsymbol{x}_i\|$, respectively. As can be seen, the first n_l subsystems are linear, with Lipschitz nonlinearities, while the following n_t subsystems are of TS type. For such hybrid systems, the following theorem has been formulated by Xu et al. (2006):

Theorem 6.2. *(Xu et al., 2006) The system described by* (6.7) *and* (6.8) *is asymptotically stable if there exist* $P_l = P_l^T > 0$, *such that*

$$\begin{aligned} &\begin{pmatrix} \mathcal{H}(P_l A_l) + (\mu_l^f)^2 I + (n_l - 1)(\mu_{li}^f)^2 I + (n_l + n_t - 1) I & P_l \\ P_l & -M_l^{-1} \end{pmatrix} < 0 \\ &M_l = \sum_{i=1, i \neq l}^{n_s} (I + D_{li} D_{li}^T) \end{aligned} \tag{6.9}$$

---

[2] For solving the LMI problems in this chapter, the SeDuMi solver within the Yalmip toolbox (Löfberg, 2004) was used.

[3] All values are given rounded to two decimal places.

[4] Since the nonlinear system is exactly represented by the fuzzy model on a compact set, in order to estimate the domain of attraction, the outermost level of the Lyapunov function has to be computed.

for $l = 1, \ldots, n_l$, $i = 1, 2, \ldots, m_l$, *and, furthermore, there exist* $P_l = P_l^T > 0$ *such that*

$$\begin{pmatrix} \mathcal{H}(P_l A_{li}) + n_l(\mu_{li}^f)^2 I + (n_l + n_t - 1)I & M_l \\ M_l^T & -I \end{pmatrix} < 0$$
$$M_l = P_l \left(A_{li1}\; A_{li2}\; \cdots\; A_{li(l-1)}\; A_{li(l+1)}\; \cdots\; A_{li(n_l+n_t)}\right) \tag{6.10}$$

for $l = n_l + 1, \ldots, n_l + n_t$, $i = 1, 2, \ldots, m_l$.

The approach of Xu et al. (2006) is useful when one has to establish the stability of a nonlinear system that can be well approximated by a linear model, and also reduces the computational complexity that would be needed to establish the stability of TS systems. A shortcoming of the method is that it is assumed that the state variables of the independent subsystems can be separated in the linear model, i.e., in the linear model, one cannot have multiplications between states belonging to different subsystems. The method is illustrated on the following example.

Example 6.2. Consider a distributed system composed of two subsystems and the corresponding interconnection terms, as follows:

$$\begin{aligned} \dot{\boldsymbol{x}}_1 &= A_1 \boldsymbol{x}_1 + \boldsymbol{f}_1(\boldsymbol{x}_1) + D_{12}\boldsymbol{x}_2 \\ \dot{\boldsymbol{x}}_2 &= \boldsymbol{A}_2(\boldsymbol{x}_2)\boldsymbol{x}_2 + \boldsymbol{f}_{21}^{\mathrm{m}}(\boldsymbol{x}_1, \boldsymbol{x}_2)\boldsymbol{x}_1 \end{aligned} \tag{6.11}$$

where

$$\boldsymbol{x}_1 = (x_1\; x_2)^T \qquad \boldsymbol{x}_2 = (x_3\; x_4)^T$$
$$A_1 = \begin{pmatrix} -1 & -1 \\ -1 & -5 \end{pmatrix} \qquad \boldsymbol{f}_1(\boldsymbol{x}_1) = \begin{pmatrix} x_2 + x_2^2 \\ 5x_2 + x_2^3 \end{pmatrix} \qquad D_{12} = \begin{pmatrix} 0 & 1/5 \\ 0 & 0 \end{pmatrix}$$
$$\boldsymbol{A}_2(\boldsymbol{x}_2) = \begin{pmatrix} -4 & 2 + 2x_3^2 \\ -2x_4^2 & -1 \end{pmatrix} \qquad \boldsymbol{f}_{21}^{\mathrm{m}}(\boldsymbol{x}_1, \boldsymbol{x}_2) = \begin{pmatrix} 1/5 & 0 \\ -1/5 & 2x_2^4 \end{pmatrix}$$
$$x_i \in [-1, 1] \quad i = 1, 2, 3, 4$$

This interconnected system is asymptotically stable, provable with a centralized common quadratic Lyapunov function.

The first subsystem is linear, the second one is nonlinear. To obtain a TS system of the form (6.8), the sector nonlinearity approach is used for the second subsystem. Four local models are obtained, with the matrices being

$$A_{21} = \begin{pmatrix} -4 & 2 \\ -2 & -1 \end{pmatrix} \quad A_{22} = \begin{pmatrix} -4 & 2 \\ 0 & -1 \end{pmatrix} \quad A_{23} = \begin{pmatrix} -4 & 4 \\ -2 & -1 \end{pmatrix} \quad A_{24} = \begin{pmatrix} -4 & 4 \\ 0 & -1 \end{pmatrix}$$

The interconnection terms are bounded, as $\|\boldsymbol{f}_{12}(\boldsymbol{x}_2)\| = 0$, and consequently $\mu_{12}^f = 0$, $\|\boldsymbol{f}_1(\boldsymbol{x}_1)\| \le \mu_1^f \|\boldsymbol{x}_1\|$, with $\mu_1^f = 2.83$, and $\|\boldsymbol{f}_{21}^{\mathrm{m}}(\boldsymbol{x}_1, \boldsymbol{x}_2)\boldsymbol{x}_1\| \le \mu_{21}^f \|\boldsymbol{x}_1\|$, for $x_i \in [-1, 1]$, $i = 1, 2, 3, 4$, with $\mu_{21} = 0.133$.

Solving the LMIs (6.9) and (6.10), one obtains $P_1 = \begin{pmatrix} 1.38 & -0.1 \\ -0.1 & 0.62 \end{pmatrix}$ and $P_2 = \begin{pmatrix} 0.47 & -0.09 \\ -0.09 & 1.44 \end{pmatrix}$, and thereby the stability of the distributed system is established. □

The results presented so far have been derived for TS systems where the individual subsystems have linear consequents, and the conditions have been established using common quadratic Lyapunov functions. An approach for distributed TS systems with affine consequents, based on piecewise Lyapunov functions (Johansson et al., 1999) has been developed by Zhang et al. (2006).

To present this result, consider a distributed system consisting of n_s interconnected subsystems, where each subsystem l, $l = 1, \ldots, n_s$, is represented by the affine TS system and the interconnection term as

$$\boldsymbol{x}_l = \sum_{i=1}^{m_l} w_{li}(\boldsymbol{z}_l)(A_{li}\boldsymbol{x}_l + a_{li}) + \sum_{j=1, j\neq l}^{n_s} D_{lj}\boldsymbol{x}_j \tag{6.12}$$

This approach is an extension of the result by Johansson et al. (1999) to distributed TS systems. Similarly to the result of Johansson et al. (1999), the analysis relies on dividing the state-space of each individual subsystem into polyhedral partitions. Let L_{l0} denote the set of indices of the partitions of the l-th subsystem that contain the origin, L_{l1} the set of indices of the partitions that do not contain it, X_{li} the region corresponding to L_{li}, and M_{li} the index set of the rules of the lth subsystem that are active in the region X_{li}.

For each subsystem, a piecewise quadratic Lyapunov function is used, such that each part of the Lyapunov function is valid only in one region. In order to parameterize the piecewise Lyapunov function, matrices F_{li}, $\bar{F}_{li}$ and E_{li}, $\bar{E}_{li}$ are constructed (see (Johansson et al., 1999)) that characterize the boundaries across the regions. Then, the following result has been formulated by Zhang et al. (2006).

Theorem 6.3. *(Zhang et al., 2006) The distributed fuzzy system* (6.12) *is asymptotically stable, if there exist $T_l = T_l^T > 0$, $\epsilon_l > 0$, symmetric matrices P_{li}, $i \in L_{l0}$, $\bar{P}_{li}$, $i \in L_{l1}$, $l = 1, \ldots, n_s$, and symmetric matrices U_{li}, $i \in L_{l1}$, and W_{lij}, $i \in L_{l0}$, $l = 1, \ldots, n_s$, $j \in M_{li}$, with positive entries, such that*

$$\begin{aligned}
&P_{li} = F_{li}^T T_i F_{li} \quad i \in L_{l0}\\
&\bar{P}_{li} = \bar{F}_{li}^T T_i \bar{F}_{li} \quad i \in L_{l1}\\
&P_{li} - E_{li}^T U_{li} E_{li} > 0 \quad i \in L_{l0}\\
&\begin{pmatrix} \mathcal{H}(P_{li}A_{lj}) + E_{li}^T W_{lij} E_{li} + \sum_{k=1,k\neq l}\epsilon_k n_{lk} D_{kl}^T D_{kl} & \sqrt{\sum_{k=1,k\neq l}^{n_s} n_{kl}P_{li}} \\ \sqrt{\sum_{k=1,k\neq l}^{n_s} n_{kl}P_{li}} & -\epsilon_l I \end{pmatrix} < 0\\
&\quad i \in L_{l0}
\end{aligned}$$

$$\bar{P}_{li} - \bar{E}_{li}^T U_{li} \bar{E}_{li} > 0 \quad i \in L_{l1}$$

$$\begin{pmatrix} \mathcal{H}(\bar{P}_{li}\bar{A}_{lj}) + \bar{E}_{li}^T W_{lij} \bar{E}_{li} + \sum_{k=1,k\neq l} \epsilon_k n_{lk} \bar{D}_{kl}^T \bar{D}_{kl} & \sqrt{\sum_{k=1,k\neq l}^{n_s} n_{kl} P_{li}} \\ \sqrt{\sum_{k=1,k\neq l}^{n_s} n_{kl} P_{li}} & -\epsilon_l I \end{pmatrix} < 0$$

$$i \in L_{l1}$$

$$\textit{where } n_{kl} = \begin{cases} 1, & \textit{if } D_{kl} \neq 0 \\ 0, & \textit{otherwise} \end{cases}$$

A shortcoming of this method is that only linear interconnection terms among the subsystems are considered. Moreover, the analysis itself, although it concerns distributed systems, is not distributed, as it has to be performed at the same time in parallel for all the subsystems. The method is illustrated on the following example.

Example 6.3. Consider the following distributed system composed of two subsystems and the corresponding interconnection terms:

$$\begin{aligned} \dot{\boldsymbol{x}}_1 &= \boldsymbol{A}_1(\boldsymbol{x}_1)\boldsymbol{x}_1 + D_{12}\boldsymbol{x}_2 \\ \dot{\boldsymbol{x}}_2 &= \boldsymbol{A}_2(\boldsymbol{x}_2)\boldsymbol{x}_2 + D_{21}\boldsymbol{x}_1 \end{aligned} \tag{6.13}$$

where

$$\boldsymbol{x}_1 = (x_1 \; x_2)^T \qquad \boldsymbol{x}_2 = (x_3 \; x_4)^T$$

$$\boldsymbol{A}_1(\boldsymbol{x}_1) = \begin{pmatrix} -1 & x_2 \\ -1 & -5 + 2x_2^2 \end{pmatrix} \qquad \boldsymbol{A}_2(\boldsymbol{x}_2) = \begin{pmatrix} -4 & 2 + 2x_3^2 \\ -2x_4^2 & -1 \end{pmatrix}$$

$$D_{12} = \begin{pmatrix} 0 & 1/5 \\ 1 & 0 \end{pmatrix} \qquad D_{21} = \begin{pmatrix} 1/5 & 0 \\ -1/5 & 0 \end{pmatrix}$$

$$x_i \in [-1, 1] \quad i = 1, 2, 3, 4$$

This interconnected system is asymptotically stable, provable with a centralized common quadratic Lyapunov function.

To obtain TS models of the subsystems, the sector nonlinearity approach is used on the two individual subsystems. For each subsystem, 4 local models are obtained, with the matrices of the first subsystem:

$$A_{11} = \begin{pmatrix} -1 & -1 \\ -1 & -5 \end{pmatrix} \quad A_{12} = \begin{pmatrix} -1 & -1 \\ -1 & -3 \end{pmatrix} \quad A_{13} = \begin{pmatrix} -1 & 1 \\ -1 & -5 \end{pmatrix} \quad A_{14} = \begin{pmatrix} -1 & 1 \\ -1 & -3 \end{pmatrix}$$

and the matrices of the second subsystems:

$$A_{21} = \begin{pmatrix} -4 & 2 \\ -2 & -1 \end{pmatrix} \quad A_{22} = \begin{pmatrix} -4 & 2 \\ 0 & -1 \end{pmatrix} \quad A_{23} = \begin{pmatrix} -4 & 4 \\ -2 & -1 \end{pmatrix} \quad A_{24} = \begin{pmatrix} -4 & 4 \\ 0 & -1 \end{pmatrix}$$

Since the fuzzy model is obtained using the sector nonlinearity approach, there is only one region for each subsystem, where all the rules of the subsystem are active.

Thus, the conditions of Theorem 6.3 are reduced to finding $P_l = P_l^T > 0, l = 1, 2,$ so that

$$\begin{pmatrix} \mathcal{H}(P_l A_{li}) + \sum_{k=1,k\neq l} \epsilon_k n_{kl} D_{kl}^T D_{kl} & \sqrt{\sum_{k=1,k\neq l}^{n_s} n_{kl} P_l} \\ \sqrt{\sum_{k=1,k\neq l}^{n_s} n_{kl} P_l} & -\epsilon_l I \end{pmatrix} < 0 \tag{6.14}$$
$$l = 1, 2 \qquad i = 1, 2, 3, 4$$

where $n_{kl} = \begin{cases} 1, & \text{if } D_{kl} \neq 0 \\ 0, & \text{otherwise} \end{cases}$

Solving (6.14) for the two subsystems yields

$$P_1 = \begin{pmatrix} 0.66 & 0 \\ 0 & 0.34 \end{pmatrix} \qquad P_2 = \begin{pmatrix} 0.48 & -0.03 \\ -0.03 & 1.20 \end{pmatrix}$$

$\epsilon_1 = 1.4$, $\epsilon_2 = 2.52$, and with this the stability of the interconnected system is established. □

One particular type of TS systems that have been extensively investigated both in stability analysis and in (robust) control are uncertain TS fuzzy systems. In case of a distributed uncertain TS system, each subsystem is described as

$$\dot{\boldsymbol{x}}_l = \sum_{i=1}^{m_l} w_{li}(\boldsymbol{z}_l)(A_{li} + \Delta A_{li})\boldsymbol{x}_l + \sum_{i=1,i\neq l}^{n_s} (B_{li} + \Delta B_{li})\boldsymbol{x}_i \tag{6.15}$$

with $\Delta A_{li} = D_{li} F_{li}(t) E_{li}$, $\Delta B_{li} = \bar{D}_{li} \bar{F}_{li}, \bar{E}_{li}$, where A_{li}, B_{li}, D_{li}, E_{li}, $\bar{D}_{li}$, and $\bar{E}_{li}$ are known constant matrices and F_{li} and $\bar{F}_{li}$ are time-varying matrices with bounded norms $F_{li} F_{li}^T \leq \mu_{li} I$ and $\bar{F}_{li} \bar{F}_{li}^T \leq \bar{\mu}_{li} I$.

Similarly to the previously presented results, the interconnection terms are assumed linear, with a Lipschitz nonlinearity. For such uncertain distributed TS systems, the following result have been formulated by Liu and Zhang (2005):

Theorem 6.4. *The distributed system* (6.15) *is asymptotically stable if there exist common positive definite matrices P_l and positive constants a_l, $l = 1, \ldots, n_s$, such that*

$$\begin{pmatrix} X_{li} & \sqrt{\mu_{li}} m_{li} P_l D_{il} & Y_l \\ \sqrt{\mu_{li}} m_{li} D_{il}^T P_l & -a_l I & 0 \\ Y_l^T & 0 & -I \end{pmatrix} < 0 \qquad i = 1, 2, \ldots, m_l$$

$$X_{li} = \mathcal{H}(P_l A_{li}) + a_l m_{li} E_{il}^T E_{il} + \sum_{k=1,k\neq l}^{n_s} (n_{lk} B_{kl}^T P_k B_{kl} + n_{kl} P_l + \bar{n}_{lk} \bar{E}_{kl}^T \bar{E}_{kl})$$
$$i = 1, 2, \ldots, m_l$$
$$Y_l = P_l[\sqrt{\bar{\mu}_{l1}} \bar{n}_{l1} \bar{D}_{l1} \ \ldots \ \sqrt{\bar{\mu}_{ll-1}} \bar{n}_{ll-1} \bar{D}_{ll-1}$$
$$\sqrt{\bar{\mu}_{ll+1}} \bar{n}_{ll+1} \bar{D}_{ll+1} \ \ldots \ \sqrt{\bar{\mu}_{ln_s}} \bar{n}_{ln_s} \bar{D}_{ln_s}] \tag{6.16}$$

where

$$n_{kl} = \begin{cases} 1, & \text{if } B_{kl} \neq 0 \\ 0, & \text{otherwise} \end{cases}$$

$$m_{li} = \begin{cases} 1, & \text{if there exists } t > 0 \text{ such that } D_{li} F_{li}(t) E_{li} \neq 0 \\ 0, & \text{otherwise} \end{cases}$$

$$\bar{n}_{kl} = \begin{cases} 1, & \text{if there exists } t > 0 \text{ such that } \bar{D}_{lk} \bar{F}_{lk}(t) \bar{E}_{lk} \neq 0 \\ 0, & \text{otherwise} \end{cases}$$

The conditions of Theorem 6.4, similarly to those of Theorem 6.3, are not distributed, in the sense that in order to establish the stability of the distributed system, the conditions have to be verified for all the subsystems at the same time, in parallel. The method is illustrated on the following example.

Example 6.4. Consider a distributed system composed of two subsystems and the corresponding interconnection terms, as follows:

$$\begin{aligned} \dot{\boldsymbol{x}}_1 &= \boldsymbol{A}_1(\boldsymbol{x}_1)\boldsymbol{x}_1 + \boldsymbol{f}_{12}^{\mathrm{m}}(\boldsymbol{x}_1, \boldsymbol{x}_2)\boldsymbol{x}_2 \\ \dot{\boldsymbol{x}}_2 &= \boldsymbol{A}_2(\boldsymbol{x}_2)\boldsymbol{x}_2 + \boldsymbol{f}_{21}^{\mathrm{m}}(\boldsymbol{x}_1, \boldsymbol{x}_2)\boldsymbol{x}_1 \end{aligned} \tag{6.17}$$

where

$$\boldsymbol{x}_1 = (x_1\ x_2)^T \qquad \boldsymbol{x}_2 = (x_3\ x_4)^T$$

$$\boldsymbol{A}_1(\boldsymbol{x}_1) = \begin{pmatrix} -1 & x_2 \\ -1 & -5 + 2x_2^2 \end{pmatrix} \qquad \boldsymbol{A}_2(\boldsymbol{x}_2) = \begin{pmatrix} -4 & 2 + 2x_3^2 \\ -2x_4^2 & -1 \end{pmatrix}$$

$$\boldsymbol{f}_{12}^{\mathrm{m}}(\boldsymbol{x}_1, \boldsymbol{x}_2) = \begin{pmatrix} 0 & 1/5 \\ -x_1^2/15 & 0 \end{pmatrix} \qquad \boldsymbol{f}_{21}^{\mathrm{m}}(\boldsymbol{x}_1, \boldsymbol{x}_2) = \begin{pmatrix} 1/5 & 0 \\ -1/5 & 2x_2^4/15 \end{pmatrix}$$

$$x_i \in [-1, 1] \quad i = 1,\ 2,\ 3,\ 4$$

The TS models of the subsystems are obtained by using the sector nonlinearity approach and the matrices of the first subsystem are

$$A_{11} = \begin{pmatrix} -1 & -1 \\ -1 & -5 \end{pmatrix} \quad A_{12} = \begin{pmatrix} -1 & -1 \\ -1 & -3 \end{pmatrix} \quad A_{13} = \begin{pmatrix} -1 & 1 \\ -1 & -5 \end{pmatrix} \quad A_{14} = \begin{pmatrix} -1 & 1 \\ -1 & -3 \end{pmatrix}$$

while the matrices of the second subsystems are

$$A_{21} = \begin{pmatrix} -4 & 2 \\ -2 & -1 \end{pmatrix} \quad A_{22} = \begin{pmatrix} -4 & 2 \\ 0 & -1 \end{pmatrix} \quad A_{23} = \begin{pmatrix} -4 & 4 \\ -2 & -1 \end{pmatrix} \quad A_{24} = \begin{pmatrix} -4 & 4 \\ 0 & -1 \end{pmatrix}$$

The interconnection terms are nonlinear and they can be expressed as $\boldsymbol{f}_{12}(\boldsymbol{x}_1, \boldsymbol{x}_2)\boldsymbol{x}_2 = (B_{12} + \Delta B_{12})\boldsymbol{x}_2$, with $B_{12} = \begin{pmatrix} 0 & 1/5 \\ 0 & 0 \end{pmatrix}$, $\Delta B_{12} = \begin{pmatrix} 0 & 0 \\ -x_1^2/15 & 0 \end{pmatrix}$, with $\|\Delta B_{12}\| \leq 1/15$, and $\boldsymbol{f}_{21}(\boldsymbol{x}_1, \boldsymbol{x}_2)\boldsymbol{x}_1 = (B_{21} + \Delta B_{21})\boldsymbol{x}_2$, with $B_{21} = \begin{pmatrix} 1/5 & 0 \\ -1/5 & 0 \end{pmatrix}$, $\Delta B_{21} = \begin{pmatrix} 0 & 0 \\ 0 & 2x_2^4/15 \end{pmatrix}$, with $\|\Delta B_{21}\| \leq 2/15$.

With these values, the conditions (6.16) are feasible, and therefore the stability of the distributed system is established. □

Results for stability analysis of distributed TS systems have been extended to stabilization and control of distributed TS systems (Akar and Özgüner, 2000; Tseng and Chen, 2001; Hsiao and Hwang, 2002; Wang et al., 2005; Tong and Zhang, 2008), including robust control (Tseng and Chen, 2001; Hsiao et al., 2005b; Wang and Tong, 2006; Liu et al., 2008; Dhbaibi et al., 2009), control of TS systems with time delays due to the interconnections (Hsiao et al., 2005a; Hua et al., 2005), and decentralized adaptive control (Chiang and Kuo, 2002; Chiang and Wang, 2003; Hua et al., 2005; Chiang, 2006; Chien and Er, 2006; Chiang and Lu, 2007; Wang et al., 2009). However, since all these results concern control design, they are not presented here. The interested reader is referred to the appropriate references.

The results that have been presented so far concern distributed systems where the structure of the system is fixed and known. Next, a method is presented for sequential analysis of distributed TS systems whose structure is not fixed, i.e., subsystems may be added or removed. A sequential analysis has the advantage that subsystems may be added to or removed from the distributed system on-line. This is not possible in case of the methods presented so far.

6.2.2 *Sequential Stability Analysis*

For sequential analysis, without loss of generality, two subsystems are considered, which are coupled through their states. It is assumed that one subsystem has already been proven stable, and a bound on the derivative of the Lyapunov function is available. In this way, the stability of the interconnected system is determined by the stability of the other subsystem and the interconnection term. After the stability of the interconnected system is established, the whole system can be considered as a stable subsystem, to which new subsystems can again be connected.

Consider a TS system (in fact one subsystem) given as:

$$\dot{\boldsymbol{x}}_2 = \sum_{i=1}^{m'} w_i'(\boldsymbol{z}')(A_{2i}\boldsymbol{x}_2) \tag{6.18}$$

Due to the addition of a new subsystem, in general both the membership functions and the local matrices change. In this chapter, we consider only the case when the membership functions change, assuming that the local matrices remain the same. Such an assumption holds for material flow systems, traffic networks, etc., where the addition of a new subsystem does not change the individual dynamics of the existing subsystems. This assumption is not required if a parallel analysis is performed, as the whole interconnected system is given prior to the analysis.

Assumption 6.1. The state matrices of the existing subsystem, A_{2i}, do not change by the addition of the new subsystem.

The restrictiveness of Assumption 6.1 largely depends on how the fuzzy model is obtained. For instance, consider the original system (6.18). If, after adding the new subsystem, the dynamics changes to

$$\dot{\boldsymbol{x}}_2 = \sum_{i=1}^{m'} w'_i(\boldsymbol{z}')(A_{2i}\boldsymbol{x}_2) + \mathbf{A}(\boldsymbol{x}_1, \boldsymbol{x}_2)\boldsymbol{x}_1$$

with $\mathbf{A}$ a smooth nonlinear matrix function that may depend on both $\boldsymbol{x}_1$ and $\boldsymbol{x}_2$, then using the sector nonlinearity approach (see Section 2.3.1), the local models of the original subsystem will remain the same (in fact they are repeated in several rules of the centralized system), although the membership functions will change, as shown in what follows. Using the sector nonlinearity approach, the interconnection term $\mathbf{A}(\boldsymbol{x}_1, \boldsymbol{x}_2)\boldsymbol{x}_1$ can be represented by the TS fuzzy system $\sum_{j=1}^{m_n} w_j(\boldsymbol{z}_n)(A_{2ji}\boldsymbol{x}_1)$, with normalized membership functions w_j, $j = 1, 2, \ldots, m_n$. Then,

$$\begin{aligned}
\dot{\boldsymbol{x}}_2 &= \sum_{i=1}^{m'} w'_i(\boldsymbol{z}')(A_{2i}\boldsymbol{x}_2) + \mathbf{A}(\boldsymbol{x}_1, \boldsymbol{x}_2)\boldsymbol{x}_1 \\
&= \sum_{i=1}^{m'} w'_i(\boldsymbol{z}')(A_{2i}\boldsymbol{x}_2) + \sum_{j=1}^{m_n} h_j(\boldsymbol{z}_n)(A_{21j}\boldsymbol{x}_1) \\
&= \sum_{i=1}^{m'} w'_i(\boldsymbol{z}') \sum_{j=1}^{m_n} h_j(\boldsymbol{z}_n)(A_{2i}\boldsymbol{x}_2 + A_{21j}\boldsymbol{x}_1) \\
&= \sum_{i=1}^{m'} \sum_{j=1}^{m_n} w'_i(\boldsymbol{z}') h_j(\boldsymbol{z}_n)(A_{2i}\boldsymbol{x}_2 + A_{21j}\boldsymbol{x}_1) \\
&= \sum_{i=1}^{m' \cdot m_n} w_i(\boldsymbol{z})(A_{2i}\boldsymbol{x}_2 + A_{21i}\boldsymbol{x}_1)
\end{aligned}$$

If the local models are obtained using Taylor series expansion (see Section 2.3.2), then this assumption becomes more restrictive.

If the global system is known prior to the analysis, and no new subsystems are added, then Assumption 6.1 is not needed.

Example 6.5. Consider the distributed system described in Example 6.1. If both subsystems are given, together with the interconnection terms, the fuzzy models can be derived as in Example 6.1, and Assumption 6.1 is not needed.

However, depending on how the distributed system is constructed, it is possible that at a certain moment only the second subsystem exists, i.e., the system is simply

$$\dot{\boldsymbol{x}}_2 = \boldsymbol{A}_2(\boldsymbol{x}_2)\boldsymbol{x}_2 \qquad \boldsymbol{A}_2(\boldsymbol{x}_2) = \begin{pmatrix} -4 & 2 + x_3^2 \\ -2x_4^2 & -1 \end{pmatrix}$$

for which a fuzzy model can be written as

$$\dot{\boldsymbol{x}}_2 = \sum_{i=1}^{4} w_{2i}^{'}(\boldsymbol{z}_2) A_{2i} \boldsymbol{x}_2$$

with A_{2i} being the matrices given in Example 6.1.

To this subsystem, the new subsystem given by

$$\dot{\boldsymbol{x}}_1 = \boldsymbol{A}_1(\boldsymbol{x}_1)\boldsymbol{x}_1$$

is added. The connections between the subsystems are realized by the interconnection terms $\boldsymbol{f}_{12}^{\mathrm{m}}(\boldsymbol{x}_1, \boldsymbol{x}_2)\boldsymbol{x}_2$ and $\boldsymbol{f}_{21}^{\mathrm{m}}(\boldsymbol{x}_1, \boldsymbol{x}_2)\boldsymbol{x}_1$ such that the interconnected system can be described as

$$\begin{aligned}
\dot{\boldsymbol{x}}_1 &= \boldsymbol{A}_1(\boldsymbol{x}_1)\boldsymbol{x}_1 + \boldsymbol{f}_{12}^{\mathrm{m}}(\boldsymbol{x}_1, \boldsymbol{x}_2)\boldsymbol{x}_2 \\
\dot{\boldsymbol{x}}_2 &= \boldsymbol{A}_2(\boldsymbol{x}_2)\boldsymbol{x}_2 + \boldsymbol{f}_{21}^{\mathrm{m}}(\boldsymbol{x}_1, \boldsymbol{x}_2)\boldsymbol{x}_1
\end{aligned}$$

Since in the interconnection term $\boldsymbol{f}_{21}(\boldsymbol{x}_1, \boldsymbol{x}_2)\boldsymbol{x}_1$ there is one nonlinearity, the new fuzzy model

$$\dot{\boldsymbol{x}}_2 = \boldsymbol{A}_2(\boldsymbol{x}_2)\boldsymbol{x}_2 + \boldsymbol{f}_{21}^{\mathrm{m}}(\boldsymbol{x}_1, \boldsymbol{x}_2)\boldsymbol{x}_1$$

i.e., the second subsystem together with the interconnection term, can be written as

$$\begin{aligned}
\dot{\boldsymbol{x}}_2 &= \sum_{i=1}^{4} w_{2i}^{'}(\boldsymbol{z}_2) A_{2i}\boldsymbol{x}_2 + \sum_{j=1}^{2} w_{21j}^{'}(\boldsymbol{z}_{21}) A_{21j}\boldsymbol{x}_1 \\
&= \sum_{i=1}^{4}\sum_{j=1}^{2} w_{2i}^{'}(\boldsymbol{z}_2) w_{21j}^{'}(\boldsymbol{z}_{21})(A_{2i}\boldsymbol{x}_2 + A_{21j}\boldsymbol{x}_1)
\end{aligned}$$

As can be seen, in this case the already available matrices A_{2i} do not change: they are repeated in the interconnected system for each nonlinearity in the interconnection term $\boldsymbol{f}_{21}^{\mathrm{m}}(\boldsymbol{x}_1, \boldsymbol{x}_2)\boldsymbol{x}_2$. □

When the new subsystem is added, and Assumption 6.1 is satisfied, the whole system, i.e., the subsystem added (with states $\boldsymbol{x}_1$), the existing subsystem (with states $\boldsymbol{x}_2$), and the interconnection terms are expressed together as

$$\begin{aligned}
\dot{\boldsymbol{x}}_1 &= \sum_{i=1}^{m} w_i(\boldsymbol{z})(A_{1i}\boldsymbol{x}_1 + A_{12i}\boldsymbol{x}_2) \\
\dot{\boldsymbol{x}}_2 &= \sum_{i=1}^{m} w_i(\boldsymbol{z})(A_{2i}\boldsymbol{x}_2 + A_{21i}\boldsymbol{x}_1)
\end{aligned} \tag{6.19}$$

The structure of system (6.19) is then presented in Figure 6.1.

For this system, the following stability conditions have been formulated (Lendek et al., 2008):

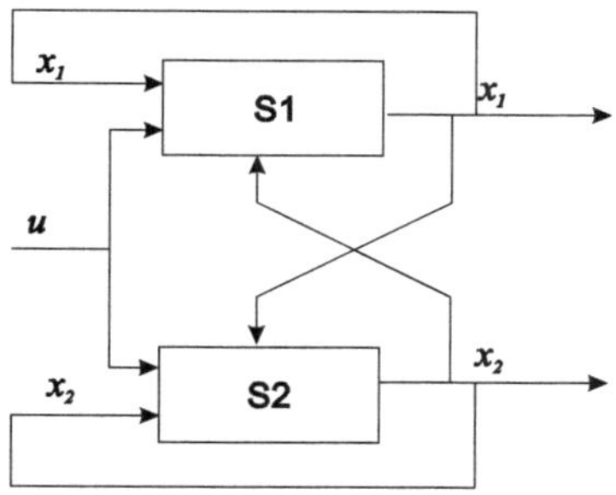

Fig. 6.1 Two subsystems coupled through their states.

Theorem 6.5. *The system* (6.19) *is asymptotically stable, if there exist* $P_1 = P_1^T > 0$, $P_2 = P_2^T > 0$, $Q_1 = Q_1^T > 0$, *and* $Q_2 = Q_2^T > 0$, *so that for* $i = 1, 2, \ldots, m$,

$$\begin{aligned}
&\mathcal{H}(P_1 A_{1i}) < -2Q_1 \\
&\mathcal{H}(P_2 A_{2i}) < -2Q_2 \\
&\lambda_{\min}(Q_1) \geq \max_i \|P_1 A_{12i}\| \\
&\frac{\lambda_{\min}(\mathcal{H}(P_1 A_{1i} + Q_1))}{\max_i \|P_1 A_{12i}\|} > \frac{\max_i \|A_{21i}^T P_2\|^2}{\lambda_{\min}(Q_2)\lambda_{\min}(\mathcal{H}(P_2 A_{2i} + Q_2))}
\end{aligned}$$

where $\lambda_{\min}(.)$ *denotes the eigenvalue with the smallest absolute magnitude.*

Proof: System (6.19) can be seen as the cascaded system

$$\begin{aligned}
\dot{\boldsymbol{x}}_1 &= \sum_{i=1}^{m} w_i(\boldsymbol{z})(A_{1i}\boldsymbol{x}_1) \\
\dot{\boldsymbol{x}}_2 &= \sum_{i=1}^{m} w_i(\boldsymbol{z})(A_{2i}\boldsymbol{x}_2 + A_{21i}\boldsymbol{x}_1)
\end{aligned} \tag{6.20}$$

with an extra feedback term given by $\sum_{i=1}^{m} w_i(\boldsymbol{z})A_{12i}\boldsymbol{x}_2$. As it has been established in Chapter 5, according to Theorem 5.3, system (6.20) is exponentially stable, if there exist $P_1 = P_1^T > 0$, $P_2 = P_2^T > 0$, $Q_1 = Q_1^T > 0$, and $Q_2 = Q_2^T > 0$, so that for $i = 1, 2, \ldots, m$, it holds that

$$\begin{aligned}
\mathcal{H}(P_1 A_{1i}) &< -2Q_1 \\
\mathcal{H}(P_2 A_{2i}) &< -2Q_2
\end{aligned} \tag{6.21}$$

In order to make the step from the stable cascaded system to the analysis of the distributed system, a Lyapunov function is needed. One way of constructing the Lyapunov function using P_1 and P_2 is by considering the function

$V_c = \boldsymbol{x}^T \text{diag}(\alpha P_1,\, P_2)\boldsymbol{x}$. This choice allows one to determine $\alpha \in \mathcal{R}^+$ so that $\dot{V}_c < -2\boldsymbol{x}^T Q\boldsymbol{x}$, with $Q = \text{diag}(\alpha Q_1,\, Q_2)$:

$$\dot{V}_c = \sum_{i=1}^{m} w_i(\boldsymbol{z})\boldsymbol{x}^T \begin{pmatrix} \alpha\mathcal{H}(P_1A_{1i}) & A_{21i}^T P_2 \\ P_2A_{21i} & \mathcal{H}(P_2A_{2i}) \end{pmatrix} \boldsymbol{x}$$

Then, $\dot{V}_c < -2\boldsymbol{x}^T Q\boldsymbol{x}$, if

$$\begin{pmatrix} \alpha\mathcal{H}(P_1A_{1i}) & {A_{21}}_i^T P_2 \\ P_2A_{21i} & \mathcal{H}(P_2A_{2i}) \end{pmatrix} < -2\begin{pmatrix} \alpha Q_1 & 0 \\ 0 & Q_2 \end{pmatrix}$$

for $i = 1,\, 2,\, \ldots,\, m$, or

$$\begin{pmatrix} \alpha\mathcal{H}(P_1A_{1i} + Q_1) & {A_{21}}_i^T P_2 \\ P_2A_{21i} & \mathcal{H}(P_2A_{2i} + Q_2) \end{pmatrix} < 0$$

for $i = 1,\, 2,\, \ldots,\, m$. Using the Schur complement, one has

$$\alpha\mathcal{H}(P_1A_{1i} + Q_1) - (A_{21i}^T P_2)(\mathcal{H}(P_2A_{2i} + Q_2))^{-1} P_2A_{21i} < 0$$

for $i = 1,\, 2,\, \ldots,\, m$, which is true if α is chosen such that

$$\alpha > \frac{1}{\lambda_{\min}(\mathcal{H}(P_1A_{1i} + Q_1))} \cdot \frac{\max_i \|A_{21i}^T P_2\|^2}{\lambda_{\min}(\mathcal{H}(P_2A_{2i} + Q_2))} \tag{6.22}$$

for $i = 1,\, 2,\, \ldots,\, m$, where $\lambda_{\min}(.)$ denotes the eigenvalue with the smallest absolute magnitude. Now, consider the full system (6.19). Using the above constructed V_c as a candidate Lyapunov function for (6.19), one obtains

$$\begin{aligned} \dot{V}_c &= \sum_{i=1}^{m} w_i(\boldsymbol{z})\boldsymbol{x}^T \left[\begin{pmatrix} \alpha\mathcal{H}(P_1A_{1i}) & A_{21i}^T P_2 \\ P_2A_{21i} & \mathcal{H}(P_2A_{2i}) \end{pmatrix} + \begin{pmatrix} 0 & \alpha P_1A_{12i} \\ \alpha A_{12i}^T P_1 & 0 \end{pmatrix} \right] \boldsymbol{x} \\ &< -2\boldsymbol{x}^T \begin{pmatrix} \alpha Q_1 & 0 \\ 0 & Q_2 \end{pmatrix} \boldsymbol{x} + 2\boldsymbol{x}^T \alpha \max_i \|P_1A_{12i}\| I\boldsymbol{x} \\ &< -2\boldsymbol{x}^T \begin{pmatrix} \alpha(Q_1 - \max_i \|P_1A_{12i}\| I) & 0 \\ 0 & Q_2 - \alpha\max_i \|P_1A_{12i}\| I \end{pmatrix} \boldsymbol{x} \end{aligned}$$

which leads to the conditions

$$\lambda_{\min}(Q_1) > \max_i \|P_1A_{12i}\| \tag{6.23}$$

$$\lambda_{\min}(Q_2) > \alpha\max_i \|P_1A_{12i}\| \tag{6.24}$$

for $i = 1,\, 2,\, \ldots,\, m$.

Combining (6.22) and (6.24), such an α exists, and V_c is a Lyapunov function for the whole system if

$$\frac{\lambda_{\min}(Q_2)}{\max_i \|P_1 A_{12i}\|} > \frac{\max_i \|A_{21i}^T P_2\|^2}{\lambda_{\min}(\mathcal{H}(P_1 A_{1i} + Q_1))\lambda_{\min}(\mathcal{H}(P_2 A_{2i} + Q_2))}$$

or

$$\frac{\lambda_{\min}(\mathcal{H}(P_1 A_{1i} + Q_1))}{\max_i \|P_1 A_{12i}\|} > \frac{\max_i \|A_{21i}^T P_2\|^2}{\lambda_{\min}(Q_2)\lambda_{\min}(\mathcal{H}(P_2 A_{2i} + Q_2))}$$

for $i = 1, 2, \ldots, m$. With this, Theorem 6.5 is proven. □

Remark: If $A_{12i} = 0$, for $i = 1, 2, \ldots, m$, or $A_{21i} = 0$, for $i = 1, 2, \ldots, m$, then based on Theorem 5.3, the system (6.19) is stable if the individual subsystems are stable, and the last two conditions are not required.

A shortcoming of Theorem 6.5 is that the conditions are not LMIs. However, LMI conditions, which, when satisfied ensure the conditions of Theorem 6.5 can be formulated using the following two-step procedure. Note that the following conditions are more conservative than those of Theorem 6.5.

Algorithm 6.1. Sequential stability analysis

1. Assume that the existing system

$$\dot{\boldsymbol{x}}_2 = \sum_{i=1}^{m} w_i(\boldsymbol{z}) A_{2i} \boldsymbol{x}$$

is already proven to be stable using a quadratic Lyapunov function and therefore P_2 and Q_2 such that $\mathcal{H}(P_2 A_{2i}) < -2Q_2$, have been computed. Thanks to this, when adding the new subsystem, with the interconnection terms, one can compute

$$\gamma = \frac{\max_i \|A_{21i}^T P_2\|^2}{\lambda_{\min}(Q_2)\min_i(\lambda_{\min}(\mathcal{H}(P_2 A_{2i} + Q_2))})$$

2. Now, for the added subsystem and the corresponding interconnection terms the conditions:

$$\begin{aligned}
&\mathcal{H}(P_1 A_{1i}) < -2Q_1 \\
&\lambda_{\min}(Q_1) \geq \max_i \|P_1 A_{12i}\| \\
&\lambda_{\min}(\mathcal{H}(P_1 A_{1i} + Q_1)) > \gamma \max_i \|P_1 A_{12i}\|
\end{aligned}$$

for $i = 1, 2, \ldots, m$, are satisfied if the LMIs

$$\begin{aligned}
&\mathcal{H}(P_1 A_{1i} + Q_1) < -t_1 I \\
&Q_1 > t_2 I \\
&\begin{pmatrix} t_2 I & \max_i \|A_{12i}\| P_1 \\ \max_i \|A_{12i}\| P_1 & t_2 I \end{pmatrix} > 0 \\
&\begin{pmatrix} t_1 I & \gamma \max_i \|A_{12i}\| P_1 \\ \gamma \max_i \|A_{12i}\| P_1 & t_1 I \end{pmatrix} > 0
\end{aligned} \tag{6.25}$$

for $i = 1, 2, \ldots, m$ are feasible.

Moreover, if one takes into consideration that new subsystems will be added to the whole system (6.19), the analysis of the new subsystems can be facilitated by minimizing the expression:

$$\frac{\|P_1\|^2}{\lambda_{\min}(Q_1)\min_i(\lambda_{\min}(\mathcal{H}(P_1 A_{1i}+Q_1)))}$$

which will in turn minimize the bound γ computed for the system (6.19). This can be achieved by solving the LMI-based convex problem: *find* $P_1 = P_1^T > 0$, $Q_1 = Q_1^T > 0$, *and maximize* t_1, t_2, t_3 *subject to* (6.25) *and* $P_1 < t_3 I$. □

The application of Theorem 6.5, using Algorithm 6.1 is illustrated on the following example.

Example 6.6. Consider the distributed system in Example 6.1, constructed as in Example 6.5. First the system available is the second subsystem

$$\dot{\boldsymbol{x}}_2 = \boldsymbol{A}_2(\boldsymbol{x}_2)\boldsymbol{x}_2 \qquad \boldsymbol{A}_2(\boldsymbol{x}_2) = \begin{pmatrix} -4 & 2+2x_3^2 \\ -2x_4^2 & -1 \end{pmatrix}$$

For this system a fuzzy model is written as

$$\dot{\boldsymbol{x}}_2 = \sum_{i=1}^{4} w'_{2i}(\boldsymbol{z}_2) A_{2i}\boldsymbol{x}_2$$

with

$$A_{21} = \begin{pmatrix} -4 & 2 \\ -2 & -1 \end{pmatrix} \quad A_{22} = \begin{pmatrix} -4 & 2 \\ 0 & -1 \end{pmatrix} \quad A_{23} = \begin{pmatrix} -4 & 4 \\ -2 & -1 \end{pmatrix} \quad A_{24} = \begin{pmatrix} -4 & 4 \\ 0 & -1 \end{pmatrix}$$

The stability of this system is verified by solving the LMI problem: find $P_2 = P_2^T > 0$, and $Q_2 = Q_2^T > 0$ such that $\mathcal{H}(P_2 A_{2i}) < -2Q_2$. One obtains $P_2 = \begin{pmatrix} 0.23 & -0.03 \\ -0.03 & 0.77 \end{pmatrix}$, $Q_2 = 0.247I$.

Now, the new subsystem is added, with the interconnection terms $\boldsymbol{f}_{21}^{\mathrm{m}}(\boldsymbol{x}_1, \boldsymbol{x}_2)\boldsymbol{x}_1$ and $\boldsymbol{f}_{12}^{\mathrm{m}}(\boldsymbol{x}_1, \boldsymbol{x}_2)\boldsymbol{x}_2$. The new subsystem is also written as a fuzzy model, with the matrices

$$A_{11} = \begin{pmatrix} -1 & -1 \\ -1 & -5 \end{pmatrix} \quad A_{12} = \begin{pmatrix} -1 & -1 \\ -1 & -3 \end{pmatrix} \quad A_{13} = \begin{pmatrix} -1 & 1 \\ -1 & -5 \end{pmatrix} \quad A_{14} = \begin{pmatrix} -1 & 1 \\ -1 & -3 \end{pmatrix}$$

The interconnection terms can also be expressed as fuzzy models, both having two rules, with the matrices $A_{211} = \begin{pmatrix} 1/5 & 0 \\ -1/5 & 0 \end{pmatrix}$, $A_{212} = \begin{pmatrix} 1/5 & 0 \\ -1/5 & 2/5 \end{pmatrix}$, $A_{121} = \begin{pmatrix} 0 & 1/5 \\ -1/50 & \end{pmatrix}$, and $A_{122} = \begin{pmatrix} 0 & 1/5 \\ 0 & 0 \end{pmatrix}$. With these values, γ from Step 1 of Algorithm 6.1 can be computed as $\gamma = 0.66$.

In the second step of Algorithm 6.1, the stability of the interconnected system

$$\begin{aligned}\dot{\boldsymbol{x}}_1 &= \boldsymbol{A}_1(\boldsymbol{x}_1)\boldsymbol{x}_1 + \boldsymbol{f}_{12}^{\mathrm{m}}(\boldsymbol{x}_1, \boldsymbol{x}_2)\boldsymbol{x}_2 \\ \dot{\boldsymbol{x}}_2 &= \boldsymbol{A}_2(\boldsymbol{x}_2)\boldsymbol{x}_2 + \boldsymbol{f}_{21}^{\mathrm{m}}(\boldsymbol{x}_1, \boldsymbol{x}_2)\boldsymbol{x}_1\end{aligned}$$

is analyzed, by solving (6.25). One obtains $P_1 = \begin{pmatrix} 0.68 & -0.07 \\ -0.07 & 0.32 \end{pmatrix}$, and $Q_1 = 0.22I$. With this, the stability of the interconnected system is established. □

A shortcoming of the approach at this point is that although the stability analysis of the second subsystem has been performed, and $V_\mathrm{c} = \begin{pmatrix} \alpha P_1 & 0 \\ 0 & P_2 \end{pmatrix}$ is used as a Lyapunov function, all that is known is that $\dot{V}_\mathrm{c} < 0$, i.e., a bound on the derivative of the Lyapunov function is not yet available. To continue the reasoning when the next subsystem will be added, it is desired that $\dot{V}_\mathrm{c} \leq -2\boldsymbol{x}^T Q \boldsymbol{x}$, for some $Q = Q^T > 0$. To obtain such a Q, consider the derivative of $\dot{V}_\mathrm{c}$,

$$\dot{V}_\mathrm{c} < -2\boldsymbol{x}^T \begin{pmatrix} \alpha(Q_1 - \max_i \|P_1 A_{12i}\| I) & 0 \\ 0 & Q_2 - \alpha \max_i \|P_1 A_{12i}\| I \end{pmatrix} \boldsymbol{x}$$

By imposing that

$$\begin{pmatrix} \alpha(Q_1 - \max_i \|P_1 A_{12i}\| I) & 0 \\ 0 & Q_2 - \alpha \max_i \|P_1 A_{12i}\| I \end{pmatrix} > \beta \begin{pmatrix} \alpha Q_1 & 0 \\ 0 & Q_2 \end{pmatrix}$$

where $i = 1, 2, \ldots, m$, for some arbitrary $\beta \in (0, 1)$, the following conditions are obtained

$$\begin{aligned} Q_1 - \max_i \|P_1 A_{12i}\| I &> \beta Q_1 \\ Q_2 - \alpha \max_i \|P_1 A_{12i}\| I &> \beta Q_2 \end{aligned}$$

i.e.,

$$\begin{aligned} (1-\beta) Q_1 &> \max_i \|P_1 A_{12i}\| I \\ (1-\beta) Q_2 &> \alpha \max_i \|P_1 A_{12i}\| I \end{aligned} \tag{6.26}$$

where $i = 1, 2, \ldots, m$. Combining (6.26) and the conditions of Theorem 6.5, the following corollary can be formulated.

Corollary 6.1. $V = \left(\boldsymbol{x}_1^T\ \boldsymbol{x}_2^T\right) \begin{pmatrix} \alpha P_1 & 0 \\ 0 & P_2 \end{pmatrix} \begin{pmatrix} \boldsymbol{x}_1 \\ \boldsymbol{x}_2 \end{pmatrix}$ *is a Lyapunov function for* (6.19) *and* $\dot{V} < \beta \left(\boldsymbol{x}_1^T\ \boldsymbol{x}_2^T\right) \begin{pmatrix} \alpha Q_1 & 0 \\ 0 & Q_2 \end{pmatrix} \begin{pmatrix} \boldsymbol{x}_1 \\ \boldsymbol{x}_2 \end{pmatrix}$ *for an arbitrary* $\beta \in (0, 1)$ *if, for* $i = 1, 2, \ldots, m$*:*

$$\begin{aligned}
&\mathcal{H}(P_1A_{1i}) < -2Q_1 \\
&\mathcal{H}(P_2A_{2i}) < -2Q_2 \\
&(1-\beta)\lambda_{\min}(Q_1) \geq \max_i \|P_1A_{12i}\| \\
&\frac{\lambda_{\min}(\mathcal{H}(P_1A_{1i}+Q_1))}{\max_i \|P_1A_{12i}\|} > \frac{\max_i \|A_{21i}^TP_2\|^2}{(1-\beta)\lambda_{\min}(Q_2)\lambda_{\min}(\mathcal{H}(P_2A_{2i}+Q_2))}
\end{aligned} \tag{6.27}$$

The application of the conditions of Corollary 6.1 are illustrated using the following example.

Example 6.7. Consider the distributed system in Example 6.6, where the stability of the interconnected system has been established, but a bound on the derivative of the Lyapunov function has not yet been found. This bound is needed for analyzing the stability if again a new subsystem is added. To obtain this bound, one has to solve (6.27) instead of (6.25).

Since the conditions (6.27) are nonlinear, with a fixed $\beta \in [0,\ 1]$, sufficient LMI conditions, similar to (6.25) are formulated as

$$\begin{aligned}
&\mathcal{H}(P_1A_{1i}+Q_1) < -t_1I \\
&Q_1 > t_2I \\
&\begin{pmatrix} (1-\beta)t_2I & \max_i \|A_{12i}\|P_1 \\ \max_i \|A_{12i}\|P_1 & (1-\beta)t_2I \end{pmatrix} > 0 \\
&\begin{pmatrix} (1-\beta)t_1I & \gamma\max_i \|A_{12i}\|P_1 \\ \gamma\max_i \|A_{12i}\|P_1 & (1-\beta)t_1I \end{pmatrix} > 0
\end{aligned} \tag{6.28}$$

for $i = 1, 2, \ldots, m$. In fact the LMIs (6.25) are a special case obtained for $\beta = 1$ of the conditions (6.28).

Conditions (6.28) may not be feasible for any β. For the distributed system in Example 6.6, we obtain that (6.28) is feasible for $\beta = 0.1$, but it is not feasible for instance for $\beta = 0.9$. Since the goal is to determine β as large as possible to facilitate the analysis of possible subsystems that will be added, one can consider (6.28) a BMI problem, with both β and P_1 decision variables. Solving (6.28) as a BMI problem[5] yields $P_1 = \begin{pmatrix} 0.23 & -0.03 \\ -0.03 & 0.77 \end{pmatrix}$, $\beta = 0.42$, and $Q_1 = 0.25I$. □

Recall that it was assumed that the interconnection terms or bounds on them are not known before adding a new subsystem. However, if $\gamma_k = \max_{ij} \|A_{kij}\|$, i.e., a bound on the interconnection terms is known beforehand, the analysis of the subsystems can be decoupled and the following result can be stated:

Theorem 6.6. *Given* $\gamma_1 = \max_i \|A_{12i}\|$ *and* $\gamma_2 = \max_i \|A_{21i}\|$, $i = 1, 2, \ldots, m$, *the distributed system* (6.19) *is exponentially stable, if there exist* $P_1 = P_1^T > 0$, $P_2 = P_2^T > 0$, $Q_1 = Q_1^T > 0$, $Q_2 = Q_2^T > 0$, *so that*

[5] For solving BMI problems, *Penbmi* (Kočvara and Stingl, 2008) has been used.

$$
\begin{aligned}
&\mathcal{H}(P_1 A_{1i}) < -2Q_1 \\
&\mathcal{H}(P_2 A_{2i}) < -2Q_2 \\
&\lambda_{\min}(\mathcal{H}(P_1 A_{1i} + Q_1)) > \lambda_{\min}(Q_1) \\
&\lambda_{\min}(\mathcal{H}(P_2 A_{2i} + Q_2)) > \lambda_{\min}(Q_2) \\
&\lambda_{\min}(Q_1) \geq \gamma_1 \|P_1\| \\
&\lambda_{\min}(Q_2) \geq \gamma_2 \|P_2\|
\end{aligned}
\tag{6.29}
$$

for $i = 1, 2, \ldots, m$, *where* $\lambda_{\min}(.)$ *is the eigenvalue with the smallest absolute magnitude.*

Proof: The last condition of Theorem 6.5 is

$$
\frac{\lambda_{\min}(\mathcal{H}(P_1 A_{1i} + Q_1))}{\max_i \|P_1 A_{12i}\|} > \frac{\max_i \|A_{21i}^T P_2\|^2}{\lambda_{\min}(Q_2)\lambda_{\min}(\mathcal{H}(P_2 A_{2i} + Q_2))}
$$

for $i = 1, 2, \ldots, m$, which can be rewritten as

$$
\begin{aligned}
\lambda_{\min}(\mathcal{H}(P_1 A_{1i} + Q_1))\lambda_{\min}(Q_2)\lambda_{\min}(\mathcal{H}(P_2 A_{2i} + Q_2)) > \\
\max_i \|P_1 A_{12i}\| \max_i \|A_{21i}^T P_2\|^2
\end{aligned}
$$

for $i = 1, 2, \ldots, m$. The third condition of Theorem 6.5 already states that for $i = 1, 2, \ldots, m$

$$
\lambda_{\min}(Q_1) \geq \max_i \|P_1 A_{12i}\| \tag{6.30}
$$

If Q_2 is similarly restricted, i.e., the condition

$$
\lambda_{\min}(Q_2) \geq \max_i \|P_2 A_{21i}\| \tag{6.31}
$$

$i = 1, 2, \ldots, m$ is imposed, then the last condition of Theorem 6.5 becomes

$$
\begin{aligned}
&\lambda_{\min}(\mathcal{H}(P_1 A_{1i} + Q_1))\lambda_{\min}(Q_2)\lambda_{\min}(\mathcal{H}(P_2 A_{2i} + Q_2)) > \lambda_{\min}(Q_1)\lambda_{\min}^2(Q_2) \\
&\lambda_{\min}(\mathcal{H}(P_1 A_{1i} + Q_1))\lambda_{\min}(\mathcal{H}(P_2 A_{2i} + Q_2)) > \lambda_{\min}(Q_1)\lambda_{\min}(Q_2)
\end{aligned}
$$

for $i = 1, 2, \ldots, m$, which is satisfied if

$$
\begin{aligned}
&\lambda_{\min}(\mathcal{H}(P_1 A_{1i} + Q_1)) > \lambda_{\min}(Q_1) \\
&\lambda_{\min}(\mathcal{H}(P_2 A_{2i} + Q_2)) > \lambda_{\min}(Q_2)
\end{aligned}
$$

for $i = 1, 2, \ldots, m$. However, since only the bounds on the interconnection terms γ_1 and γ_2 are known, instead of (6.30) and (6.31) one has to use

$$
\begin{aligned}
&\lambda_{\min}(Q_1) \geq \gamma_1 \|P_1\| \\
&\lambda_{\min}(Q_2) \geq \gamma_2 \|P_2\|
\end{aligned}
\tag{6.32}
$$

Together with the restrictions on Q_1 and Q_2, and (6.32), the conditions expressed in Theorem 6.6 are obtained. □

Example 6.8. Consider again the distributed system of Example 6.6. To apply Theorem 6.6, one can bound the interconnection terms as $\gamma_2 = \max_i \|A_{21i}\| = 0.46$ and $\gamma_1 = \max_i \|A_{12i}\| = 0.2$, $i = 1, 2, \ldots, m$. However, similarly to the result in Example 6.4, the conditions (6.29) are not feasible. In this case, the unfeasibility of the conditions is due to the "too large" interconnection term $\boldsymbol{f}_{21}$, in fact γ_2. If this interconnection term would be such that this bound has the value $\gamma_2/2$, the stability of the interconnected system can be established using (6.29). □

The conditions of Theorem 6.6 are similar to those reported by Wang and Lin (2005). As it has already been stated, the parallel analysis has the advantage that it is less conservative when for each subsystem the strength of the interconnection terms is approximately the same and are weak in both directions, i.e., there is a weak coupling (Sandell et al., 1978) between the subsystems. A shortcoming of such approaches is that the result can only be obtained if bounds on the interconnection terms that are introduced to the system by the addition of a new subsystem are known beforehand. This condition is not needed for Theorem 6.5, as, thanks to the sequential analysis, the interconnection terms only need to be known when the subsystem that introduces them is analyzed. In fact, Theorem 6.5 and the resulting Algorithm 6.1, are comparable to methods developed for strong coupling, i.e., only one of the subsystems has to converge quickly enough so that stability is preserved. The sequential approach can also be thought of as an asymmetrical weak coupling, i.e., only one of the influences has to be weak enough for stability to be preserved.

One of the assumptions that restricts the applicability of the presented methods is that they require the stability of the individual subsystems. Since no restrictions are assumed on the structure of the interconnected system (e.g., conditions such as only the neighboring subsystems are interconnected), this is a reasonable assumption. Moreover, this shortcoming can be alleviated for instance for sequential stability analysis (Theorem 6.5) by using a full Lyapunov matrix instead of the block-diagonal one when deriving the stability conditions. However, such an approach results in even more complex nonlinear conditions. Note that this is a disadvantage mainly of interconnected systems where subsystems are added on-line. If the whole interconnected system is known prior to the analysis, this disadvantage may disappear, for instance because the addition of new subsystems changes the dynamics of the individual subsystems (e.g., two inverted pendulums connected with a spring).

On the other hand, if subsystems can be added and removed over time, for establishing the stability of the interconnected system, it is necessary that the individual subsystems are stable. If no subsystem is removed, this assumption is no longer necessary, i.e., theoretically it is allowed that some subsystems are unstable as long as the other subsystems stabilize it. Results for such systems exist for special cases of linear systems, but, in the context of TS fuzzy systems, this issue has not been addressed.

6.3 Distributed Observer Design for TS Systems

Distributed observer design has been early recognized as an important problem and attracted research interest since the 1970s. However, most of the obtained results concern linear systems.

The general approach is that first one constructs a set of observers for the independent subsystems. However, unless there is no interconnection between the subsystems, the set of local observers does not guarantee the convergence of the estimation error to zero for the interconnected system. Therefore, one either has to incorporate an appropriate compensation to account for the influence of other subsystems or determine conditions under which the collection of the individual observers is a valid observer for the distributed system. In this section, we consider distributed observer design for TS systems, and present conditions that ensure that the estimation error converges to zero.

6.3.1 General Framework

For observer design, we consider a system consisting of n_s subsystems, with each subsystem l represented by the TS model

$$\begin{aligned}\dot{\boldsymbol{x}}_l &= \sum_{i=1}^{m_l} w_i(\boldsymbol{z})\Big(A_{li}\boldsymbol{x}_l + B_{li}\boldsymbol{u}_l + a_{li} + \sum_{j=1,j\neq l}^{n_s} \boldsymbol{f}_{lij}(\boldsymbol{x}_j)\Big)\\ \boldsymbol{y}_l &= \sum_{i=1}^{m_l} w_i(\boldsymbol{z})\Big(C_{li}\boldsymbol{x}_l + c_{li} + \sum_{j=1,j\neq l}^{n_s} \boldsymbol{h}_{lij}(\boldsymbol{x}_j)\Big)\end{aligned} \tag{6.33}$$

where $\boldsymbol{x}_l$, and $\boldsymbol{u}_l$, $l = 1, 2, \ldots, n_s$ denote the state and input vectors of the lth subsystem, m_l is the number of rules in the fuzzy representation of the lth subsystem, A_{li}, B_{li}, C_{li}, a_{li}, and c_{li} are the corresponding local matrices and biases, and $\boldsymbol{f}_{lij}$ and $\boldsymbol{h}_{lij}$ denote the (usually nonlinear) interconnection terms from other subsystems. A general assumption, similarly to stability analysis, is that these interconnection terms are Lipschitz in the states, i.e., there exist μ^f_{lij} and μ^h_{lij} such that $\|\boldsymbol{f}_{lij}(\boldsymbol{x}_l)\| \leq \mu^f_{lij}\|\boldsymbol{x}_l\|$, and $\|\boldsymbol{h}_{lij}(\boldsymbol{x}_l)\| \leq \mu^h_{lij}\|\boldsymbol{x}_l\|$.

We start with presenting a result for the linear case. Observer design for TS systems in general follows the same line of reasoning.

For linear systems, in general, two main approaches have been considered. The first approach considers the case when, for some reason, the estimated values cannot be communicated. In such a case, the interconnection term is usually treated as an unknown input, and is either decoupled or estimated. While unknown input observers have attracted research interest for linear systems (Xiong and Saif, 2003; Pertew et al., 2005; Hsieh, 2009), and for centralized TS systems with linear measurements (Marx et al., 2007) to the authors' best knowledge, this setting has not been considered in the context of distributed TS systems.

For distributed linear systems, Saif and Guan (1992) proposed a decentralized observer scheme based on this approach. In the approach of Saif and Guan (1992), a local observer is designed for each subsystem that treats the influence from other subsystems as an unknown input and effectively decouples them. A similar method has been employed by Hou and Müller (1994). Note, however, that designing such an observer is not possible in all cases, even less in the case of TS systems with a nonlinear measurement model.

The second approach is when the measured or estimated variables are communicated between the subsystems that directly influence each other. Although in practice such an assumption can lead to a communication overhead, or may be even impossible to realize, in theoretical developments this is a common assumption.

For the linear case, consider the distributed system where each subsystem l, $l = 1, \ldots, n_s$, is described as

$$\begin{aligned} \dot{\boldsymbol{x}}_l &= A_l \boldsymbol{x}_l + B_l \boldsymbol{u}_l + \sum_{i=1, i \neq l}^{n_s} D_{li} \boldsymbol{x}_i \\ \boldsymbol{y}_l &= C_l \boldsymbol{x}_l \end{aligned} \tag{6.34}$$

For this system, the following observer can be used, see (Sundareshan and Elbanna, 1990) and the references therein:

$$\dot{\widehat{\boldsymbol{x}}}_l = (A_l - L_l C_l)\widehat{\boldsymbol{x}}_l + L_l \boldsymbol{y} + B_l \boldsymbol{u}_l + \sum_{i=1, i \neq l}^{n_s} D_{li} \widehat{\boldsymbol{x}}_i$$

If the subsystems are independent, i.e., $D_{li} = 0$, L_l can be determined as for instance $L_l = P_l C_l^T$, where $P_l = P_l^T > 0$ is the solution of the Riccati equation

$$\mathcal{H}(P_l A_l) - P_l C_l^T C_l P_l + Q_l = 0$$

with $Q_l = Q_l^T > 0$.

Otherwise, the error dynamics for the overall distributed system can be written as

$$\dot{\boldsymbol{e}} = (A + D - LC)\boldsymbol{e} \tag{6.35}$$

with $A = \text{diag}(A_1, \ldots, A_{n_s}), C = \text{diag}(C_1, \ldots, C_{n_s}), L = \text{diag}(L_1, \ldots, L_{n_s})$, and $D = \begin{pmatrix} 0 & D_{12} & \cdots & D_{1n_s} \\ D_{21} & 0 & \cdots & D_{2n_s} \\ \vdots & \vdots & \cdots & \vdots \\ D_{n_s1} & D_{n_s2} & \cdots & 0 \end{pmatrix}$. The observer design problem is then reduced to finding L such that the error system (6.35) is stable.

The assumption that the estimate of the variables is communicated is used for instance by Sundareshan and Elbanna (1990) for linear systems. The result obtained relies on block-diagonal Lyapunov functions and establishes that the matrix block-diagonal terms in the derivative of the Lyapunov function are dominant. Although the method presented by (Sundareshan and Elbanna, 1990) requires the transformation of each subsystem into observer canonical form, it is in effect equivalent to the extension of Theorem 6.1 to observer design, under the assumption that the estimates of the states are communicated between the subsystems.

In fact, the results presented in Section 6.2.1 can be directly extended for observer design under the assumptions that 1) the scheduling vector depends only on measured variables and 2) the estimated states are communicated between the subsystems that influence each other.

Under these assumptions, consider the distributed TS fuzzy system where each subsystem $l = 1, 2, \ldots, n_s$ is described as

$$\begin{aligned} \dot{\boldsymbol{x}}_l &= \sum_{i=1}^{m_l} w_{li}(\boldsymbol{z}_l)(A_{li}\boldsymbol{x}_l + B_{li}\boldsymbol{u}_l + a_i + \sum_{j=1, j\neq l}^{n_s} A_{lij}\boldsymbol{x}_j) \\ \boldsymbol{y} &= \sum_{i=1}^{m_l} w_{li}(\boldsymbol{z}_l)(C_{li}\boldsymbol{x}_l + c_{li}) \end{aligned} \tag{6.36}$$

For each subsystem $l = 1, 2, \ldots, n_s$, the following observer can be used

$$\begin{aligned} \dot{\widehat{\boldsymbol{x}}}_l &= \sum_{i=1}^{m_l} w_{li}(\boldsymbol{z}_l)\Big(A_{li}\widehat{\boldsymbol{x}}_l + B_{li}\boldsymbol{u}_l + a_i + \sum_{j=1, j\neq l}^{n_s} A_{lij}\widehat{\boldsymbol{x}}_j \\ &\quad + L_{li}(\boldsymbol{y}_l - \widehat{\boldsymbol{y}}_l)\Big) \\ \boldsymbol{y}_l &= \sum_{i=1}^{m_l} w_{li}(\boldsymbol{z}_l)(C_{li}\widehat{\boldsymbol{x}}_l + c_{li}) \end{aligned} \tag{6.37}$$

and the estimation error dynamics of subsystem l are expressed as

$$\dot{\boldsymbol{e}}_l = \sum_{i=1}^{m_l}\sum_{j=1}^{m_l} w_{li}(\boldsymbol{z}_l)w_{lj}(\boldsymbol{z}_l)\Big((A_{li} - L_{li}C_{lj})\boldsymbol{e}_l + \sum_{k=1, k\neq l}^{n_s} A_{lik}\boldsymbol{e}_k\Big) \tag{6.38}$$

System (6.38) is equivalent to system (6.3). Depending on the interconnection terms A_{lik}, the conditions presented in Section 6.2.1, namely Theorems 6.1–6.4 can be used.

However, the extension of the results regarding stability analysis of distributed TS systems to observer design has not been reported in the literature. Instead, parallel observer-based control design (Chiang and Kuo, 2002; Chiang and Wang, 2003; Hua et al., 2005; Chien and Er, 2006; Huang and Ho, 2007; Tseng, 2008) has been addressed.

In general, for centralized TS systems, in observer-based control, the observer and controller gains are designed using the separation principle. However, the separation principle only holds if the scheduling variables do not depend on states that are not measured. Furthermore, if a subsystem is influenced by the states of other subsystems that are not known (communicated), the separation principle cannot be used. Therefore, results for observer-based control design usually employ a two-step procedure. Another issue that in general is not taken into account is that the measurements of one subsystem may depend on the states of the other subsystems.

For distributed uncertain TS systems with time delay due to the the communication, a robust observer-based control design method has been proposed by Tong et al. (2007). In this approach the observer and controller design cannot be separated, due to the uncertainty in the system matrices. The same is valid in the case of the result by Tseng (2008): the observer and the controller have to be designed simultaneously. Without a stabilizing state-feedback controller, the convergence to zero of the estimation error cannot be guaranteed.

Distributed observer-based control design has been considered in several settings, such as tracking control (Tseng, 2008), adaptive control (Chiang and Wang, 2003; Chien and Er, 2006; Huang and Ho, 2007), robust control (Chiang and Kuo, 2002; Chiang and Wang, 2003; Chiang, 2006), control in the presence of time delay (Hua et al., 2005; Chiang, 2006; Chiang and Lu, 2007), and their combinations.

Results for discrete-time controller and observer-based controller design include, but are not limited to (Akar and Özgüner, 2000; Hsiao and Hwang, 2002). Also for discrete-time systems the observer design problem is not considered separately.

In the next two sections we present the extension of the results of Section 6.2.2 to observer design. Consider a distributed system where each subsystem is represented by a TS fuzzy model, to which new subsystems may be added online. The goal is to design an asymptotically stable observer for the whole system. Sequential design is considered, where an observer is designed for each newly added subsystem, without modifying the already existing observers, so that the overall observer is stable.

Note that it is not assumed that the subsystems are stabilized or controlled at a known value (i.e., the states are not at some known constant value). However, the estimated states are communicated among the subsystems that influence each other.

For the ease of notation and without loss of generality, similarly to Section 6.2.2, only two subsystems are considered. The observer structure is depicted in Figure 6.2.

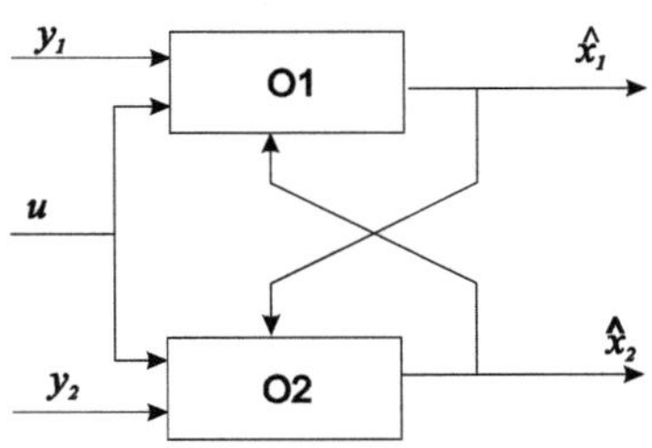

Fig. 6.2 Distributed observer for two subsystems.

Then, the fuzzy system consists of two subsystems:

$$\begin{aligned}
\dot{\boldsymbol{x}}_1 &= \sum_{i=1}^{m} w_i(\boldsymbol{z})(A_{1i}\boldsymbol{x}_1 + B_{1i}\boldsymbol{u} + A_{12i}\boldsymbol{x}_2 + a_{1i}) \\
\boldsymbol{y}_1 &= \sum_{i=1}^{m} w_i(\boldsymbol{z})(C_{11i}\boldsymbol{x}_1 + C_{12i}\boldsymbol{x}_2 + c_{1i}) \\
\dot{\boldsymbol{x}}_2 &= \sum_{i=1}^{m} w_i(\boldsymbol{z})(A_{2i}\boldsymbol{x}_2 + B_{2i}\boldsymbol{u} + A_{21i}\boldsymbol{x}_1 + a_{2i}) \\
\boldsymbol{y}_2 &= \sum_{i=1}^{m} w_i(\boldsymbol{z})(C_{22i}\boldsymbol{x}_2 + C_{21i}\boldsymbol{x}_1 + c_{2i})
\end{aligned} \tag{6.39}$$

and the observer is of the form:

$$\begin{aligned}
\dot{\widehat{\boldsymbol{x}}}_1 &= \sum_{i=1}^{m} w_i(\widehat{\boldsymbol{z}})(A_{1i}\widehat{\boldsymbol{x}}_1 + B_{1i}\boldsymbol{u} + A_{12i}\widehat{\boldsymbol{x}}_2 + a_{1i} + L_{1i}(\boldsymbol{y}_1 - \widehat{\boldsymbol{y}}_1)) \\
\widehat{\boldsymbol{y}}_1 &= \sum_{i=1}^{m} w_i(\widehat{\boldsymbol{z}})(C_{11i}\widehat{\boldsymbol{x}}_1 + C_{12i}\widehat{\boldsymbol{x}}_2 + c_{1i}) \\
\dot{\widehat{\boldsymbol{x}}}_2 &= \sum_{i=1}^{m} w_i(\widehat{\boldsymbol{z}})(A_{2i}\widehat{\boldsymbol{x}}_2 + B_{2i}\boldsymbol{u} + A_{21i}\widehat{\boldsymbol{x}}_1 + a_{2i} + L_{2i}(\boldsymbol{y}_2 - \widehat{\boldsymbol{y}}_2)) \\
\widehat{\boldsymbol{y}}_2 &= \sum_{i=1}^{m} w_i(\widehat{\boldsymbol{z}})(C_{22i}\widehat{\boldsymbol{x}}_2 + C_{21i}\widehat{\boldsymbol{x}}_1 + c_{2i})
\end{aligned} \tag{6.40}$$

The goal is to design the observer gains L_{1i}, $i = 1, 2, \ldots, m$, for each rule of the subsystem with states $\boldsymbol{x}_1$ so that (6.40) guarantees the convergence of the estimation error $\boldsymbol{x} - \widehat{\boldsymbol{x}}$ to zero, given that the gains L_{2i}, $i = 1, 2, \ldots, m$, have already been designed such that the observer

$$\begin{aligned}
\dot{\widehat{\boldsymbol{x}}}_2 &= \sum_{i=1}^{m} w_i(\widehat{\boldsymbol{z}})(A_{2i}\widehat{\boldsymbol{x}}_2 + B_{2i}\boldsymbol{u} + a_{2i} + L_{2i}(\boldsymbol{y}_2 - \widehat{\boldsymbol{y}}_2)) \\
\widehat{\boldsymbol{y}}_2 &= \sum_{i=1}^{m} w_i(\widehat{\boldsymbol{z}})(C_{22i}\widehat{\boldsymbol{x}}_2 + c_{2i})
\end{aligned}$$

guarantees the convergence to zero of the estimation error for the second subsystem without the interconnection terms:

$$\dot{\boldsymbol{x}}_2 = \sum_{i=1}^{m} w_i(\boldsymbol{z})(A_{2i}\boldsymbol{x}_2 + B_{2i}\boldsymbol{u} + a_{2i})$$
$$\boldsymbol{y}_2 = \sum_{i=1}^{m} w_i(\boldsymbol{z})(C_{22i}\boldsymbol{x}_2 + dc_{2i})$$

The system structure considered is characterized by coupling both in the states and the measurements, and is presented in Figure 6.3. In what follows, two cases are distinguished: 1) the scheduling vector depends only on measured or known variables and 2) the scheduling variables depend on states that have to be estimated.

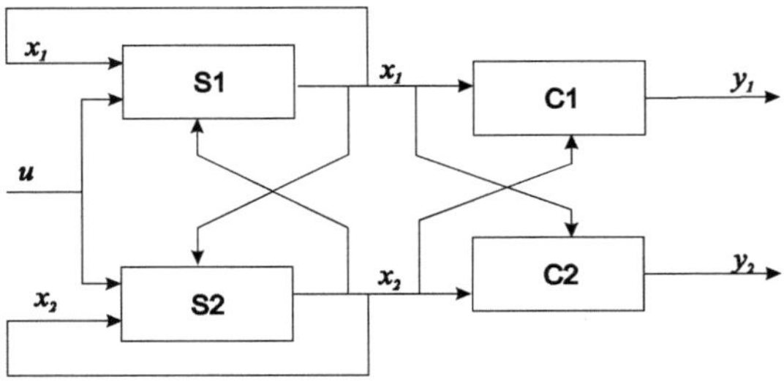

Fig. 6.3 Two subsystems coupled through their states and measurements.

6.3.2 *Sequential Design: Measured Scheduling Vector*

Assuming that the scheduling vector depends only on measured variables, and that the estimated states are communicated among the subsystems, the error systems can be expressed as:

$$\begin{aligned}\dot{\boldsymbol{e}}_1 &= \sum_{i=1}^{m}\sum_{j=1}^{m} w_i(\boldsymbol{z})w_j(\boldsymbol{z})(A_{1i}\boldsymbol{e}_1 + A_{12i}\boldsymbol{e}_2 - L_{1i}C_{1j}\boldsymbol{e}) \\ \dot{\boldsymbol{e}}_2 &= \sum_{i=1}^{m}\sum_{j=1}^{m} w_i(\boldsymbol{z})w_j(\boldsymbol{z})(A_{2i}\boldsymbol{e}_2 + A_{21i}\boldsymbol{e}_1 - L_{2i}C_{2j}\boldsymbol{e})\end{aligned} \tag{6.41}$$

where $C_{1i} = (C_{11i} \;\; C_{12i})$ and $C_{2i} = (C_{21i} \;\; C_{22i})$, or

$$\dot{\boldsymbol{e}} = \sum_{i=1}^{m}\sum_{j=1}^{m} w_i(\boldsymbol{z})w_j(\boldsymbol{z}) \begin{pmatrix} A_{1i} - L_{1i}C_{11j} & A_{12i} - L_{1i}C_{12j} \\ A_{21i} - L_{2i}C_{21j} & A_{2i} - L_{2i}C_{22j} \end{pmatrix} \boldsymbol{e} \tag{6.42}$$

Since L_{1i}, $i = 1, 2, \ldots, m$, need to be designed, a simple special case is when there exist $P_1 = P_1^T > 0$ and L_{1i}, so that $\mathcal{H}(P_1(A_{1i} - L_{1i}C_{11j})) < 0$ and $A_{12i} - L_{1i}C_{12j} = 0$, $i = 1, 2, \ldots, m$, $j = 1, 2, \ldots, m$, $\forall i \neq j : \;\; \exists \boldsymbol{z} : w_i(\boldsymbol{z})w_j(\boldsymbol{z}) \neq 0$. In this case the error system (6.42) is cascaded, further restrictions are not necessary, and the stability conditions for the estimation error dynamics can be summarized as the consequence of Theorem 5.3:

Corollary 6.2. *The error system* (6.42) *is asymptotically stable if there exist* $P_1 = P_1^T > 0$, $P_2 = P_2^T > 0$, L_{1i}, *and* L_{2i}, $i = 1, 2, \ldots, m$, *so that*

$$\begin{aligned}
&\mathcal{H}(P_1(A_{1i} - L_{1i}C_{11j})) < 0\\
&\mathcal{H}(P_2(A_{2i} - L_{2i}C_{22j})) < 0\\
&A_{12i} - L_{1i}C_{12j} = 0
\end{aligned}$$

for $i = 1, 2, \ldots, m$, $j = 1, 2, \ldots, m$, $\forall i \neq j: \ \exists \boldsymbol{z}: \ w_i(\boldsymbol{z})w_j(\boldsymbol{z}) \neq 0$.

Note that the third condition of Corollary 6.2 can rarely be satisfied if the system is not cascaded, i.e., in general it is not possible to find L_{1i} such that $A_{12i} - L_{1i}C_{12j} = 0$, $i = 1, 2, \ldots, m$, $j = 1, 2, \ldots, m$, $\forall i \neq j: \ \exists \boldsymbol{z}: \ w_i(\boldsymbol{z})w_j(\boldsymbol{z}) \neq 0$.

Therefore, the results from Section 6.2.2, in particular Theorem 6.5, are appropriately modified:

Corollary 6.3. *The error system* (6.42) *is exponentially stable, if there exist* L_{1i}, L_{2i}, $i = 1, 2, \ldots, m$, $P_1 = P_1^T > 0$, $P_2 = P_2^T > 0$, $Q_1 = Q_1^T > 0$, *and* $Q_2 = Q_2^T > 0$, *so that*

$$\begin{aligned}
&\mathcal{H}(P_1G_{1ij}) < -2Q_1\\
&\mathcal{H}(P_2G_{2ij}) < -2Q_2\\
&\lambda_{\min}(Q_1) \geq \max_{ij} \|P_1G_{12ij}\|\\
&\frac{\lambda_{\min}(\mathcal{H}(P_1G_{1ij} + Q_1))}{\max_{ij} \|P_1G_{12ij}\|} > \frac{\max_{ij} \|P_2G_{21ij}\|^2}{\lambda_{\min}(Q_2)\lambda_{\min}(\mathcal{H}(P_2G_{2ij} + Q_2))}
\end{aligned} \tag{6.43}$$

for $i = 1, 2, \ldots, m$, $j = 1, 2, \ldots, m$, $\forall i \neq j: \ \exists \boldsymbol{z}: \ w_i(\boldsymbol{z})w_j(\boldsymbol{z}) \neq 0$, *where* $G_{1ij} = A_{1i} - L_{1i}C_{11j}$, $G_{2ij} = A_{2i} - L_{2i}C_{22j}$, $G_{12ij} = A_{12i} - L_{1i}C_{12j}$, $G_{21ij} = A_{21i} - L_{2i}C_{21j}$, *and* $\lambda_{\min}$ *denotes the eigenvalue with the smallest absolute magnitude.*

Similarly to Algorithm 6.1, sufficient LMI conditions can be formulated, which, although more conservative than the conditions of Corollary 6.3, when satisfied, ensure that the conditions of Corollary 6.3 are satisfied. These LMI conditions lead to a sequential implementation, as follows. Assume that an observer has been designed for the subsystem

$$\begin{aligned}
\dot{\boldsymbol{x}}_2 &= \sum_{i=1}^{m} w_i(\boldsymbol{z})(A_{2i}\boldsymbol{x}_2 + B_{2i}\boldsymbol{u} + a_{2i})\\
\boldsymbol{y}_2 &= \sum_{i=1}^{m} w_i(\boldsymbol{z})(C_{22i}\boldsymbol{x}_2 + c_{2i})
\end{aligned}$$

the matrices P_2, Q_2, and the gains L_{2i}, $i = 1, 2, \ldots, m$ are known, and therefore, G_{2ij} can be computed. After adding the interconnection terms, G_{21ij}, $i = 1, 2, \ldots, m$, $j = 1, 2, \ldots, m$, also the ratio

$$\gamma = \frac{\max_{ij} \|P_2 G_{21ij}\|^2}{\lambda_{\min}(Q_2)\lambda_{\min}(\mathcal{H}(P_2 G_{2ij} + Q_2))}$$

can be computed. The conditions of Corollary 6.3 are then reduced to finding L_{2i}, $i = 1, 2, \ldots, m$, $P_1 = P_1^T > 0$, $Q_1 = Q_1^T > 0$, so that for $i = 1, 2, \ldots, m$, $j = 1, 2, \ldots, m$

$$\begin{aligned}
&\mathcal{H}(P_1 G_{1ij}) < -2Q_1 \\
&\lambda_{\min}(Q_1) \geq \max_{ij} \|P_1 G_{12ij}\| \\
&\lambda_{\min}(\mathcal{H}(P_1 G_{1ij} + Q_1)) > \gamma \max_{ij} \|P_1 G_{12ij}\|
\end{aligned}$$

which are satisfied if

$$\begin{aligned}
&\mathcal{H}(P_1 G_{1ij} + Q_1) < 0 \\
&Q_1 \geq \max_{ij} \|P_1 G_{12ij}\| I \\
&\mathcal{H}(P_1 G_{1ij} + Q_1) < -\gamma \max_{ij} \|P_1 G_{12ij}\| I
\end{aligned}$$

These conditions, in turn are satisfied if the following LMIs are feasible, with the change of variables $M_i = P_1 L_{1i}$, $i = 1, 2, \ldots, m$: find L_{2i}, $i = 1, 2, \ldots, m$, $P_1 = P_1^T > 0$, $Q_1 = Q_1^T > 0$, $t_1 > 0$, $t_2 > 0$, and M_i, $i = 1, 2, \ldots, m$, so that for $i = 1, 2, \ldots, m$, $j = 1, 2, \ldots, m$ it holds that

$$\begin{aligned}
&\mathcal{H}(P_1 A_{1i} - M_i C_{1i} + Q_1) < -t_1 I \\
&Q_1 > t_2 I \\
&\begin{pmatrix} t_2 I & P_1 A_{12i} - M_i C_{12j} \\ (P_1 A_{12i} - M_i C_{12j})^T & t_2 I \end{pmatrix} > 0 \\
&\begin{pmatrix} t_1 I & P_1 \gamma A_{12i} - M_i \gamma C_{12j} \\ (P_1 \gamma A_{12i} - M_i \gamma C_{12j})^T & t_1 I \end{pmatrix} > 0
\end{aligned}$$

The steps are summarized as follows:

Algorithm 6.2. Sequential observer design.

1. For the existing observer of the subsystem

$$\begin{aligned}
\dot{\boldsymbol{x}}_2 &= \sum_{i=1}^{m} w_i(\boldsymbol{z})(A_{2i}\boldsymbol{x}_2 + B_{2i}\boldsymbol{u} + a_{2i}) \\
\boldsymbol{y}_2 &= \sum_{i=1}^{m} w_i(\boldsymbol{z})(C_{22i}\boldsymbol{x}_2 + c_{2i})
\end{aligned}$$

compute

$$\bar{\gamma} = \frac{\|P_2\|^2}{\lambda_{\min}(Q_2)\lambda_{\min}(\mathcal{H}(P_2 G_{2ij} + Q_2))}$$

2. When the new subsystem and corresponding interconnection terms are added, compute $\gamma = \bar{\gamma} \max_{ij} \|G_{21ij}\|^2$. To design the observer for this subsystem, solve the LMI problem: find L_{2i}, $i = 1, 2, \ldots, m$, $P_1 = P_1^T > 0$, $Q_1 = Q_1^T > 0$, $t_1 > 0$, $t_2 > 0$ M_i, $i = 1, 2, \ldots, m$, so that for $i = 1, 2, \ldots, m$, $j = 1, 2, \ldots, m$

$$\begin{aligned}
&\mathcal{H}(P_1 A_{1i} - M_i C_{1i} + Q_1) < -t_1 I \\
&Q_1 > t_2 I \\
&\begin{pmatrix} t_2 I & P_1 A_{12i} - M_i C_{12j} \\ (P_1 A_{12i} - M_i C_{12j})^T & t_2 I \end{pmatrix} > 0 \\
&\begin{pmatrix} t_1 I & P_1 \gamma A_{12i} - M_i \gamma C_{12j} \\ (P_1 \gamma A_{12i} - M_i \gamma C_{12j})^T & t_1 I \end{pmatrix} > 0
\end{aligned}$$

The application of Algorithm 6.2 is illustrated on the following example.

Example 6.9. Consider a distributed system consisting of two subsystems as follows:

$$\begin{aligned}
\dot{\boldsymbol{x}}_1 &= \begin{pmatrix} -x_1 & 1 \\ x_1^2 & -3 \end{pmatrix} \boldsymbol{x}_1 + \begin{pmatrix} 1 \\ 0 \end{pmatrix} u_1 + \begin{pmatrix} 3 \\ 4 \end{pmatrix} + \begin{pmatrix} 0 & 1 \\ 1 & 0 \end{pmatrix} \boldsymbol{x}_2 \\
\boldsymbol{y}_1 &= (1\ 0)\, \boldsymbol{x}_1 \\
\dot{\boldsymbol{x}}_2 &= \begin{pmatrix} -2x_3^2 + 3 & x_3 \\ -2 & -1 \end{pmatrix} \boldsymbol{x}_2 + \begin{pmatrix} 3 \\ x_3 \end{pmatrix} u_2 + \begin{pmatrix} 1 \\ 2 \end{pmatrix} + \begin{pmatrix} x_1 & 0 \\ 0 & 1 \end{pmatrix} \boldsymbol{x}_1 \\
\boldsymbol{y}_2 &= (1\ 0)\, \boldsymbol{x}_2
\end{aligned} \tag{6.44}$$

where $\boldsymbol{x}_1 = (x_1\ x_2)^T$, $\boldsymbol{x}_2 = (x_3\ x_4)^T$, $x_i \in [-1, 1]$, $i = 1, 2, 3, 4$, $u_1, u_2 \in \mathbb{R}$. Our goal is to design an observer observer for this distributed system.

An exact TS fuzzy representation of this system is obtained using the sector nonlinearity approach. The scheduling variables are chosen as x_1 and x_1^2 for the first subsystem and x_1, x_3, x_3^2 for the second subsystem. Since x_1 and x_3 are measured, the scheduling vectors of the subsystems do not depend on states that have to be estimated.

Note that in the actual computation of the observer gains only the local state matrices take part, i.e., the input matrices and the affine terms do not influence the computation of the gains. Then, for the second independent subsystem (i.e., the second subsystem without the interconnection terms), four distinct local state matrices can be determined:

$$A_{21} = \begin{pmatrix} 3 & -1 \\ -2 & -1 \end{pmatrix} \quad A_{22} = \begin{pmatrix} 3 & 1 \\ -2 & -1 \end{pmatrix} \quad A_{23} = \begin{pmatrix} 1 & -1 \\ -2 & -1 \end{pmatrix} \quad A_{24} = \begin{pmatrix} 1 & 1 \\ -2 & -1 \end{pmatrix}$$

For this subsystem, the observer designed has the gains

$$L_{21} = \begin{pmatrix} 684.41 \\ -3.00 \end{pmatrix} \quad L_{22} = \begin{pmatrix} 684.41 \\ -1.00 \end{pmatrix} \quad L_{23} = \begin{pmatrix} 679.85 \\ -3.00 \end{pmatrix} \quad L_{24} = \begin{pmatrix} 679.85 \\ -1.00 \end{pmatrix}$$

the Lyapunov matrix is $P_2 = 0.1I$, and $Q_2 = 0.05I$. The value of $\bar{\gamma}$ from the first step of Algorithm 6.2 is obtained as $\bar{\gamma} = 1.0612$.

Now, the new subsystem, i.e., the subsystem with state variables x_1 and x_2 is considered. Again, four distinct state matrices are obtained:

$$A_{11} = \begin{pmatrix} 1 & 1 \\ 0 & -3 \end{pmatrix} \quad A_{12} = \begin{pmatrix} 1 & 1 \\ 1 & -3 \end{pmatrix} \quad A_{13} = \begin{pmatrix} -1 & 1 \\ 0 & -3 \end{pmatrix} \quad A_{14} = \begin{pmatrix} -1 & 1 \\ 1 & -3 \end{pmatrix}$$

The interconnection term from this subsystem to the existing subsystem is $\begin{pmatrix} 0 & 1 \\ 1 & 0 \end{pmatrix} \boldsymbol{x}_2$. Since the interconnection is realized only through the states, i.e., the measurements of each subsystem concern only its own states, $\max_{ij} \|G_{21ij}\|^2 = \max_{ij} \|A_{21ij}\|^2 = 1$, and consequently $\gamma = 1.0612$.

Solving the LMIs from the second step of Algorithm 6.2, we obtain

$$L_{11} = \begin{pmatrix} 4.81 \\ 0.10 \end{pmatrix} \quad L_{12} = \begin{pmatrix} 4.81 \\ 0.20 \end{pmatrix} \quad L_{13} = \begin{pmatrix} 4.61 \\ 0.10 \end{pmatrix} \quad L_{14} = \begin{pmatrix} 4.61 \\ 0.20 \end{pmatrix}$$

and with this, all the observer gains have been computed. □

Similarly to Theorem 6.5, and Algorithm 6.1, the shortcoming of the conditions at this point is that the Lyapunov function for both subsystems and a bound on its derivative have not yet been determined. To overcome this, the following corollary can be formulated, similarly to Corollary 6.1:

Corollary 6.4. $V = (\boldsymbol{e}_1^T\ \boldsymbol{e}_2^T) \begin{pmatrix} \alpha P_1 & 0 \\ 0 & P_2 \end{pmatrix} \begin{pmatrix} \boldsymbol{e}_1 \\ \boldsymbol{e}_2 \end{pmatrix}$ *is a Lyapunov function for* (6.42) *and* $\dot{V} < (\boldsymbol{e}_1^T\ \boldsymbol{e}_2^T)\, \beta \begin{pmatrix} \alpha Q_1 & 0 \\ 0 & Q_2 \end{pmatrix} \begin{pmatrix} \boldsymbol{e}_1 \\ \boldsymbol{e}_2 \end{pmatrix}$ *for an arbitrary* $\beta \in (0, 1)$ *if there exist* L_{1i}, L_{2i}, $i = 1, 2, \ldots, m$, $P_1 = P_1^T > 0$, $P_2 = P_2^T > 0$, $Q_1 = Q_1^T > 0$, *and* $Q_2 = Q_2^T > 0$, *so that*

$$\begin{aligned}
&\mathcal{H}(P_1 G_{1ij}) < -2Q_1 \\
&\mathcal{H}(P_2 G_{2ij}) < -2Q_2 \\
&(1-\beta)\lambda_{\min}(Q_1) \geq \max_i \|P_1 G_{12ij}\| \\
&\frac{\lambda_{\min}(\mathcal{H}(P_1 G_{1ij} + Q_1))}{\max_{ij} \|P_1 G_{12ij}\|} > \frac{\max_{ij} \|P_2 G_{21ij}\|^2}{(1-\beta)\lambda_{\min}(Q_2)\lambda_{\min}(\mathcal{H}(P_2 G_{2ij} + Q_2))}
\end{aligned} \tag{6.45}$$

for $i = 1, 2, \ldots, m$, $j = 1, 2, \ldots, m$, $\forall i \neq j : \quad \exists \boldsymbol{z} : \quad w_i(\boldsymbol{z}) w_j(\boldsymbol{z}) \neq 0$, *where* $G_{1ij} = A_{1i} - L_{1i}C_{11j}$, $G_{2ij} = A_{2i} - L_{2i}C_{22j}$, $G_{12ij} = A_{12i} - L_{1i}C_{12j}$, $G_{21ij} = A_{21i} - L_{2i}C_{21j}$, *and* $\lambda_{\min}$ *denotes the eigenvalue with the smallest absolute magnitude.*

The application of the conditions of Corollary 6.4 are illustrated on the following example.

Example 6.10. Consider the distributed system in Example 6.9. In Example 6.9, the observers have been designed for the interconnected system, but a bound on the the derivative of the Lyapunov function has not yet been found. To determine this bound, (6.45) is solved instead of (6.43).

Since the conditions (6.45) are nonlinear, they are transformed into conditions similar to (6.28):

$$
\begin{aligned}
&\mathcal{H}(P_1 G_{1ij} + Q_1) < -t_1 I \\
&Q_1 > t_2 I \\
&\begin{pmatrix} (1-\beta)t_2 I & \max_i \|G_{12ij}\| P_1 \\ \max_i \|G_{12ij}\| P_1 & (1-\beta)t_2 I \end{pmatrix} > 0 \\
&\begin{pmatrix} (1-\beta)t_1 I & \gamma \max_i \|G_{12ij}\| P_1 \\ \gamma \max_i \|G_{12ij}\| P_1 & (1-\beta)t_1 I \end{pmatrix} > 0
\end{aligned}
\tag{6.46}
$$

for $i = 1, 2, \ldots, m, j = 1, 2, \ldots, m$, with $\beta \in [0, 1]$ arbitrarily chosen. However, since (6.28) may not be feasible for any β, and the goal is to determine β as large as possible to facilitate the analysis of possible subsystems that will be added, (6.46) is considered a BMI problem, with both β and P_1 decision variables. Solving (6.46) yields the observer gains

$$
L_{11} = \begin{pmatrix} 1280.2 \\ 0.1 \end{pmatrix} \quad L_{12} = \begin{pmatrix} 1280.2 \\ 0.2 \end{pmatrix} \quad L_{13} = \begin{pmatrix} 1280.0 \\ 0.1 \end{pmatrix} \quad L_{14} = \begin{pmatrix} 1280.0 \\ 0.2 \end{pmatrix}
$$

with $P_1 = 0.5I$, $\beta = 0.29$, and $Q_1 = 1.5 \cdot 10^6 I$. □

Algorithm 6.2 is useful if no bound on the interconnection terms is known before the subsystem is added. If a bound on A_{12i}, A_{21i}, C_{21i}, C_{12i}, $i = 1, 2, \ldots, m$ is known beforehand, the design can be decoupled starting with the first subsystem, by analyzing the last condition of Corollary 6.3. Although LMI conditions are obtained, the following manipulations introduce conservativeness.

Imposing a condition similar to that of the third condition of Corollary 6.3 to the second subsystem, i.e., for $i = 1, 2, \ldots, m$, $j = 1, 2, \ldots, m$, $\forall i \neq j : \ \exists \boldsymbol{z} : w_i(\boldsymbol{z}) w_j(\boldsymbol{z}) \neq 0$

$$
\lambda_{\min}(Q_2) \geq \max_{ij} \|P_2 G_{21ij}\|
$$

one obtains

$$
\frac{\max_{ij} \|P_2 G_{21ij}\|^2}{\lambda_{\min}(Q_2)\lambda_{\min}(\mathcal{H}(P_2 G_{2ij} + Q_2))} \leq \frac{\max_{ij} \|P_2 G_{21ij}\|}{\lambda_{\min}(\mathcal{H}(P_2 G_{2ij} + Q_2))}
$$

where $\lambda_{\min}(\mathcal{H}(P_2 G_{2ij} + Q_2))$ denotes the eigenvalue with the minimum absolute magnitude of all the matrices $(\mathcal{H}(P_2 G_{2ij} + Q_2))$,

$i = 1, 2, \ldots, m$, $j = 1, 2, \ldots, m$. This expression is similar to that of the reciprocal of the first part of the fourth condition of Corollary 6.3, i.e.,

$$\frac{\lambda_{\min}(\mathcal{H}(P_1 G_{1ij} + Q_1))}{\max_{ij} \|P_1 G_{12ij}\|}$$

By imposing for both subsystems

$$\frac{\lambda_{\min}(\mathcal{H}(P_k G_{kij} + Q_k))}{\max_{ij} \|P_k T_{kpij}\|} > 1$$

where T_{kpij}, $k = 1, 2$, $p = 1, 2$, $k \neq p$, $i = 1, 2, \ldots, m$, $j = 1, 2, \ldots, m$, is the interconnection term from subsystem p influencing the subsystem k, $T_{kpij} = A_{kpi} - L_{ki} C_{kpj}$, the conditions are decoupled. This result can be summarized as:

Corollary 6.5. *The error system* (6.42) *is exponentially stable, if there exist* L_{kij}, $k = 1, 2$, $i = 1, 2, \ldots, m$, $j = 1, 2, \ldots, m$, $P_k = P_k^T > 0$, *and* $Q_k = Q_k^T > 0$ *so that*

$$\begin{aligned} &\mathcal{H}(P_k G_{kij}) < -2Q_k \\ &\lambda_{\min}(Q_k) \geq \max_{ijp} \|P_k T_{kpij}\| \\ &\lambda_{\min}(\mathcal{H}(P_k G_{kij} + Q_k)) > \max_{ijp} \|P_k T_{kpij}\| \end{aligned} \tag{6.47}$$

for $i = 1, 2, \ldots, m$, $j = 1, 2, \ldots, m$, $k = 1, 2$, $p = 1, 2$, $k \neq p$, $\forall i \neq j : \exists \boldsymbol{z} : \; w_i(\boldsymbol{z}) w_j(\boldsymbol{z}) \neq 0$, *where* $G_{kij} = A_{ki} - L_{ki} C_{kj}$, $T_{kpij} = A_{kpi} - L_{ki} C_{kpj}$ *is the interconnection term that influences the subsystem* k, L_{ki}, $i = 1, 2, \ldots, m$ *are the observer gains of the* k*th subsystem, and* $\lambda_{\min}$ *denotes the eigenvalue with the smallest absolute magnitude.*

The above conditions are not LMIs. By imposing that

$$\lambda_{\min}(\mathcal{H}(P_k G_{kij} + Q_k)) > \lambda_{\min}(Q_k)$$

for $i = 1, 2, \ldots, m$, $j = 1, 2, \ldots, m$, $k = 1, 2$, and that $t_{k\mathrm{m}} I \leq Q_k \leq t_{k\mathrm{M}} I$, conditions (6.47) are satisfied if

$$\begin{aligned} &t_{k\mathrm{m}} I \leq Q_k \leq t_{k\mathrm{M}} I \\ &\mathcal{H}(P_k G_{kij} + Q_k) < -t_{k\mathrm{M}} I \\ &t_{k\mathrm{m}} I \geq \max_{ijp} \|P_k T_{kpij}\| \end{aligned} \tag{6.48}$$

for $i = 1, 2, \ldots, m$, $j = 1, 2, \ldots, m$, $k = 1, 2$, $p = 1, 2$, $k \neq p$. Note that conditions (6.48) are only sufficient, and therefore more conservative than (6.47).

Recall, that the interconnection term T_{kpij} is in fact $T_{kpij} = A_{kpi} - L_{ki} C_{kpj}$, $k = 1, 2, p = 1, 2, k \neq p$, i.e., the interconnection term in the error dynamics. However, only the bounds on the interconnection terms in the subsystems are known, i.e., $\mu_{Ak} = \max_{pi} \|A_{kpi}\|$ and $\mu_{Ck} = \max_{pi} \|C_{kpi}\|$, where k is the number of the current subsystem, $k = 1, 2$. To be able to use these bounds, let Q_k be the sum of two positive definite matrices, $Q_k = Q_{kA} + Q_{kC}$ that satisfy

$$\begin{aligned} Q_{kA} &\geq \mu_{Ak} \|P_k\| I \\ Q_{kC} &\geq \mu_{Ck} \max_i \|P_k L_{ki}\| I \end{aligned}$$

The above conditions may be expressed as LMIs:

$$\begin{aligned} &Q_{kA} \geq t_{k1} I \\ &Q_{kC} \geq t_{k2} I \\ &\begin{pmatrix} t_{k1} I & \mu_{Ak} P_k \\ \mu_{Ak} P_k & t_{k1} I \end{pmatrix} \geq 0 \\ &\begin{pmatrix} t_{k2} I & \mu_{Ck} M_{ki} \\ \mu_{Ck} M_{ki}^T & t_{k2} I \end{pmatrix} \geq 0 \end{aligned}$$

where $M_{ki} = P_k L_{ki}$.

With the decoupled LMI conditions, the result is summarized as:

Theorem 6.7. *The error system* (6.42) *is exponentially stable, if there exist* M_{ki}, $i = 1, 2, \ldots, m$, $P_k = P_k^T > 0$, $Q_k = Q_k^T$, $t_{k1} > 0$, $t_{k2} > 0$, $t_{k\mathrm{M}} > 0$, *and* $t_{k\mathrm{m}} > 0$, $k = 1, 2$, *so that*

$$\begin{aligned} &t_{k\mathrm{m}} I \leq Q_k \leq t_{k\mathrm{M}} I \\ &\mathcal{H}(P_k G_{kij} + Q_k) < -t_{k\mathrm{M}} I \\ &t_{k\mathrm{m}} I \geq Q_{kA} + Q_{kC} \\ &Q_{kA} \geq t_{k1} I \\ &Q_{kC} \geq t_{k2} I \\ &\begin{pmatrix} t_{k1} I & \mu_{Ak} P_k \\ \mu_{Ak} P_k & t_{k1} I \end{pmatrix} \geq 0 \\ &\begin{pmatrix} t_{k2} I & \mu_{Ck} M_{ki} \\ \mu_{Ck} M_{ki}^T & t_{k2} I \end{pmatrix} \geq 0 \end{aligned} \tag{6.49}$$

for $i = 1, 2, \ldots, m$, $j = 1, 2, \ldots, m$, $\forall i \neq j : \quad \exists \boldsymbol{z} : \quad w_i(\boldsymbol{z}) w_j(\boldsymbol{z}) \neq 0$, $k = 1, 2$.

Note that this result can only be used if a bound on the interconnection terms is known a priori. Also, the conditions of Theorem 6.7 are more conservative than those of Corollary 6.3.

The observer design using the conditions of Theorem 6.7 is illustrated on the following example.

Example 6.11. Consider again the distributed system in Example 6.9, with the TS fuzzy representation obtained in Example 6.9. For both subsystems, we have $\mu_{A1} = \mu_{A2} = 1$, and $\mu_{C1} = \mu_{C2} = 0$. Therefore, the last condition of (6.49) is satisfied for any $t_{k2} \geq 0$, and the LMI problem that has to be solved is *to find* M_{ki},

$i = 1, 2, \ldots, m$, $P_k = P_k^T > 0$, $Q_k = Q_k^T$, $t_{k1} > 0$, $t_{k2} > 0$, $t_{k\mathrm{M}} > 0$, $t_{k\mathrm{m}} > 0$, $k = 1, 2$, *so that*

$$
\begin{aligned}
&t_{k\mathrm{m}} I \leq Q_k \leq t_{k\mathrm{M}} I \\
&\mathcal{H}(P_k G_{kij} + Q_k) < -t_{k\mathrm{M}} I \\
&t_{k\mathrm{m}} I \geq Q_{kA} + Q_{kC} \\
&Q_{kA} \geq t_{k1} I \\
&Q_{kC} \geq t_{k2} I \\
&\begin{pmatrix} t_{k1} I & \mu_{Ak} P_k \\ \mu_{Ak} P_k & t_{k1} I \end{pmatrix} \geq 0
\end{aligned}
$$

for $i = 1, 2, \ldots, m$, $j = 1, 2, \ldots, m$, $\forall i \neq j : \quad \exists \boldsymbol{z} : \quad w_i(\boldsymbol{z}) w_j(\boldsymbol{z}) \neq 0$, $k = 1, 2$.

Solving conditions (6.49) for the subsystem with state variables x_1 and x_2 yields the observer gains

$$
L_{11} = \begin{pmatrix} 0.88 \\ 0.14 \end{pmatrix} \quad L_{12} = \begin{pmatrix} 0.88 \\ 0.24 \end{pmatrix} \quad L_{13} = \begin{pmatrix} 0.68 \\ 0.14 \end{pmatrix} \quad L_{14} = \begin{pmatrix} 0.68 \\ 0.23 \end{pmatrix}
$$

However, the LMI referring to the other subsystem, with state variables x_3 and x_4, is infeasible for $\mu_{A21} = 1$, and can only be solved, if the bound on the interconnection term is lower. For the bound $\mu_{A2} = 0.5$, the observer gains are obtained as

$$
L_{21} = \begin{pmatrix} 7.42 \\ -2.27 \end{pmatrix} \quad L_{22} = \begin{pmatrix} 7.42 \\ -1.72 \end{pmatrix} \quad L_{23} = \begin{pmatrix} 5.42 \\ -2.27 \end{pmatrix} \quad L_{24} = \begin{pmatrix} 5.42 \\ -1.72 \end{pmatrix} \quad \square
$$

6.3.3 Sequential Design: Estimated Scheduling Vector

Consider now the case when the scheduling vector depends on the states to be estimated. For the simplicity of notation, only the case when the measurement matrices are common for all the rules of a subsystem is presented. If the measurement matrices are different for each rule, the observers can be designed in a similar fashion, although the conditions become more complex.

The error dynamics (similarly to Section 6.3.2) are expressed as:

$$
\begin{aligned}
\dot{\boldsymbol{e}}_1 =& \sum_{i=1}^{m} w_i(\hat{\boldsymbol{z}})(A_{1i}\boldsymbol{e}_1 + A_{12i}\boldsymbol{e}_2 - L_{1i}C_1\boldsymbol{e}) \\
&+ \sum_{i=1}^{m} (w_i(\boldsymbol{z}) - w_i(\hat{\boldsymbol{z}}))(A_{1i}\boldsymbol{x}_1 + B_{1i}\boldsymbol{u} + A_{12i}\boldsymbol{x}_2) \\
\dot{\boldsymbol{e}}_2 =& \sum_{i=1}^{m} w_i(\hat{\boldsymbol{z}})(A_{2i}\boldsymbol{e}_2 + A_{21i}\boldsymbol{e}_1 - L_{2i}C_2\boldsymbol{e}) \\
&+ \sum_{i=1}^{m} (w_i(\boldsymbol{z}) - w_i(\hat{\boldsymbol{z}}))(A_{2i}\boldsymbol{x}_2 + B_{2i}\boldsymbol{u} + A_{21i}\boldsymbol{x}_1)
\end{aligned}
\tag{6.50}
$$

or

$$\begin{aligned}\dot{\boldsymbol{e}} &= \sum_{i=1}^{m} w_i(\widehat{\boldsymbol{z}}) \begin{pmatrix} A_{1i} - L_{1i}C_{11} & A_{12i} - L_{1i}C_{12} \\ A_{21i} - L_{2i}C_{21} & A_{2i} - L_{2i}C_{22} \end{pmatrix} \boldsymbol{e} \\ &+ \sum_{i=1}^{m} (w_i(\boldsymbol{z}) - w_i(\widehat{\boldsymbol{z}})) \begin{pmatrix} A_{1i}\boldsymbol{x}_1 + B_{1i}\boldsymbol{u} + A_{12i}\boldsymbol{x}_2 \\ A_{2i}\boldsymbol{x}_2 + B_{2i}\boldsymbol{u} + A_{21i}\boldsymbol{x}_1 \end{pmatrix}\end{aligned} \tag{6.51}$$

In case of a centralized observer design in general it is assumed that

$$\Delta = \sum_{i=1}^{m} (w_i(\boldsymbol{z}) - w_i(\widehat{\boldsymbol{z}})) \begin{pmatrix} A_{1i}\boldsymbol{x}_1 + B_{1i}\boldsymbol{u} + A_{12i}\boldsymbol{x}_2 \\ A_{2i}\boldsymbol{x}_2 + B_{2i}\boldsymbol{u} + A_{21i}\boldsymbol{x}_1 \end{pmatrix}$$

is Lipschitz in $\boldsymbol{e}$, i.e., $\|\Delta\| \leq \mu\|\boldsymbol{e}\|$, for some $\mu > 0$. This condition can also be formulated as $\Delta = F\boldsymbol{e}$, with F an uncertainty, such that $\|F\| \leq \mu$. In distributed observer design, the estimation error for the already existing subsystem is

$$\begin{aligned}\dot{\boldsymbol{e}}_2 &= \sum_{i=1}^{m} w_i(\hat{\bar{\boldsymbol{z}}})(A_{2i}\boldsymbol{e}_2 - L_{2i}C_{22}\boldsymbol{e}_2) \\ &+ \sum_{i=1}^{m} (w_i(\bar{\boldsymbol{z}}) - w_i(\hat{\bar{\boldsymbol{z}}}))(A_{2i}\boldsymbol{x}_2 + B_{2i}\boldsymbol{u})\end{aligned} \tag{6.52}$$

where $\bar{\boldsymbol{z}}$ depends *only* on the states of this subsystem. For this subsystem, there already exists a condition on the model-observer mismatch, i.e.,

$$\|\bar{\Delta}\| = \|\sum_{i=1}^{m} (w_i(\bar{\boldsymbol{z}}) - w_i(\hat{\bar{\boldsymbol{z}}}))(A_{2i}\boldsymbol{x}_2 + B_{2i}\boldsymbol{u})\| \leq \mu_2\|\boldsymbol{e}_2\|$$

When a new subsystem is introduced, both $\boldsymbol{z}$ and Δ change. In order to keep the symmetry and to have condition similar to that of centralized observer design, one can require that Δ is expressed as

$$\sum_{i=1}^{m} (w_i(\boldsymbol{z}) - w_i(\widehat{\boldsymbol{z}})) \begin{pmatrix} A_{1i}\boldsymbol{x}_1 + B_{1i}\boldsymbol{u} + A_{12i}\boldsymbol{x}_1 \\ A_{2i}\boldsymbol{x}_2 + B_{2i}\boldsymbol{u} + A_{21i}\boldsymbol{x}_1 \end{pmatrix} = \begin{pmatrix} F_1 & F_{12} \\ F_{21} & F_2 \end{pmatrix} \boldsymbol{e} \tag{6.53}$$

and that the uncertainties are bounded

$$\begin{aligned}\|F_{12}\| &\leq \mu_{12} \\ \|F_1\| &\leq \mu_1 \\ \|F_{21}\| &\leq \mu_{21} \\ \|F_2\| &\leq \mu_2\end{aligned} \tag{6.54}$$

With the considerations above, the following Proposition is formulated:

Proposition 6.1. *The error system* (6.51) *is asymptotically stable, if there exist* $P_1 = P_1^T > 0$, $P_2 = P_2^T > 0$, $Q_1 = Q_1^T > 0$, $Q_2 = Q_2^T > 0$, L_{1i}, *and* L_{2i}, $i = 1, 2, \ldots, m$ *so that* (6.53) *and* (6.54) *are satisfied and*

$$
\begin{aligned}
&\mathcal{H}(P_2(G_{2i} + F_2)) < -2Q_2 \\
&\mathcal{H}(P_1 G_{1i}) < -2Q_1 \\
&\lambda_{\min}(\mathcal{H}(Q_1 + P_1 F_1)) > \max_i \|P_1(G_{12i} + F_{12})\| \\
&\frac{\lambda_{\min}(\mathcal{H}(P_1 G_{1i} + Q_1))}{\max_i \|P_1(G_{12i} + F_{12})\|} > \frac{\max_i \|P_2(G_{21i} + F_{21})\|^2}{\lambda_{\min}(Q_2)\lambda_{\min}(\mathcal{H}(P_2(G_{2i} + F_2) + Q_2))}
\end{aligned}
$$

for $i = 1, 2, \ldots, m$, *where* $G_{1i} = A_{1i} - L_{1i}C_{11}$, $G_{2i} = A_{2i} - L_{2i}C_{21}$, $G_{12i} = A_{12i} - L_{1i}C_{12}$, $G_{21i} = A_{21i} - L_{2i}C_{21}$, *and* $\lambda_{\min}$ *denotes the eigenvalue with the smallest absolute magnitude.*

Proof: Similarly to the proof of Theorem 6.5, one can see system (6.51) as the cascaded system

$$
\begin{aligned}
\dot{\boldsymbol{e}}_{\mathrm{c}} = &\sum_{i=1}^{m} w_i(\widehat{\boldsymbol{z}}) \begin{pmatrix} (A_{1i} - L_{1i}C_{11})\boldsymbol{e}_{1\mathrm{c}} \\ A_{2i}\boldsymbol{e}_{2\mathrm{c}} + A_{21i}\boldsymbol{e}_{1\mathrm{c}} - L_{2i}C_2\boldsymbol{e}_{\mathrm{c}} \end{pmatrix} \\
&+ \sum_{i=1}^{m} (w_i(\boldsymbol{z}) - w_i(\widehat{\boldsymbol{z}})) \begin{pmatrix} 0 \\ A_{2i}\boldsymbol{x}_2 + B_{2i}\boldsymbol{u} + A_{21i}\boldsymbol{x}_1 \end{pmatrix}
\end{aligned}
\tag{6.55}
$$

and an extra feedback term. System (6.55) is asymptotically stable, if the conditions of Theorem 5.10 are satisfied. Moreover, exponential stability can also be ensured by using somewhat more conservative conditions: if there exist $P_1 = P_1^T > 0$, $P_2 = P_2^T > 0$, $Q_1 = Q_1^T > 0$, $Q_2 = Q_2^T > 0$, $\mu_2 \geq 0$, $\mu_{21} \geq 0$, F_2, and F_{21} so that

$$
\begin{aligned}
&\mathcal{H}(P_1 G_{1i}) < -2Q_1 \quad i = 1, 2, \ldots, m \\
&\sum_{i=1}^{m} (w_i(\boldsymbol{z}) - w_i(\widehat{\boldsymbol{z}}))(A_{2i}\boldsymbol{x}_2 + B_{2i}\boldsymbol{u} + A_{21i}\boldsymbol{x}_1) = \begin{pmatrix} F_{21} & F_2 \end{pmatrix} \boldsymbol{e}_{\mathrm{c}} \\
&\|F_{21}\| \leq \mu_{21} \\
&\|F_2\| \leq \mu_2 \\
&\mathcal{H}(P_2(G_{2i} + F_2)) < -2Q_2 \quad i = 1, 2, \ldots, m
\end{aligned}
$$

with $G_{1i} = A_{1i} - L_{1i}C_{11}$ and $G_{2i} = A_{2i} - L_{2i}C_{21}$.

The condition $\mathcal{H}(P_2(G_{2i} + F_2)) < -2Q_2$ ensures that the already existing error system is exponentially stable. Moreover, when the new (error) subsystem is connected, the bound on F_2 should not change, i.e., although the new subsystem may influence the membership functions, it should not influence the model-observer mismatch of the second subsystem.

The conditions above also ensure that there exists $\alpha \in \mathcal{R}^+$ so that

$$
V_{\mathrm{c}} = \boldsymbol{e}_{\mathrm{c}}^T \mathrm{diag}(\alpha P_1, P_2)\boldsymbol{e}_{\mathrm{c}}
$$

is a Lyapunov function for (6.55) and $\dot{V}_c < -2\boldsymbol{e}_c^T Q \boldsymbol{e}_c$, with $Q = \text{diag}(\alpha Q_1, Q_2)$ and $G_{21i} = A_{21i} - L_{2i}C_{21}$. To show this, consider the Lyapunov function

$$V_c = \boldsymbol{e}_c^T \begin{pmatrix} \alpha P_1 & 0 \\ 0 & P_2 \end{pmatrix} \boldsymbol{e}_c$$

The derivative is:

$$\dot{V}_c = \sum_{i=1}^{m} w_i(\hat{\boldsymbol{z}}) \boldsymbol{e}_c^T \mathcal{H} \begin{pmatrix} \alpha P_1 G_{1i} & 0 \\ P_2(G_{21i} + F_{21}) & P_2(G_{2i} + F_2) \end{pmatrix} \boldsymbol{e}_c$$

Then, $\dot{V}_c < -2\boldsymbol{e}_c^T Q \boldsymbol{e}_c$ if

$$\begin{pmatrix} \alpha\mathcal{H}(P_1 G_{1i}) & (G_{21i} + F_{21})^T P_2 \\ P_2(G_{21i} + F_{21}) & \mathcal{H}(P_2(G_{2i} + F_2)) \end{pmatrix} < -2 \begin{pmatrix} \alpha Q_1 & 0 \\ 0 & Q_2 \end{pmatrix}$$

for $i = 1, 2, \ldots, m$, which amounts to

$$\begin{pmatrix} \alpha\mathcal{H}(P_1 G_{1i} + Q_1) & (G_{21i} + F_{21})^T P_2 \\ P_2(G_{21i} + F_{21}) & \mathcal{H}(P_2(G_{2i} + F_2) + Q_2) \end{pmatrix} < 0$$

for $i = 1, 2, \ldots, m$.

Using the Schur complement, one obtains

$$\begin{aligned} &\alpha\mathcal{H}(P_1 G_{1i} + Q_1) \\ &\quad - (G_{21i} + F_{21})^T P_2 (\mathcal{H}(P_2(G_{2i} + F_2) + Q_2))^{-1} P_2 (G_{21i} + F_{21}) < 0 \end{aligned}$$

which is satisfied by any α chosen such that for $i = 1, 2, \ldots, m$

$$\alpha > \frac{1}{\lambda_{\min}(\mathcal{H}(P_1 G_{1i} + Q_1))} \cdot \frac{\max_i \|P_2(G_{21i} + F_{21})\|^2}{\lambda_{\min}(\mathcal{H}(P_2(G_{2i} + F_2) + Q_2))} \tag{6.56}$$

Now, consider the full error system (6.51), together with the assumptions

$$\begin{aligned} &\sum_{i=1}^{m} (w_i(\boldsymbol{z}) - w_i(\hat{\boldsymbol{z}}))(A_{1i}\boldsymbol{x}_1 + B_{1i}\boldsymbol{u} + A_{12i}\boldsymbol{x}_1) = \begin{pmatrix} F_1 & F_{12} \end{pmatrix} \boldsymbol{e} \\ &\|F_{12}\| \leq \mu_{12} \\ &\|F_1\| \leq \mu_1 \end{aligned} \tag{6.57}$$

These assumptions, combined with the assumption that

$$\begin{aligned} &\sum_{i=1}^{m} (w_i(\boldsymbol{z}) - w_i(\hat{\boldsymbol{z}}))(A_{2i}\boldsymbol{x}_2 + B_{2i}\boldsymbol{u} + A_{21i}\boldsymbol{x}_1) = \begin{pmatrix} F_{21} & F_2 \end{pmatrix} \boldsymbol{e}_c \\ &\|F_{21}\| \leq \mu_{21} \\ &\|F_2\| \leq \mu_2 \end{aligned} \tag{6.58}$$

are effectively equivalent to those that would be used in the centralized design (see Theorem 4.5).

By using the above constructed $V = V_c$ as a candidate Lyapunov function for (6.51), one obtains:

$$\begin{aligned}
\dot{V} &= \sum_{i=1}^{m} w_i(\hat{z}) e^T \left(\begin{pmatrix} \alpha\mathcal{H}(P_1G_{1i}) & G_{21i}^T P_2 \\ P_2G_{21i} & \mathcal{H}(P_2G_{2i}) \end{pmatrix} + \begin{pmatrix} 0 & \alpha P_1 G_{12i} \\ \alpha G_{12i}^T P_1 & 0 \end{pmatrix} \right) e \\
&+ e^T \begin{pmatrix} \alpha\mathcal{H}(P_1F_1) & \alpha P_1 F_{12} \\ \alpha(P_1F_{12})^T & 0 \end{pmatrix} e \\
&< -e^T \mathcal{H} \begin{pmatrix} \alpha(Q_1 + P_1F_1) & 0 \\ 0 & Q_2 \end{pmatrix} e + 2e^T (\alpha \max_i \|P_1(G_{12i} + F_{12})\|) I e \\
&< -e^T \begin{pmatrix} \alpha\mathcal{H}(Q_1 + P_1F_1 - \alpha \max_i \|P_1(G_{12i} + F_{12})\| I) & 0 \\ 0 & \mathcal{H}(Q_2 - \alpha \max_i \|P_1(G_{12i} + F_{12})\| I) \end{pmatrix} e
\end{aligned}$$

leading to the conditions

$$\lambda_{\min}(\mathcal{H}(Q_1 + P_1F_1)) > \max_i \|P_1(G_{12i} + F_{12})\| \tag{6.59}$$

$$\lambda_{\min}(Q_2) > \alpha \max_i \|P_1(G_{12i} + F_{12})\| \tag{6.60}$$

for $i = 1, 2, \ldots, m$. Combining (6.56) and (6.60), we get that such an α exists, and $V = V_c$ is a Lyapunov function if

$$\frac{\lambda_{\min}(Q_2)}{\max_i \|P_1(G_{12i} + F_{12})\|} > \frac{\max_i \|P_2(G_{21i} + F_{21})\|^2}{\lambda_{\min}(\mathcal{H}(P_1G_{1i} + Q_1))\lambda_{\min}(\mathcal{H}(P_2(G_{2i} + F_2) + Q_2))}$$

for $i = 1, 2, \ldots, m$, or

$$\frac{\lambda_{\min}(\mathcal{H}(P_1G_{1i} + Q_1))}{\max_i \|P_1(G_{12i} + F_{12})\|} > \frac{\max_i \|P_2(G_{21i} + F_{21})\|^2}{\lambda_{\min}(Q_2)\lambda_{\min}(\mathcal{H}(P_2(G_{2i} + F_2) + Q_2))}$$

for $i = 1, 2, \ldots, m$. With this, the proof is concluded. □

If the scheduling vector depends on states to be estimated, a cascaded error system can only be obtained in special cases. As in Section 6.3.2, the conditions of Proposition 6.1 can be implemented in a two-step algorithm, similarly to Algorithm 6.2. A decoupled design, similar to that given in Theorem 6.7 is also possible, if bounds on the interconnection terms are known in advance.

The decentralized observer design for the case when the scheduling vector depends on the unmeasured state variables is illustrated on the following example.

Example 6.12. Consider a decentralized system, composed of four subsystems, with their interconnection terms as presented in Figure 6.4.

The order in which the subsystems are added to the distributed system is as follows: first only Subsystem 1 exists, to which Subsystem 2 is added, then Subsystem 3 is connected to the system composed of Subsystem 1 and Subsystem 2, and finally,

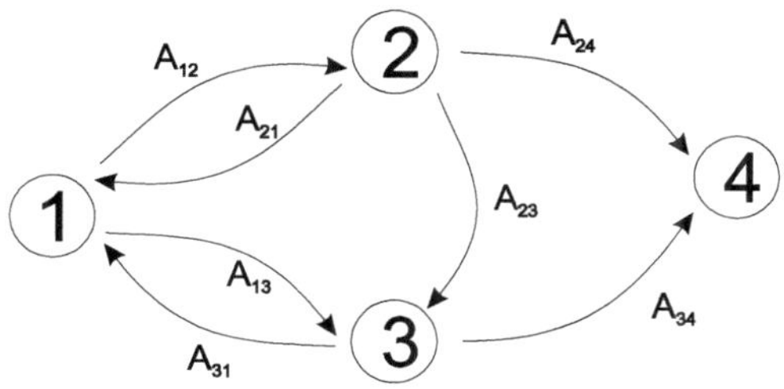

Fig. 6.4 Four subsystems with their interconnections.

Subsystem 4 is connected to the existing system. The individual subsystems and the interconnections are described as:

1. Subsystem 1: The scheduling variable z_1 is a measured variable, with the membership functions presented in Figure 6.5.

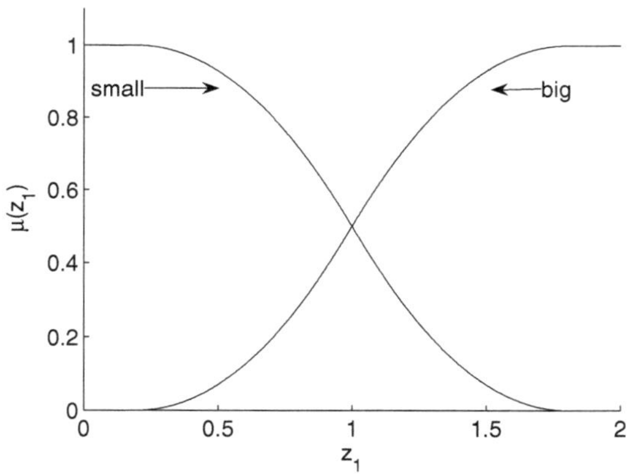

Fig. 6.5 Membership functions of z_1.

Model rule 1:
If z_1 is small then

$$\dot{\boldsymbol{x}}_1 = \begin{pmatrix} -1 & 1 \\ 2 & 3 \end{pmatrix} \boldsymbol{x}_1 + \begin{pmatrix} 1 \\ 2 \end{pmatrix} u$$

$$y_1 = (1 \;\; 0)\boldsymbol{x}_1$$

Model rule 2:
If z_1 is big then

$$\dot{\boldsymbol{x}}_1 = \begin{pmatrix} 2 & 3 \\ 0 & -4 \end{pmatrix} \boldsymbol{x}_1 + \begin{pmatrix} 1 \\ 2 \end{pmatrix} u$$

$$y_1 = (1 \;\; 0)\boldsymbol{x}_1$$

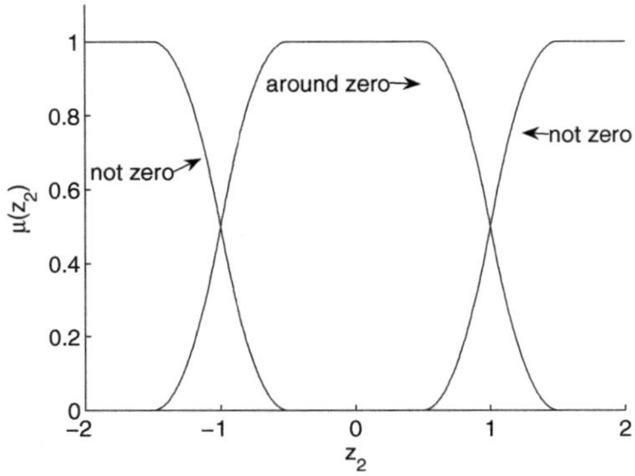

Fig. 6.6 Membership function of z_2.

2. Subsystem 2: The scheduling variable z_2 is x_{22}, a state to be estimated, with the membership functions presented in Figure 6.6. The states are bounded, $x_{21}, x_{22} \in [-2,\ 2]$.

   Model rule 1:
   If z_2 is not zero then

$$\dot{\boldsymbol{x}}_2 = \begin{pmatrix} -0.5 & 1.5 \\ 0 & -1 \end{pmatrix} \boldsymbol{x}_2 + \begin{pmatrix} 0.1 \\ 0.1 \end{pmatrix}$$
$$\boldsymbol{y}_2 = \begin{pmatrix} 1 & 10 \\ 0.1 & 0 \end{pmatrix} \boldsymbol{x}_2$$

   Model rule 2:
   If z_2 is around zero then

$$\dot{\boldsymbol{x}}_2 = \begin{pmatrix} 0.5 & 3 \\ 0 & 1 \end{pmatrix} \boldsymbol{x}_2 + \begin{pmatrix} 0 \\ 0.1 \end{pmatrix}$$
$$\boldsymbol{y}_2 = \begin{pmatrix} 1 & 10 \\ 0.1 & 0 \end{pmatrix} \boldsymbol{x}_2$$

   The interconnection terms are as follows:
   If z_1 is small and x_{22} is around zero then: $A_{12} = \begin{pmatrix} 0.1 & 0.8 \\ 0.5 & 0 \end{pmatrix}$, $A_{21} = \begin{pmatrix} 0.2 & 0.3 \\ 0.1 & 0 \end{pmatrix}$.
   If z_1 is big and x_{22} is around zero then: $A_{12} = \begin{pmatrix} -0.3 & 0.1 \\ 0.2 & 0.3 \end{pmatrix}$, $A_{21} = \begin{pmatrix} -0.2 & -0.3 \\ 0.1 & 0 \end{pmatrix}$.
   Otherwise there is no direct connection between the subsystems 1 and 2.
3. Subsystem 3: The scheduling variable z_3 is an exogenous measured variable, with the same membership functions as z_1.

Model rule 1:
If z_3 is small then

$$\dot{\boldsymbol{x}}_3 = \begin{pmatrix} 1 & 2 \\ 3 & -3 \end{pmatrix} \boldsymbol{x}_3 + \begin{pmatrix} 3 \\ 4 \end{pmatrix} u$$
$$y_3 = (1 \ \ 2)\boldsymbol{x}_3$$

Model rule 2:
If z_3 is big then

$$\dot{\boldsymbol{x}}_3 = \begin{pmatrix} -2 & 0 \\ 2 & 2 \end{pmatrix} \boldsymbol{x}_3 + \begin{pmatrix} 3 \\ 4 \end{pmatrix} u$$
$$y_3 = (1 \ \ 2)\boldsymbol{x}_3$$

The interconnection terms are as follows:
If z_3 is big then: $A_{13} = \begin{pmatrix} 0.4 & 0.3 \\ 0.8 & 0 \end{pmatrix}$.
If z_1 is big and z_3 is small then: $A_{31} = \begin{pmatrix} 0.2 & -0.3 \\ 0.1 & 0 \end{pmatrix}$.
if z_2 is around zero then: $A_{32} = \begin{pmatrix} -0.2 & 0.1 \\ 0 & -0.1 \end{pmatrix}$.
Otherwise, there is no connection between the subsystems 1 and 3 and between 2 and 3, respectively.

4. Subsystem 4: The scheduling variable z_4 depends on x_4, $z_4 = x_{41} + x_{42} + 4$, and the membership functions are $w_1(z_4) = 0.125(x_{41} + x_{42} + 4)$ (corresponding to "z_4 is small"), and $w_2(z_4) = 1 - w_1(z_4)$ ("z_4 is big"). The states are bounded, $x_{41},\ x_{41} \in [-2,\ 2]$ and the input is bounded, $u \in [-0.5, 0.5]$.

Model rule 1:
If z_4 is small then

$$\dot{\boldsymbol{x}}_4 = \begin{pmatrix} -2 & 0 \\ 2 & -3 \end{pmatrix} \boldsymbol{x}_4 + \begin{pmatrix} 3 \\ 2 \end{pmatrix} u$$
$$y_4 = (1 \ \ 0)\boldsymbol{x}_4$$

Model rule 2:
If z_4 is big then

$$\dot{\boldsymbol{x}}_4 = \begin{pmatrix} -1 & 3 \\ 2 & -1 \end{pmatrix} \boldsymbol{x}_4 + \begin{pmatrix} 1 \\ 0 \end{pmatrix} u$$
$$y_4 = (1 \ \ 0)\boldsymbol{x}_4$$

The interconnection terms are as follows:
If rule 1 is active then: $A_{42} = \begin{pmatrix} 0.1 & 0.5 \\ 0.8 & 1 \end{pmatrix}$.
If rule 2 is active then: $A_{43} = \begin{pmatrix} 0.2 & 0.5 \\ -0.5 & 0 \end{pmatrix}$.
Otherwise, there is no connection between the subsystems 1 and 4, 2 and 4, and 3 and 4, respectively.

The observers are designed sequentially, as the subsystems are added, based on the conditions of Corollary 6.3 and Proposition 6.1. First, an observer is designed

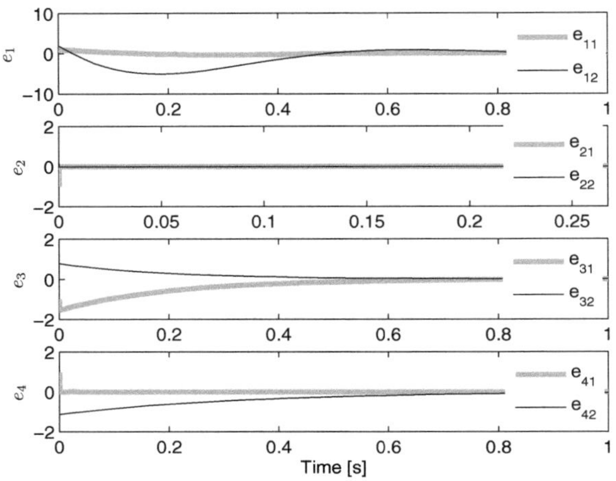

Fig. 6.7 Estimation error for the subsystems using distributed observers.

for Subsystem 1, without further conditions. Second, an observer is designed for Subsystem 2, taking into account the interconnection terms with Subsystem 1. Third, an observer is designed for Subsystem 3, considering the interconnection terms with Subsystems 1 and 2. Finally, an observer is designed for Subsystem 4. Note that the group of subsystems 1-2-3 and subsystem 4 are in cascade, and therefore, based on stability conditions for cascaded systems, it is sufficient that the independent observers are stable.

A typical error trajectory can be seen in Figure 6.7. The system was simulated using the *ode23tb* method (trapezoidal rule with second order backward difference formula) for solving ordinary differential equations of Matlab. This particular trajectory was computed for a randomly generated input and (where applicable) a random scheduling vector, with the true initial state being $(1\ \ 2\ \ -1\ \ 3\ \ -1\ \ 1\ \ 1\ \ -1)^T$ and the estimated initial states being zero. As expected, the error converges asymptotically to zero. Note that the error for the second subsystem converges very fast. This is because of two reasons: 1) the scheduling variable of this subsystem is a state to be estimated, and the observer has to be robust enough to handle the model mismatch and 2) at the same time the observer has to comply with the restrictions imposed by the interconnections to the first subsystem.

If the subsystems are added sequentially, and a bound on the interconnection terms is not known before the subsystem is added, one has to redesign a centralized observer each time a subsystem is added. A model of the centralized system can be obtained by taking all possible combinations of the interconnected subsystems and an observer can be designed for this system. This means that both the number of rules and the dimension of the LMI problem to be solved increase in every step: for the first subsystem, 2 LMIs of dimension 2 have to be solved, when the second subsystem is added, 4 LMIs of dimension 4, for the third subsystem 8 LMIs of dimension 6, and finally, when the fourth is added, 16 LMIs of dimension 8 need to be solved. □

6.4 Summary

Many physical systems, such as power systems, communication networks, economic systems, and traffic networks are composed of lower-dimensional subsystems that are interconnected. In this chapter, the stability of such decentralized systems has been studied for the case when the subsystems are represented as TS fuzzy systems. We have reviewed parallel and sequential methods for analyzing the stability of and design observers for TS fuzzy subsystems. The presented approaches reduce the dimension of the problem to be solved, by analyzing the stability of the overall system based on the individual subsystems and the strength of the interconnection terms. The same holds for observer design. Observers can be designed for the individual subsystems in parallel or sequentially. Such a design has the advantage that the observer does not need to be redesigned every time a subsystem is added or removed.

A shortcoming of the methods presented is that they rely on the existence of a common quadratic Lyapunov function. Moreover, the derivation of LMI conditions, although facilitating the easier design, introduce conservativeness.

Chapter 7
Adaptive Observers for TS Systems

Many processes change over time or due to the influence of unknown inputs that cannot be measured. In some cases, the change is considerable and may be due to the degradation of parts of the system, actuator faults, or disturbances that should be detected as soon as possible. Therefore, this change has to be considered when designing observers and controllers. In this chapter, methods to design adaptive observers that simultaneously estimate the states and unknown inputs of systems represented by TS fuzzy models are presented.

7.1 Introduction

Adaptive observers are in general defined as observers that simultaneously estimate the states and unknown parameters of a dynamic system, by processing the measurements online. Adaptive parameter estimators or observers in general fall into two categories (Sheikholeslam, 1995): 1) input-output identification methods and 2) model-based algorithms that use an observer. In this chapter, we address the second category, that is, model-based adaptive observers.

Although in general the application of adaptive observers concerns the estimation of states and parameters, in this book we also refer to unknown input observers as being adaptive observers. In fact, in this context, the unknown inputs are considered to be time-varying parameters. In what follows, we briefly review adaptive observers for linear and nonlinear dynamic systems.

Starting from the 1970s, the problem of state estimation in the presence of unknown inputs or unknown parameters has attracted significant research interest (Landau, 1979; Narendra and Annaswamy, 1989; Sastry and Bodson, 1989). However, these early results concern mainly linear time-invariant, single-input single-output systems, and their application for nonlinear dynamic systems is not straightforward. When applying adaptive state and input observers designed for linear systems to a nonlinear system, such as the approach proposed in (Xiong and Saif, 2003), the observer can only be used in a small neighborhood of the linearization point. In general, observers designed for linear systems are rarely able to estimate the states and the inputs of the nonlinear system.

Zs. Lendek et al.: Stability and Observer Design Using TS Models, STUDFUZZ 262, pp. 149–182.
springerlink.com

For nonlinear systems, a general approach, both in adaptive controller and observer design is to assume that the system is SISO and is in or can be transformed into observer canonical form (Marino and Tomei, 1995; Marino et al., 2001; Park et al., 2001; Park and Park, 2003; Wang and Luoh, 2004; Tong et al., 2004; Wang and Chai, 2005; Park et al., 2005; Ho et al., 2005; Hyun et al., 2006). Although by this transformation the physical meaning of the state variables is lost, it facilitates the observer design. By using a quadratic Lyapunov function, ensuring strictly positive real conditions, the Kalman-Yakubovic-Popov lemma can be applied and the adaptive laws are deduced from the Lyapunov synthesis. A shortcoming of these observers is that they do not incorporate prior information and cannot be used when physical states have to be estimated, or when a model is not in a canonical form. Robust versions of these adaptive observers have also been derived for systems affected by a bounded disturbance, by adding a robustness term (Park et al., 2001; Park and Park, 2003; Park et al., 2005; Wang and Chai, 2005; Labiod and Guerra, 2007). In many cases, when using both an observer and a controller, the robustness term is incorporated in the controller instead, to deal with the approximation error and disturbances (Tong et al., 2004; Ho et al., 2005).

A special class of adaptive TS models that can be considered is the class of so-called evolving fuzzy models (Angelov and Filev, 2004b). However, these models are in general identified from data, and result in an input-output description, not a state-space formulation. The interested reader is referred to (Angelov and Filev, 2003, 2004b,a; Angelov et al., 2008; Angelov and Zhou, 2008).

Results for MIMO systems include high-gain observers (Zhang and Xu, 2001), special observer canonical forms of the system (Zhu and Pagilla, 2003; Wang and Luoh, 2004), linearly parameterized neural networks (Ruiz Vargas and Hemerly, 2001; Hovakimyan et al., 2002), and observers based on a known linear part of the model (Ha and Trinh, 2004).

Several methods (Marino et al., 2001; Pertew et al., 2005, 2006) exist that estimate the states and parameters of nonlinear systems composed of a known linear part and a Lipschitz nonlinearity affected by unknown inputs. For instance, (Ha and Trinh, 2004) provided an estimation method for a class of nonlinear systems, where the known part of the nonlinearity is Lipschitz in the states and inputs. This method relies on an assumption related to a rank condition on a matrix composed of the direct feedthrough term and of a distribution matrix of the unknown nonlinear terms affected by the unknown inputs.

The design of observers in the presence of unknown inputs is an important problem, since in many cases not all the inputs are known (Xiong and Saif, 2003; Pertew et al., 2005, 2006). For instance, in machine tool and manipulator applications, the cutting force exerted by the tool or the exerting force/torque of the robot is needed, but it is very difficult or expensive to measure (Corless and Tu, 1998; Ha and Trinh, 2004). Load estimation in e.g., electricity distribution networks (Sheldrake, 2005), or wind turbines (Li and Chen, 2005) is necessary for proper planning and operation. In biomechanics, the myoskeletal system can be regarded as a dynamic system, where segment positions and trajectories are the system outputs and joint torques are the non-measurable inputs (Guelton et al., 2008b). In traffic control, time-varying parameters have to be tracked, which can be regarded as unknown inputs (Wang and Papageorgiou, 2005). In chaotic systems, for chaos synchronization and secure communication, one has to estimate not only the state, but also the input signal (Liao and Huang, 1999).

Adaptive input observers have also been used for fault detection, even when all inputs are known (see (Frank, 1990) and the references therein). In fact, the class of adaptive observers has received considerable interest in fault detection and identification, where the unknown inputs represent the effect of actuator faults or plant components and its presence has to be detected as soon as possible. However, these methods usually concern linear systems (Zhang et al., 2005) and only detect the fault, but do not estimate it (Marx et al., 2007). A method for TS systems in descriptor form has been proposed by Marx et al. (2007) to estimate the states in the presence of unknown inputs. This method is based either on decoupling the unknown inputs, or on attenuating their effects on the states. If decoupling is possible, the states are correctly estimated, and single faults can be isolated by using a bank of observers. However, the faults (unknown inputs) cannot be reconstructed. For the case when the decoupling of all the unknown inputs is not possible, Marx et al. (2007) also proposed a method to attenuate their effect on the states.

In this chapter, methods to design observers that estimate the states and the unknown inputs of TS fuzzy systems are presented. First, the general framework is introduced, starting from adaptive observers for linear systems. Afterwards, adaptive observers for TS fuzzy systems are considered. Two types of unknown inputs are considered. First, inputs that are or can be approximated by polynomial functions of time, are studied. Such inputs can be, for instance, biases in the model or ramp inputs acting on the model. This approach is in fact a generalization of several results that consider (approximately) constant unknown inputs. The second type of input considered is uncertainty in the model dynamics.

7.2 Unknown Input Estimation

Adaptive observers have already been investigated in the 1970s, for linear systems influenced by unknown inputs, with the goal that both the states of the system and the unknown inputs are estimated. A complete analysis of input observability and input reconstruction for linear systems has been provided by Hou and Patton (1998).

The early results concern linear systems of the form

$$\begin{aligned}\dot{\boldsymbol{x}} &= A\boldsymbol{x} + B\boldsymbol{d}\\ \boldsymbol{y} &= C\boldsymbol{x} + D\boldsymbol{d}\end{aligned}$$

where $\boldsymbol{d}$ denotes the unknown input, and A, B, C, and D are the system matrices. For such systems, Hostetter and Meditch (1973) proposed the observer

$$\begin{aligned}\dot{\boldsymbol{\xi}} &= F\boldsymbol{\xi} + G\boldsymbol{y}\\ \chi &= L\boldsymbol{y} + M\boldsymbol{\xi}\end{aligned}$$

where $\chi = R \begin{pmatrix} \widehat{\boldsymbol{x}} \\ \hat{\boldsymbol{d}} \end{pmatrix}$, i.e., it is the estimate of a linear transformation of the states unknown inputs, and the matrices F, G, L, and M are determined (see Hostetter and Meditch (1973)) such that the estimation error $\boldsymbol{e} = \boldsymbol{x} - \widehat{\boldsymbol{x}}$ converges to zero.

Although the method of Hostetter and Meditch (1973) can only be used for SISO systems in canonical forms, they also proposed that the observer should be extended with the derivatives of the unknown input. That is, if the unknown input is not constant, but time-varying, its derivatives can be included in the vector χ, and a so-called 'k-observer', in fact a Luenberger observer for this augmented system can be used. With this observer both the states and the unknown inputs can are estimated.

For nonlinear systems, the problem of estimating both the states and the unknown inputs of the system is motivated in part by machine tool and manipulator applications. There are many situations when an observer is required to estimate the cutting force of a machine tool or the exerting force/torque of a robotic manipulator. This problem has been addressed by Corless and Tu (1998), where the nonlinear part, a state-dependent and time-varying function, is also the unknown input.

Corless and Tu (1998) considered systems described by

$$\begin{aligned} \dot{\boldsymbol{x}} &= A\boldsymbol{x} + B\boldsymbol{f}(t, \boldsymbol{x}) \\ \boldsymbol{y} &= C\boldsymbol{x} \end{aligned}$$

where $\boldsymbol{f}(t, \boldsymbol{x})$ is considered an unknown input, i.e., $\boldsymbol{d} \triangleq \boldsymbol{f}(t, \boldsymbol{x})$, and $\text{rank}(CB) = \text{rank}(B)$, i.e., the number of measurements is necessarily greater than or equal to the number of inputs. Moreover, an initial estimate $\boldsymbol{f}_0(t, \boldsymbol{x})$ of $\boldsymbol{f}(t, \boldsymbol{x})$ is assumed to be available. The goal is to estimate both the state vector $\boldsymbol{x}$ and the unknown input $\boldsymbol{d}$.

Under the condition that $\text{rank}(B) = n_d$, i.e., B has full column rank, where n_d denotes the number of unknown inputs, and $\boldsymbol{d}$ is such that

$$\begin{aligned} &\|\boldsymbol{d} - \boldsymbol{f}_0(t, \widehat{\boldsymbol{x}})\| \le \beta_1 + \kappa_1 \|\boldsymbol{x} - \widehat{\boldsymbol{x}}\| \\ &\left\| \frac{d\boldsymbol{d}}{dt} - \frac{\partial \boldsymbol{f}_0}{\partial t}(t, \widehat{\boldsymbol{x}}) - \frac{\partial \boldsymbol{f}_0}{\partial t}(t, \widehat{\boldsymbol{x}})\dot{\widehat{\boldsymbol{x}}} \right\| \le \beta_2 + \kappa_{21} \|\boldsymbol{x} - \widehat{\boldsymbol{x}}\| + \kappa_{22} \|\dot{\boldsymbol{x}} - \dot{\widehat{\boldsymbol{x}}}\| \end{aligned}$$

holds, $\forall t, \boldsymbol{x}, \dot{\boldsymbol{x}}$, where β_1, β_2, κ_1, κ_{21}, and κ_{22} are known non-negative constants, Corless and Tu (1998) proposed the observer

$$\begin{aligned} \dot{\widehat{\boldsymbol{x}}} &= A\widehat{\boldsymbol{x}} + B\widehat{\boldsymbol{d}} + L(C\widehat{\boldsymbol{x}} - \boldsymbol{y}) \\ \widehat{\boldsymbol{d}} &= \boldsymbol{f}_0(t, \widehat{\boldsymbol{x}}) - \gamma G(C\widehat{\boldsymbol{x}} - \boldsymbol{y}) \end{aligned}$$

where the matrices L, G, and $P = P^T > 0$ are determined such that

$$\begin{aligned} &\mathcal{H}(P(A + LC)) < 0 \\ &B^T P = GC \end{aligned}$$

hold.

The estimator proposed by Corless and Tu (1998) is actually a high-gain observer, and it exists only under very strict conditions. Although it does not achieve asymptotic stability, the unknown inputs can be estimated to any degree of accuracy, by increasing γ. Moreover, this estimator does not require differentiation of the measured output, which would be problematic when the measurements are corrupted by noise.

The result of Corless and Tu (1998) has been extended and the conditions relaxed by Xiong and Saif (2003), by showing that the bound on the derivative of the unknown input is not necessary. Xiong and Saif (2003) consider the dynamic system

$$\begin{aligned}\dot{\boldsymbol{x}} &= A\boldsymbol{x} + B\boldsymbol{u} + G\boldsymbol{d}\\ \boldsymbol{y} &= C\boldsymbol{x}\end{aligned}$$

with the G and C matrices having full (column) rank, where $\boldsymbol{u}$ is a known input and $\boldsymbol{d}$ is an unknown input. The proposed observer is

$$\begin{aligned}\dot{\boldsymbol{z}} &= F\boldsymbol{z} + L\boldsymbol{y} + TB\boldsymbol{u} + TG\widehat{\boldsymbol{d}}\\ \widehat{\boldsymbol{d}} &= \gamma(W\boldsymbol{y} - N\boldsymbol{z})\end{aligned}$$

and guarantees that $\lim_{t\to\infty}(\boldsymbol{z} - T\boldsymbol{x}) \leq \epsilon$, for an arbitrary $\epsilon > 0$, under the conditions, that

$$\begin{aligned}&FT - TA + LC = 0\\ &N = (TG)^T P\\ &NT = G^T T^T PT = WC\\ &\operatorname{rank}(TG) = \operatorname{rank}(G) = n_d\end{aligned}$$

where F is a stable matrix, and $PF + F^T P = -Q$, $Q = Q^T > 0$. Moreover, if $\boldsymbol{d}$ is constant, then the estimates converge asymptotically to the true values.

A class of systems that has been extensively investigated comprises systems of the form

$$\begin{aligned}\dot{\boldsymbol{x}} &= A\boldsymbol{x} + \phi(\boldsymbol{x}, \boldsymbol{u}) + B\boldsymbol{f}(\boldsymbol{x}, \boldsymbol{u})\boldsymbol{\theta}\\ \boldsymbol{y} &= C\boldsymbol{x}\end{aligned}$$

where $B \in \mathbb{R}^{n_x \times n_d}$, and $\boldsymbol{\theta}$ are unknown parameters to be estimated.

For such systems, under the conditions that there exists a matrix $P = P^T > 0$ such that $B^T P = C_1$, with C_1 being a linear combination of the rows of C, that the nonlinear functions ϕ and $\boldsymbol{f}$ are Lipschitz continuous in the states, i.e., there exist γ_1 and γ_2 such that

$$\begin{aligned}\|\phi(\boldsymbol{x}, \boldsymbol{u}) - \phi(\widehat{\boldsymbol{x}}, \boldsymbol{u})\| &\leq \gamma_1 \|\boldsymbol{x} - \widehat{\boldsymbol{x}}\|\\ \|\boldsymbol{f}(\boldsymbol{x}, \boldsymbol{u}) - \boldsymbol{f}(\widehat{\boldsymbol{x}}, \boldsymbol{u})\| &\leq \gamma_2 \|\boldsymbol{x} - \widehat{\boldsymbol{x}}\|\end{aligned}$$

that the parameter vector $\boldsymbol{\theta}$ is bounded, i.e., $\|\boldsymbol{\theta}\| \leq \gamma_3$, for some $\gamma_3 > 0$, Rajamani and Hedrick (1995) proposed the observer

$$\begin{aligned}\dot{\widehat{\boldsymbol{x}}} =& A\widehat{\boldsymbol{x}} + \phi(\widehat{\boldsymbol{x}}, \boldsymbol{u}) + B\boldsymbol{f}(\widehat{\boldsymbol{x}}, \boldsymbol{u})\widehat{\boldsymbol{\theta}} + L(\boldsymbol{y} - C\widehat{\boldsymbol{x}})\\ \dot{\widehat{\boldsymbol{\theta}}} =& \varphi \boldsymbol{f}(\widehat{\boldsymbol{x}}, \boldsymbol{u})^T(\boldsymbol{y} - C\widehat{\boldsymbol{x}})\end{aligned}$$

Then, the estimated states converge asymptotically to the true values, if the gain L is chosen such that

$$\begin{aligned}\gamma_1 + \gamma_2\gamma_3\|B\| \leq \frac{\lambda_{\min}(Q)}{2\lambda_{\max}(P)}\\ \mathcal{H}(P(A - LC)) = -Q\end{aligned}$$

where $Q = Q^T > 0$ and $\varphi > 0$ are arbitrarily chosen, and $\lambda_{\min}$ and $\lambda_{\max}$ denote the smallest and largest eigenvalue, respectively.

A consequence of the results of Rajamani and Hedrick (1995) is, that in a similar way, adaptive observers can be designed for systems of the form

$$\begin{aligned}\dot{\boldsymbol{x}} =& A\boldsymbol{x} + \phi(\boldsymbol{y}, \boldsymbol{u}) + B\boldsymbol{f}(\boldsymbol{y}, \boldsymbol{u})\boldsymbol{\theta}\\ \boldsymbol{y} =& C\boldsymbol{x}\end{aligned}$$

a type of systems that have been extensively investigated by Marino (1990); Marino and Tomei (1995); Zhang and Xu (2001). For such systems, Marino and Tomei (1995) proposed an observer that guarantees an arbitrary rate of exponential convergence provided that the system is persistently excited, while Zhang and Xu (2001) proposed high-gain observers.

Ha and Trinh (2004) considered the dynamic system

$$\begin{aligned}\dot{\boldsymbol{x}} =& A\boldsymbol{x} + B\boldsymbol{d} + \boldsymbol{f}(\boldsymbol{x}, \boldsymbol{d}, \boldsymbol{y})\\ \boldsymbol{y} =& C\boldsymbol{x} + D\boldsymbol{d}\end{aligned}$$

where $\boldsymbol{d}$ is an unknown input (of dimension n_d) and the vector function $\boldsymbol{f}$ is composed of an unknown vector function $\boldsymbol{f}_u$ and a known vector function $\boldsymbol{f}_L$, i.e., it can be written as

$$\boldsymbol{f}(\boldsymbol{x}, \boldsymbol{d}, \boldsymbol{y}) = \boldsymbol{f}_L(\boldsymbol{x}, \boldsymbol{d}, \boldsymbol{y}) + W\boldsymbol{f}_u(\boldsymbol{x}, \boldsymbol{d}, \boldsymbol{y})$$

where W has full column rank n_f that is the dimension of $\boldsymbol{f}_u$. It is assumed that 1) the function $\boldsymbol{f}_L$ is Lipschitz in $\boldsymbol{x}$ and $\boldsymbol{d}$, i.e., there exists $\gamma > 0$ so that

$$\|\boldsymbol{f}_L(\boldsymbol{x}, \boldsymbol{d}, \boldsymbol{y}) - \boldsymbol{f}_L(\widehat{\boldsymbol{x}}, \widehat{\boldsymbol{d}}, \boldsymbol{y})\| \leq \gamma \left\| \begin{matrix}\boldsymbol{x} - \widehat{\boldsymbol{x}}\\ \boldsymbol{d} - \widehat{\boldsymbol{d}}\end{matrix} \right\|$$

$\forall \boldsymbol{x}, \widehat{\boldsymbol{x}} \in \mathbb{R}^{n_x}$ and $\forall \boldsymbol{d}, \widehat{\boldsymbol{d}} \in \mathbb{R}^{n_d}$, and furthermore, 2)

$$\text{rank}(D\ CW) = n_d + n_f$$

In order to design the observer, the system is written as

$$E\dot{\boldsymbol{\xi}} = M\boldsymbol{\xi} + \boldsymbol{f}_L(\boldsymbol{\xi}, \boldsymbol{y}) + W\boldsymbol{f}_u(\boldsymbol{\xi}, \boldsymbol{y})$$

with $\boldsymbol{\xi} = (\boldsymbol{x}^T \ \boldsymbol{d}^T)^T$, $E = (I \ 0)$, $M = (A \ B)$, $H = (C \ D)$. The observer proposed is

$$\begin{aligned} \dot{\chi} &= N\chi + L\boldsymbol{y} + R\boldsymbol{f}_L(\hat{\boldsymbol{\xi}}, \boldsymbol{y}) \\ \hat{\boldsymbol{\xi}} &= \chi + U\boldsymbol{y} \end{aligned} \tag{7.1}$$

For this observer, Ha and Trinh (2004) formulated the following result.

Theorem 7.1. *(Ha and Trinh, 2004) The estimation error dynamics obtained by using the observer* (7.1) *are asymptotically stable, if there exist* $P = P^T > 0$, X, Y, $\delta_1 > 0$, *and* $\delta_2 > 0$ *such that*

$$\begin{pmatrix} \mathcal{H}(P\phi + Y\Psi - XH) + \gamma^2(\delta_1 + \delta_2)I & PJ & YG \\ J^T P & -\delta_1 I & 0 \\ G^T Y^T & 0 & -\delta_2 I \end{pmatrix} < 0 \tag{7.2}$$

where

$$\begin{aligned} R &= J + ZG & Z &= P^{-1}Y \\ U &= V + ZK & & \\ N &= \phi + Z\Psi - FH & F &= P^{-1}X \\ L &= F + NU & & \\ \phi &= JM & J &= [I \ 0]S^{\dagger}[I \ 0]^T \\ S &= \begin{pmatrix} E & W \\ H & 0 \end{pmatrix} & \phi &= GM \\ G &= (I - S^{\dagger}S)[I \ 0]^T & V &= [I \ 0]S^{\dagger}[0 \ I]^T \\ K &= (I - S^{\dagger}S)[0 \ I]^T & S^{\dagger} &= (S^T S)^{-1}S^T \end{aligned}$$

where the superscript † *denotes the Moore-Penrose pseudoinverse.*

To increase the possibility that (7.2) is feasible, the Lipschitz constant γ should be as small as possible.

For TS fuzzy systems, unknown input observers have been developed for systems in descriptor form. One of these, developed by Marx et al. (2007), involves the estimation of the states by decoupling or when decoupling is not possible, attenuating the unknown inputs. However, this observer is not able to estimate the unknown input, and therefore is not presented here.

An unknown input observer that is able to estimate both the states and the unknown input for TS fuzzy systems in descriptor form has been developed by Guelton et al. (2008b). Consider the TS system in the descriptor form

$$\begin{aligned} \sum_{k=1}^{m_e} \nu_k(\boldsymbol{z})E_k\dot{\boldsymbol{x}} &= \sum_{i=1}^{m} w_i(\boldsymbol{z})(A_i\boldsymbol{x} + M_i\boldsymbol{d}) \\ \boldsymbol{y} &= \sum_{i=1}^{m} w_i(\boldsymbol{z})C_i\boldsymbol{x} \end{aligned} \tag{7.3}$$

where $\boldsymbol{d}$ is the vector of unknown inputs, E_k are regular matrices, and ν_k, $k = 1, 2, \ldots, m_\mathrm{e}$, are normalized membership functions.

By considering the unknown inputs as states, and under the assumption that $\dot{\boldsymbol{d}} = 0$, the system (7.3) can be reformulated as

$$\begin{aligned}\sum_{k=1}^{m_\mathrm{e}} \nu_k(\boldsymbol{z}) E_k^\mathrm{e} \dot{\boldsymbol{x}}^\mathrm{e} &= \sum_{i=1}^{m} w_i(\boldsymbol{z}) A_i^\mathrm{e} \boldsymbol{x}^\mathrm{e} \\ \boldsymbol{y} &= \sum_{i=1}^{m} w_i(\boldsymbol{z}) C_i^\mathrm{e} \boldsymbol{x}^\mathrm{e}\end{aligned} \tag{7.4}$$

with $\boldsymbol{x}^\mathrm{e} = (\boldsymbol{x}\ \boldsymbol{d})^T$, $E_k^\mathrm{e} = \begin{pmatrix} E_k & 0 \\ 0 & I \end{pmatrix}$, $A_k^\mathrm{e} = \begin{pmatrix} A_i & M_i \\ 0 & 0 \end{pmatrix}$, and $C_i^\mathrm{e} = \begin{pmatrix} C_i & 0 \end{pmatrix}$. Now, (7.4) can be written as

$$\begin{aligned}E^* \dot{\boldsymbol{x}}^* &= \sum_{k=1}^{m_e} \sum_{i=1}^{m} \nu_k(\boldsymbol{z}) w_i(\boldsymbol{z}) (A_{ik}^* \boldsymbol{x}^*) \\ \boldsymbol{y} &= \sum_{i=1}^{m} w_i(\boldsymbol{z}) C_i^* \boldsymbol{x}^*\end{aligned} \tag{7.5}$$

with $\boldsymbol{x}^* = (\boldsymbol{x}^\mathrm{e}\ \dot{\boldsymbol{x}}^\mathrm{e})^T$, $E^* = \begin{pmatrix} I & 0 \\ 0 & 0 \end{pmatrix}$, $A_{ik}^* = \begin{pmatrix} 0 & I \\ A_i^\mathrm{e} & -E_k^\mathrm{e} \end{pmatrix}$, and $C_i^* = \begin{pmatrix} C_i^\mathrm{e} & 0 \end{pmatrix}$.

The observer proposed by Guelton et al. (2008b) for (7.5) is

$$\begin{aligned}E^* \dot{\widehat{\boldsymbol{x}}}^* &= \sum_{k=1}^{m_\mathrm{e}} \sum_{i=1}^{m} \nu_k(\boldsymbol{z}) w_i(\boldsymbol{z}) (A_{ik}^* \widehat{\boldsymbol{x}}^*) + \sum_{k=1}^{m_\mathrm{e}} \sum_{i=1}^{m} \nu_k(\boldsymbol{z}) w_i(\boldsymbol{z}) K_{ik}^* (\boldsymbol{y} - \widehat{\boldsymbol{y}}) \\ \widehat{\boldsymbol{y}} &= \sum_{i=1}^{m} w_i(\boldsymbol{z}) C_i \widehat{\boldsymbol{x}}^*\end{aligned} \tag{7.6}$$

For this observer, the following result has been formulated:

Theorem 7.2. *(Guelton et al., 2008b) The estimation error obtained by using the observer* (7.6) *converges to zero, if there exist matrices P_3 and P_4, $P_1 = P_1^T > 0$, K_{ki}, $k = 1, 2, \ldots, m_e$, $i = 1, 2, \ldots, m$, such that*

$$\begin{aligned}\Gamma_{ii}^k &< 0 \\ \Gamma_{ij}^k + \Gamma_{ji}^k &< 0\end{aligned}$$

where

$$\Gamma_{ij}^k = \begin{pmatrix} A_i^T P_3 + P_3^T A_i - C_i^T K_{jk}^T P_3 - P_3^T K_{jk} C_i & X_{ijk}^T \\ X_{ijk} & -E_k^T P_4 - P_4^T E_k \end{pmatrix}$$

for $i = 1, 2, \ldots, m$, $j = i+1, i+2, \ldots, m$, $k = 1, 2, \ldots, m_e$, with $X_{ijk} = P_1 - E_k^T P_3 + P_4^T A_i - P_4^T K_{jk} C_i$.

In (Guelton et al., 2008b), the observer (7.6) has been applied to the estimation of joint torques in a human stance. However, this observer has two shortcomings. First, it is assumed that the unknown inputs are constants, or slowly varying, that is, $\dot{\boldsymbol{d}} \approx 0$. Second, it is implicitly assumed that the scheduling vectors are known (measured), as they can be used in the observer.

In the sequel, we consider similar adaptive unknown input observers, for unknown inputs that are polynomial functions of time, for TS fuzzy systems in the classical (non-descriptor) form. The observers are designed, similarly to the results presented so far, based on the already identified model, such that, together with an appropriate update law for estimating the unknown inputs, they ensure the convergence of the estimation error.

The TS fuzzy system considered is of the form

$$\begin{aligned}\dot{\boldsymbol{x}} &= \sum_{i=1}^{m} w_i(\boldsymbol{z})(A_i\boldsymbol{x} + B_i\boldsymbol{u} + M_i\boldsymbol{d} + a_i)\\ \boldsymbol{y} &= \sum_{i=1}^{m} w_i(\boldsymbol{z})(C_i\boldsymbol{x} + c_i)\end{aligned} \tag{7.7}$$

where A_i, B_i, M_i, a_i, C_i, c_i, $i = 1, 2, \ldots, m$, are the known matrices and biases of the ith local model, and the vector $\boldsymbol{d}$ is an unknown input. This input can represent disturbances acting on the process, effects of uncertain dynamics, or actuator faults. Two types of unknown inputs are distinguished. The first type of unknown input considered is an unstructured input that can be approximated by a polynomial function that varies over time. The second type of unknown input considered is unmodelled dynamics. In both cases, the goal is to design an observer that simultaneously estimates both the state vector $\boldsymbol{x}$ and the unknown input $\boldsymbol{d}$. To ensure the observability of the unknown inputs from the available measurements, in the sequel, it is assumed that

Assumption 7.1. *The matrices M_i, $i = 1, 2, \ldots, m$ have full column rank, and* $\operatorname{rank}(CM_i) = \operatorname{rank}(M_i)$, *$i = 1, 2, \ldots, m$.*

Given that our goal is to estimate both the states and the unknown inputs, this assumption is not restrictive.

The observer considered is of the form:

$$\begin{aligned}\dot{\widehat{\boldsymbol{x}}} &= \sum_{i=1}^{m} w_i(\widehat{\boldsymbol{z}})(A_i\widehat{\boldsymbol{x}} + B_i\boldsymbol{u} + a_i + M_i\widehat{\boldsymbol{d}} + L_i(\boldsymbol{y} - \widehat{\boldsymbol{y}}))\\ \widehat{\boldsymbol{y}} &= \sum_{i=1}^{m} w_i(\widehat{\boldsymbol{z}})(C_i\widehat{\boldsymbol{x}} + c_i)\\ \dot{\widehat{\boldsymbol{d}}} &= \boldsymbol{f}^{\mathrm{u}}(\widehat{\boldsymbol{d}}, \boldsymbol{w}(\widehat{\boldsymbol{z}}), \widehat{\boldsymbol{x}}, \boldsymbol{y})\end{aligned} \tag{7.8}$$

where L_i, $i = 1, 2, \ldots, m$, are the gain matrices to be designed for each rule, and $\boldsymbol{f}^{\mathrm{u}}$, the update law for $\boldsymbol{d}$, should be determined so that the estimation errors $\boldsymbol{x} - \widehat{\boldsymbol{x}}$ and $\boldsymbol{d} - \widehat{\boldsymbol{d}}$ converge asymptotically to zero, or to a neighborhood of zero.

Next to the two types of unknown inputs considered, two cases, depending on whether or not the scheduling vector depends on the states to be estimated are distinguished. The observer design is considered in both cases.

7.3 Estimation of Unknown Polynomial Inputs

In this section we consider the case when the unknown input is or can be approximated by a polynomial function in time. Such inputs may represent biases in the model, time-varying disturbances acting on the process, degradation in time, or even failure of actuators. In what follows, conditions to design a fuzzy observer and bounds on the estimation error are presented.

To design the observer, consider the TS fuzzy system of the form (7.7), where $\boldsymbol{d}$ is an unknown input that is a function of time so that there exists $p \in \mathbb{N}$ so that $\boldsymbol{d}^{(p)} = 0$, or $\|\boldsymbol{d}^{(p)}\| \leq \mu_{\boldsymbol{d}}$, for some $\mu_{\boldsymbol{d}} > 0$, i.e., the unknown input is $p-1$-th order polynomial function or can be approximated by a p-th order polynomial function of time. It is assumed that the states, the unknown input $\boldsymbol{d}$, and the derivatives of $\boldsymbol{d}$ are observable from $\boldsymbol{y}$.

7.3.1 *Measured Scheduling Vector*

First, consider the case when the scheduling variables are known (measured), and therefore their values can be directly used in the observer. Moreover, assume that the unknown input is indeed a $(p-1)$th order polynomial function of time, i.e., $\boldsymbol{d}^{(p)} = 0$. For this case, the following result holds.

Theorem 7.3. *The estimation error* $\mathbf{e} = \boldsymbol{x} - \widehat{\boldsymbol{x}}$ *obtained by using the observer*

$$\begin{aligned}
\dot{\widehat{\boldsymbol{x}}} &= \sum_{i=1}^{m} w_i(\boldsymbol{z})[A_i\widehat{\boldsymbol{x}} + B_i\boldsymbol{u} + L_i(\boldsymbol{y} - \widehat{\boldsymbol{y}}) + M_i\widehat{\boldsymbol{d}} + a_i] \\
\widehat{\boldsymbol{y}} &= \sum_{i=1}^{m} w_i(\boldsymbol{z})(C_i\widehat{\boldsymbol{x}} + c_i) \\
\widehat{\boldsymbol{d}}^{(p)} &= \sum_{i=1}^{m} w_i(\boldsymbol{z})\Lambda_i^p(\boldsymbol{y} - \widehat{\boldsymbol{y}}) \\
\widehat{\boldsymbol{d}}^{(k)} &= \sum_{i=1}^{m} w_i(\boldsymbol{z})(\Lambda_i^k(\boldsymbol{y} - \widehat{\boldsymbol{y}}) + \widehat{\boldsymbol{d}}^{(k+1)}) \\
& \textit{for } k = 1, \ldots, p-1
\end{aligned} \tag{7.9}$$

is exponentially stable if there exist $P = P^T > 0$, L_i, Λ_i^k, $i = 1, 2, \ldots, m$, $k = 1, 2, \ldots, p$, *so that*

$$\mathcal{H}\left(P\begin{pmatrix} A_i - L_iC_j + A_j - L_jC_i & M_i + M_j & 0 & \cdots & 0 \\ -\Lambda_i^1 C_j - \Lambda_j^1 C_i & 0 & 2I & \cdots & 0 \\ \vdots & \vdots & \vdots & \ddots & \vdots \\ -\Lambda_i^{p-1} C_j - \Lambda_j^{p-1} C_i & 0 & 0 & \cdots & 2I \\ -\Lambda_i^{p} C_j - \Lambda_j^{p} C_i & 0 & 0 & \cdots & 0 \end{pmatrix}\right) < 0 \tag{7.10}$$

for $i = 1, 2, \ldots, m$, $j = i+1, i+2, \ldots, m$, $\forall i < j:\ \exists \boldsymbol{z}:\ w_i(\boldsymbol{z})w_j(\boldsymbol{z}) \neq 0$.

Proof: An extended error system, containing both the state error and the derivatives of the input error $\overline{\boldsymbol{d}} = \boldsymbol{d} - \widehat{\boldsymbol{d}}$, can be expressed as:

$$\dot{\boldsymbol{e}}_a = \begin{pmatrix} \dot{\boldsymbol{e}} \\ \dot{\overline{\boldsymbol{d}}} \\ \ddot{\overline{\boldsymbol{d}}} \\ \vdots \\ \overline{\boldsymbol{d}}^{(p)} \end{pmatrix} = \sum_{i=1}^{m} w_i(\boldsymbol{z})w_i(\boldsymbol{z}) \begin{pmatrix} A_i - L_iC_i & M_i & 0 & \cdots & 0 \\ -\Lambda_i^1 C_i & 0 & I & \cdots & 0 \\ \vdots & \vdots & \vdots & \ddots & \vdots \\ -\Lambda_i^{p-1} C_i & 0 & 0 & \cdots & I \\ -\Lambda_i^{p} C_i & 0 & 0 & \cdots & 0 \end{pmatrix} \begin{pmatrix} \boldsymbol{e} \\ \overline{\boldsymbol{d}} \\ \vdots \\ \overline{\boldsymbol{d}}^{(p-1)} \end{pmatrix}$$

$$+ \sum_{i=1}^{m} w_i(\boldsymbol{z}) \sum_{\substack{j=1 \\ j>i}}^{m} w_j(\boldsymbol{z})$$

$$\cdot \begin{pmatrix} A_i - L_iC_j + A_j - L_jC_i & M_i + M_j & 0 & \cdots & 0 \\ -\Lambda_i^1 C_j - \Lambda_j^1 C_i & 0 & 2I & \cdots & 0 \\ \vdots & \vdots & \vdots & \ddots & \vdots \\ -\Lambda_i^{p-1} C_j - \Lambda_j^{p-1} C_i & 0 & 0 & \cdots & 2I \\ -\Lambda_i^{p} C_j - \Lambda_j^{p} C_i & 0 & 0 & \cdots & 0 \end{pmatrix} \begin{pmatrix} \boldsymbol{e} \\ \overline{\boldsymbol{d}} \\ \vdots \\ \overline{\boldsymbol{d}}^{(p-1)} \end{pmatrix}$$

Using a quadratic Lyapunov function $V = \boldsymbol{e}_a^T P \boldsymbol{e}_a$ for the extended error vector $\dot{\boldsymbol{e}}_a$, its derivative is expressed as:

$$\dot{V} = \sum_{i=1}^{m} w_i(\boldsymbol{z})w_i(\boldsymbol{z}) \boldsymbol{e}_a^T \mathcal{H}\left(P \begin{pmatrix} A_i - L_iC_i & M_i & 0 & \cdots & 0 \\ -\Lambda_i^1 C_i & 0 & I & \cdots & 0 \\ \vdots & \vdots & \vdots & \ddots & \vdots \\ -\Lambda_i^{p-1} C_i & 0 & 0 & \cdots & I \\ -\Lambda_i^{p} C_i & 0 & 0 & \cdots & 0 \end{pmatrix}\right) \boldsymbol{e}_a$$

$$+ \sum_{i=1}^{m} w_i(\boldsymbol{z}) \sum_{\substack{j=1 \\ j>i}}^{m} w_j(\boldsymbol{z})$$

$$\cdot\, \boldsymbol{e}_a^T \mathcal{H}\left(P \begin{pmatrix} A_i - L_iC_j + A_j - L_jC_i & M_i + M_j & 0 & \cdots & 0 \\ -\Lambda_i^1 C_j - \Lambda_j^1 C_i & 0 & 2I & \cdots & 0 \\ \vdots & \vdots & \vdots & \ddots & \vdots \\ -\Lambda_i^{p-1} C_j - \Lambda_j^{p-1} C_i & 0 & 0 & \cdots & 2I \\ -\Lambda_i^{p} C_j - \Lambda_j^{p} C_i & 0 & 0 & \cdots & 0 \end{pmatrix}\right) \boldsymbol{e}_a$$

which is negative definite if condition (7.10) is satisfied. □

Note that for simplicity, Theorem 7.3 has been formulated using the conditions of Wang et al. (1996). More relaxed conditions can be formulated using Lemmas 3.1 and 3.2.

Although the conditions of Theorem 7.3 are not LMIs, they can be transformed into LMIs using the change of variables $X_i = P\left(-L_i \; -\Lambda_i^1 \; \dots \; -\Lambda_i^{p-1} \; -\Lambda_i^p\right)^T$. The observer design using the conditions of Theorem 7.3 is illustrated using the following example.

Example 7.1. Consider the nonlinear dynamic system

$$\begin{aligned}\dot{\boldsymbol{x}} &= \begin{pmatrix} -x_1^2 & 1 & 0 \\ 0 & -1 & 1 \\ 1 & x_2 & -3 \end{pmatrix} \begin{pmatrix} x_1 \\ x_2 \\ x_3 \end{pmatrix} + \begin{pmatrix} 1 & 0 \\ 0 & 1 \\ 0 & 0 \end{pmatrix} \begin{pmatrix} d_1 \\ d_2 \end{pmatrix} \\ \boldsymbol{y} &= \begin{pmatrix} 1 & 0 & 0 \\ 0 & 1 & 0 \end{pmatrix} \begin{pmatrix} x_1 \\ x_2 \\ x_3 \end{pmatrix}\end{aligned} \tag{7.11}$$

with $x_1,\, x_2,\, x_3 \in [-5, 5]$, where $\boldsymbol{d} = \left(d_1 \; d_2\right)^T$ is an unknown input.

This system can be exactly represented (using the sector nonlinearity approach) by a 4-rule fuzzy system with the local matrices

$$A_1 = \begin{pmatrix} -25 & 1 & 0 \\ 0 & -1 & 1 \\ 1 & -5 & -3 \end{pmatrix} \qquad A_2 = \begin{pmatrix} -25 & 1 & 0 \\ 0 & -1 & 1 \\ 1 & 5 & -3 \end{pmatrix}$$

$$A_3 = \begin{pmatrix} 0 & 1 & 0 \\ 0 & -1 & 1 \\ 1 & -5 & -3 \end{pmatrix} \qquad A_4 = \begin{pmatrix} 0 & 1 & 0 \\ 0 & -1 & 1 \\ 1 & 5 & -3 \end{pmatrix}$$

$$M_i = \begin{pmatrix} 1 & 0 \\ 0 & 1 \\ 0 & 0 \end{pmatrix} \qquad C_i = \begin{pmatrix} 1 & 0 & 0 \\ 0 & 1 & 0 \end{pmatrix} \qquad i = 1,\, 2,\, 3,\, 4$$

and membership functions $w_1 = \eta_0^1 \eta_0^2$, $w_2 = \eta_0^1 \eta_1^2$, $w_3 = \eta_1^1 \eta_0^2$, $w_4 = \eta_1^1 \eta_1^2$, where $\eta_0^1 = \frac{x_1^2}{25}$, $\eta_0^2 = \frac{5 - x_2}{10}$, $\eta_1^1 = 1 - \eta_0^1$, and $\eta_1^2 = 1 - \eta_0^2$. Note that the scheduling variables are $z_1 = x_1$ and $z_2 = x_2$, which are both measured.

For this example the input $\boldsymbol{d}$ is assumed to be a second order polynomial function of time. To design the observer, the conditions (7.10) are transformed into LMIs using the change of variables $X_i = P\left(-L_i \; -\Lambda_i^1 \; \dots \; -\Lambda_i^{p-1} \; -\Lambda_i^p\right)^T$. Then, taking into account that the measurement matrix C_i is common for all the rules, the LMIs to be solved are *find* $P = P^T > 0$, X_i, $i = 1, 2, \ldots, 4$, *such that*

$$\mathcal{H}\left(P\begin{pmatrix} A_i & M_i & 0 & \cdots & 0 \\ 0 & 0 & I & \cdots & 0 \\ \vdots & \vdots & \vdots & \ddots & \vdots \\ 0 & 0 & 0 & \cdots & I \\ 0 & 0 & 0 & \cdots & 0 \end{pmatrix} - X_i\begin{pmatrix} C & 0 & \cdots & 0 \end{pmatrix}\right) < 0$$

for $i = 1, 2, \ldots, 4$.

The observer gains are found as[1]

$$L_1 = 10^3\begin{pmatrix} 0.21 & 0.03 \\ 0.00 & 0.07 \\ 0.00 & -0.01 \\ 1.60 & 0.20 \\ -0.01 & 0.42 \\ 4.54 & 0.51 \\ -0.07 & 1.13 \\ 5.93 & 0.53 \\ -0.12 & 1.45 \\ 3.01 & 0.20 \\ -0.05 & 0.73 \end{pmatrix} \quad L_2 = 10^3\begin{pmatrix} 0.25 & 0.04 \\ 0.017 & 0.03 \\ -0.00 & 0.01 \\ 1.93 & 0.26 \\ 0.08 & 0.22 \\ 5.46 & 0.70 \\ 0.16 & 0.60 \\ 7.11 & 0.84 \\ 0.16 & 0.77 \\ 3.58 & 0.39 \\ 0.09 & 0.39 \end{pmatrix}$$

$$L_3 = 10^3\begin{pmatrix} 0.07 & 0.03 \\ 0.01 & 0.07 \\ 0.00 & -0.01 \\ 0.49 & 0.23 \\ 0.02 & 0.42 \\ 1.39 & 0.61 \\ 0.05 & 1.13 \\ 1.81 & 0.67 \\ 0.06 & 1.45 \\ 0.91 & 0.27 \\ 0.03 & 0.73 \end{pmatrix} \quad L_4 = 10^3\begin{pmatrix} 0.08 & 0.02 \\ 0.01 & 0.03 \\ 0.00 & 0.00 \\ 0.55 & 0.10 \\ 0.03 & 0.19 \\ 1.55 & 0.25 \\ 0.08 & 0.52 \\ 2.02 & 0.26 \\ 0.09 & 0.67 \\ 1.02 & 0.10 \\ 0.05 & 0.34 \end{pmatrix}$$

To illustrate the estimation, a trajectory[2] of the estimation error using the above observer gains is presented in Figure 7.1(a). The corresponding true and estimated unknown inputs are given in Figure 7.1(b). The true input vector is the second order polynomial

$$\boldsymbol{d} = \begin{pmatrix} 0.6t^2 + 0.6t + 0.3 \\ 0.3t^2 + 1.2t + 0.6 \end{pmatrix}$$

[1] Throughout the chapter, all numerical values are rounded to two decimal places.

[2] Throughout this chapter, for numerical integration, the *ode45* Matlab function was used.

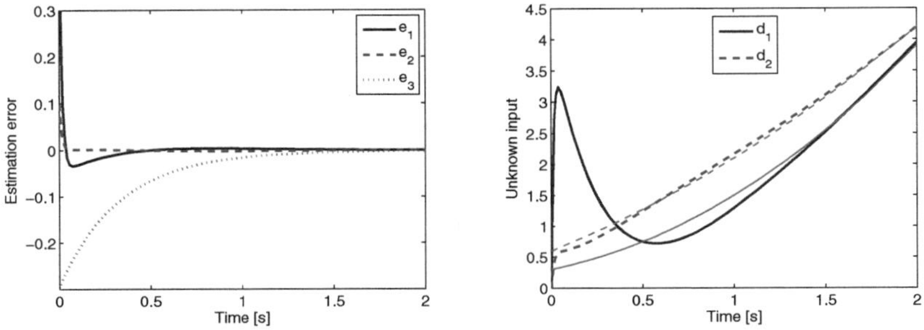

(a) Estimation error for the states.

(b) The unknown inputs (thin line) and their estimate (thick line).

Fig. 7.1 Simulation results for Example 7.1.

The true initial state vector was $(0.5\ 0.1\ \ -0.3)^T$, and the estimated initial states were $(0\ 0\ 0)^T$. As can be seen, both the state and the input estimates converge to the true values. □

Remark: In order to design observers with a desired convergence rate α, Corollary 4.2 can be combined with Theorem 7.3. Then, the following result can be formulated.

Corollary 7.1. *The estimation error of the observer* (7.9) *converges with a rate at least α if there exists $P = P^T > 0$, L_i, Λ_i^k, $i = 1, 2, \ldots, m$, $k = 1, 2, \ldots, p$, so that*

$$\mathcal{H}\left(P\begin{pmatrix} A_i - L_iC_j + A_j - L_jC_i & M_i + M_j & 0 & \cdots & 0 \\ -\Lambda_i^1 C_j - \Lambda_j^1 C_i & 0 & 2I & \cdots & 0 \\ \vdots & \vdots & \vdots & \ddots & \vdots \\ -\Lambda_i^{p-1} C_j - \Lambda_j^{p-1} C_i & 0 & 0 & \cdots & 2I \\ -\Lambda_i^{p} C_j - \Lambda_j^{p} C_i & 0 & 0 & \cdots & 0 \end{pmatrix}\right) + 4\alpha P < 0 \tag{7.12}$$

for $i = 1, 2, \ldots, m$, $j = i+1, i+2, \ldots, m$, $\forall i < j :\ \exists \boldsymbol{z} :\ w_i(\boldsymbol{z})w_j(\boldsymbol{z}) \neq 0.$

The design of an observer with a desired error convergence rate is illustrated using the following example.

Example 7.2. Consider the nonlinear system of Example 7.1. To maximize the convergence rate of the observer, the following general eigenvalue problem has to be solved:

maximize α, such that the following LMI is feasible: find $P = P^T > 0$, X_i, $i = 1, 2, 3, 4$, so that

$$\mathcal{H}\left(P\begin{pmatrix} A_i & M_i & 0 & \cdots & 0 \\ 0 & 0 & I & \cdots & 0 \\ \vdots & \vdots & \vdots & \ddots & \vdots \\ 0 & 0 & 0 & \cdots & I \\ 0 & 0 & 0 & \cdots & 0 \end{pmatrix} - X_i \begin{pmatrix} C & 0 & \cdots & 0 \end{pmatrix}\right) + 2\alpha P < 0$$

for $i = 1, 2, 3, 4$.

The observer gains are obtained as

$$L_1 = 10^3 \begin{pmatrix} 0.09 & -0.00 \\ 0.00 & 0.02 \\ -0.01 & 0.04 \\ 2.02 & -0.04 \\ 0.05 & 0.32 \\ 16.91 & -0.38 \\ 0.22 & 3.45 \\ 70.85 & -1.59 \\ 0.49 & 14.70 \\ 118.42 & -2.66 \\ -1.15 & 31.21 \end{pmatrix} \quad L_2 = 10^3 \begin{pmatrix} 0.09 & -0.00 \\ 0.00 & 0.02 \\ -0.01 & 0.05 \\ 2.01 & -0.08 \\ 0.04 & 0.32 \\ 16.86 & -0.71 \\ 0.13 & 3.45 \\ 70.63 & -2.98 \\ 0.11 & 14.70 \\ 118.06 & -4.98 \\ -1.95 & 31.26 \end{pmatrix}$$

$$L_3 = 10^3 \begin{pmatrix} 0.02 & 1-0.00 \\ 0.00 & 0.02 \\ 0.00 & 0.04 \\ 0.35 & -0.07 \\ 0.02 & 0.32 \\ 3.02 & -0.65 \\ 0.23 & 3.44 \\ 12.69 & -2.72 \\ 0.90 & 14.68 \\ 21.26 & -4.53 \\ 1.52 & 31.21 \end{pmatrix} \quad L_4 = 10^3 \begin{pmatrix} 0.02 & -0.00 \\ 0.00 & 0.02 \\ 0.00 & 0.05 \\ 0.35 & -0.08 \\ 0.03 & 0.32 \\ 3.02 & -0.72 \\ 0.26 & 3.44 \\ 12.71 & -3.01 \\ 1.05 & 14.68 \\ 21.29 & -5.01 \\ 1.83 & 31.21 \end{pmatrix}$$

and $\alpha = 4.02$. A trajectory of the estimation error and the true and estimated unknown inputs, for the same setting as in Example 7.1 is given in Figures 7.2(a) and 7.2(b). As can be seen, the estimates converge faster than in Example 7.1, but the overshoot is also larger. □

In many cases, the unknown input acting on the system is not polynomial. However, the input is in general bounded or it is possible to determine a bound on some derivative of it. Therefore, assume that there exists $p \in \mathbb{N}$ so that $\boldsymbol{d}^{(p)}$, the p-th derivative, is bounded by a known constant, i.e., $\|\boldsymbol{d}^{(p)}\| < \mu_{\boldsymbol{d}}$, and $\boldsymbol{d}^{(j)}$, $j = 1, 2, \ldots, p$ are observable from $\boldsymbol{y}$. In this case, although the estimation error does not converge to zero, it is bounded, and an upper bound can be computed as follows.

In the case when the scheduling vector does not depend on unmeasured states, the error system can be written as:

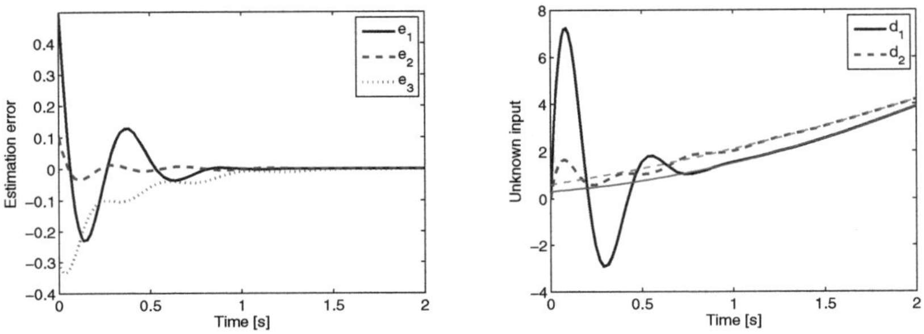

(a) Estimation errors for the states.

(b) The unknown inputs (thin line) and their estimate (thick line).

Fig. 7.2 Simulation results for Example 7.2.

$$
\begin{aligned}
\dot{\boldsymbol{e}}_a =& \sum_{i=1}^{m} w_i(\boldsymbol{z}) w_i(\boldsymbol{z}) \begin{pmatrix} A_i - L_i C_i & M_i & 0 & \cdots & 0 \\ -\Lambda_i^1 C_i & 0 & I & \cdots & 0 \\ \vdots & \vdots & \vdots & \ddots & \vdots \\ -\Lambda_i^{p-1} C_i & 0 & 0 & \cdots & I \\ -\Lambda_i^p C_i & 0 & 0 & \cdots & 0 \end{pmatrix} \boldsymbol{e}_a + \begin{pmatrix} 0 \\ 0 \\ 0 \\ \vdots \\ \boldsymbol{d}^{(p)} \end{pmatrix} \\
&+ \sum_{i=1}^{m} w_i(\boldsymbol{z}) \sum_{\substack{j=1 \\ j>i}}^{m} w_j(\boldsymbol{z}) \\
&\cdot \begin{pmatrix} A_i - L_i C_j + A_j - L_j C_i & M_i + M_j & 0 & \cdots & 0 \\ -\Lambda_i^1 C_j - \Lambda_j^1 C_i & 0 & 2I & \cdots & 0 \\ \vdots & \vdots & \vdots & \ddots & \vdots \\ -\Lambda_i^{p-1} C_j - \Lambda_j^{p-1} C_i & 0 & 0 & \cdots & 2I \\ -\Lambda_i^p C_j - \Lambda_j^p C_i & 0 & 0 & \cdots & 0 \end{pmatrix} \boldsymbol{e}_a
\end{aligned} \tag{7.13}
$$

For the error dynamics above, the following result can be stated.

Theorem 7.4. *The error described by* (7.13)*, with* $\|\boldsymbol{d}^{(p)}\| < \mu_{\boldsymbol{d}}$*, where* $\mu_{\boldsymbol{d}} > 0$ *is a known constant, is ultimately bounded by a ball with radius*

$$
\gamma = \sqrt{\frac{\lambda_{\max}(P)}{\lambda_{\min}(P)} \frac{\lambda_{\max}(P)\mu_{\boldsymbol{d}}}{\sigma \lambda_{\min}(Q)}} \tag{7.14}
$$

where $\sigma \in (0, 1)$*, if there exist* $P = P^T > 0$*,* $Q = Q^T > 0$*,* L_i*, and* Λ_i^k*,* $i = 1, 2, \ldots, m$*,* $k = 1, 2, \ldots, p$*, so that*

$$
\mathcal{H}\left(P\begin{pmatrix} A_i - L_iC_j + A_j - L_jC_i & M_i + M_j & 0 & \cdots & 0 \\ -\Lambda_i^1 C_j - \Lambda_j^1 C_i & 0 & 2I & \cdots & 0 \\ \vdots & \vdots & \vdots & \ddots & \vdots \\ -\Lambda_i^{p-1} C_j - \Lambda_j^{p-1} C_i & 0 & 0 & \cdots & 2I \\ -\Lambda_i^p C_j - \Lambda_j^p C_i & 0 & 0 & \cdots & 0 \end{pmatrix}\right) < -4Q \qquad (7.15)
$$

for $i = 1, 2, \ldots, m$, $j = i+1, i+2, \ldots, m$, $\forall i < j: \ \exists \boldsymbol{z}: \ w_i(\boldsymbol{z})w_j(\boldsymbol{z}) \neq 0$, *where* $\lambda_{\min}$ *and* $\lambda_{\max}$ *denote the eigenvalues with the smallest and largest absolute magnitude, respectively.*

Proof: Consider a quadratic Lyapunov function $V = \boldsymbol{e}_a^T P \boldsymbol{e}_a$ for the extended error vector, and $Q = Q^T$ is a positive definite matrix such that (7.15) is satisfied. We have

$$
\begin{aligned}
\dot{V} =& \sum_{i=1}^{m} w_i(\boldsymbol{z}) w_i(\boldsymbol{z}) \boldsymbol{e}_a^T \mathcal{H}\left(P\begin{pmatrix} A_i - L_iC_i & M_i & 0 & \cdots & 0 \\ -\Lambda_i^1 C_i & 0 & I & \cdots & 0 \\ \vdots & \vdots & \vdots & \ddots & \vdots \\ -\Lambda_i^{p-1} C_i & 0 & 0 & \cdots & I \\ -\Lambda_i^p C_i & 0 & 0 & \cdots & 0 \end{pmatrix}\right)\boldsymbol{e}_a \\
&+ \sum_{i=1}^{m} w_i(\boldsymbol{z}) \sum_{\substack{j=1 \\ j>i}}^{m} w_j(\boldsymbol{z}) \\
&\boldsymbol{e}_a^T \mathcal{H}\left(P\begin{pmatrix} A_i - L_iC_j + A_j - L_jC_i & M_i + M_j & 0 & \cdots & 0 \\ -\Lambda_i^1 C_j - \Lambda_j^1 C_i & 0 & 2I & \cdots & 0 \\ \vdots & \vdots & \vdots & \ddots & \vdots \\ -\Lambda_i^{p-1} C_j - \Lambda_j^{p-1} C_i & 0 & 0 & \cdots & 2I \\ -\Lambda_i^p C_j - \Lambda_j^p C_i & 0 & 0 & \cdots & 0 \end{pmatrix}\right)\boldsymbol{e}_a \\
&+ \sum_{i=1}^{m} w_i(\boldsymbol{z}) 2\boldsymbol{e}_a^T P \left(0\ 0\ 0 \cdots \ \boldsymbol{d}^{(p)T}\right)^T \\
\leq& -2\lambda_{\min}(Q)\|\boldsymbol{e}_a\|^2 + 2\lambda_{\max}(P)\|\boldsymbol{e}_a\|\mu_{\boldsymbol{d}} \\
\leq& -2(1-\sigma)\lambda_{\min}(Q)\|\boldsymbol{e}_a\|^2 - 2(\sigma\lambda_{\min}(Q)\|\boldsymbol{e}_a\|^2 - \lambda_{\max}(P)\|\boldsymbol{e}_a\|\mu_{\boldsymbol{d}})
\end{aligned}
$$

where $\sigma \in (0, 1)$ is arbitrarily chosen. Then, $\dot{V}$ is negative definite if

$$
\sigma\lambda_{\min}(Q)\|\boldsymbol{e}_a\|^2 - \lambda_{\max}(P)\|\boldsymbol{e}_a\|\mu_{\boldsymbol{d}} > 0
$$

or

$$
\|\boldsymbol{e}_a\| > \frac{\lambda_{\max}(P)\mu_{\boldsymbol{d}}}{\sigma\lambda_{\min}(Q)}
$$

Since $\lambda_{\min}(P)\|\boldsymbol{e}_a\|^2 \leq V \leq \lambda_{\max}(P)\|\boldsymbol{e}_a\|^2$, using Theorem 4.18 of (Khalil, 2002) it can be concluded that $\|\boldsymbol{e}_a\|$ converges exponentially to a ball with radius

$$\gamma = \sqrt{\frac{\lambda_{\max}(P)}{\lambda_{\min}(P)} \frac{\lambda_{\max}(P)\mu_{\boldsymbol{d}}}{\sigma\lambda_{\min}(Q)}} \tag{7.16}$$

which is a global uniform ultimate bound on the estimation error (Khalil, 2002). □

Remark: The bound (7.16) can be minimized by using the relaxation in (Tuan et al., 2001) and solving the following optimization problem:

maximize α_1, α_2, α_3 *so that there exist* $P = P^T > 0$, L_i, Λ_i^k, $i = 1, 2, \ldots, m$, $k = 1, 2, \ldots, p$, *subject to*

$$\begin{aligned}
&\Gamma_{ij} = \mathcal{H}\left(P \begin{pmatrix} A_i - L_i C_j & M_i & 0 & \cdots & 0 \\ -\Lambda_i^1 C_j & 0 & I & \cdots & 0 \\ \vdots & \vdots & \vdots & \ddots & \vdots \\ -\Lambda_i^{p-1} C_j & 0 & 0 & \cdots & I \\ -\Lambda_i^p C_j & 0 & 0 & \cdots & 0 \end{pmatrix} \right) \\
&\Gamma_{ii} > 0 \\
&\frac{2}{m-1}\Gamma_{ii} + \Gamma_{ij} + \Gamma_{ji} < -\alpha_3 I \\
&-P > -\alpha_2 I \\
&P > \alpha_1 I
\end{aligned} \tag{7.17}$$

for $i = 1, 2, \ldots, m$, $j = 1, 2, \ldots, m$, $i \neq j$.

Note however, that the obtained bound is only an upper bound on the estimation error, and in general it is very conservative. The observer design when the unknown input is approximated by a polynomial function of time is illustrated on the following example.

Example 7.3. Consider the nonlinear system in Example 7.1, and let the unknown input be given by $\boldsymbol{d} = (\sin(0.5t) \;\; \cos(0.5t))^T$. This input is not polynomial, but it is bounded, and its derivatives are bounded. For instance, $\|\boldsymbol{d}^{(2)}\| \leq 0.35$.

To design the observer, the unknown input is approximated by a first order (linear in time) polynomial. Solving (7.17), the ultimate bound is obtained as $\gamma = 63.02$. A trajectory of the estimation error and the true and estimated unknown inputs, for the same setting as in Example 7.1 are given in Figures 7.2(a) and 7.2(b). As can be seen in the figures, the computed bound on the estimation errors is very conservative. □

The results presented so far can be considered a generalization of the design proposed by Guelton et al. (2008b) to higher order inputs for classical TS fuzzy models. In the next section, this result is extended to the case when the scheduling vector depends on states that are not measured.

7.3.2 Estimated Scheduling Vector

In this section, we consider the case when the scheduling vector depends on states that are not measured, and therefore an estimate of the scheduling vector has to be

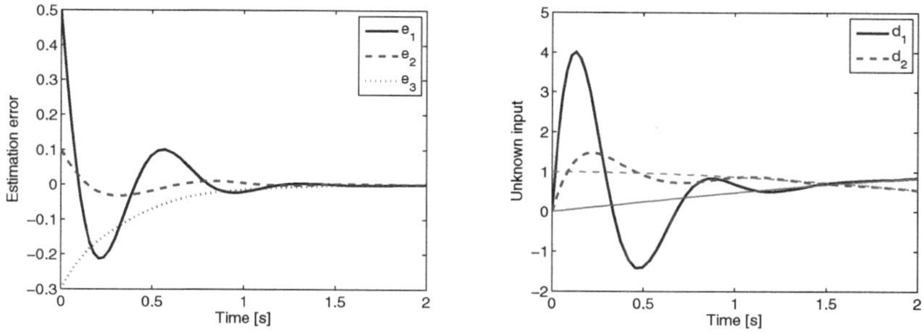

(a) Estimation errors for the states.

(b) The unknown input (thin line) and its estimate (thick line).

Fig. 7.3 Simulation results for Example 7.3.

used in the observer. For simplicity of notation, only the case when the measurement matrix is common for all rules is presented. Note, however, that if the measurement matrices are different, the observer can be designed similarly, although the conditions become more conservative.

The observer considered now is of the form

$$\begin{aligned} \dot{\widehat{\boldsymbol{x}}} &= \sum_{i=1}^{m} w_i(\widehat{\boldsymbol{z}})\Big(A_i\widehat{\boldsymbol{x}} + B_i\boldsymbol{u} + L_i(\boldsymbol{y}-\widehat{\boldsymbol{y}}) + M_i\widehat{\boldsymbol{d}} + a_i\Big) \\ \widehat{\boldsymbol{y}} &= C\widehat{\boldsymbol{x}} \\ \widehat{\boldsymbol{d}}^{(p)} &= \sum_{i=1}^{m} w_i(\widehat{\boldsymbol{z}})\Lambda_i^p(\boldsymbol{y}-\widehat{\boldsymbol{y}}) \\ \widehat{\boldsymbol{d}}^{(j)} &= \sum_{i=1}^{m} w_i(\widehat{\boldsymbol{z}})(\Lambda_i^j(\boldsymbol{y}-\widehat{\boldsymbol{y}}) + \widehat{\boldsymbol{d}}^{(j+1)}) \\ &\text{for } j = 1, \ldots, p-1 \end{aligned} \tag{7.18}$$

The extended error system becomes:

$$\begin{aligned} \dot{\boldsymbol{e}}_a = \sum_{i=1}^{m} w_i(\widehat{\boldsymbol{z}}) &\begin{pmatrix} A_i - L_iC & M_i & 0 & \cdots & 0 \\ -\Lambda_i^1 C & 0 & I & \cdots & 0 \\ \vdots & \vdots & \vdots & \ddots & \vdots \\ -\Lambda_i^{p-1} C & 0 & 0 & \cdots & I \\ -\Lambda_i^p C & 0 & 0 & \cdots & 0 \end{pmatrix} \begin{pmatrix} \boldsymbol{e} \\ \overline{\boldsymbol{d}} \\ \vdots \\ \overline{\boldsymbol{d}}^{(p-1)} \end{pmatrix} \\ + \sum_{i=1}^{m}(w_i(\boldsymbol{z}) - w_i(\widehat{\boldsymbol{z}}))&(I\ 0\ \ldots\ 0)^T \cdot (A_i\boldsymbol{x} + B_i\boldsymbol{u} + M_i\boldsymbol{d} + a_i) \end{aligned} \tag{7.19}$$

If the condition $\| \sum_{i=1}^m (w_i(\boldsymbol{z}) - w_i(\widehat{\boldsymbol{z}}))(A_i\boldsymbol{x} + B_i\boldsymbol{u} + M_i\boldsymbol{d} + a_i)\| \leq \mu\|\boldsymbol{e}\|$, for some $\mu > 0$ is satisfied, then $\| \sum_{i=1}^m (w_i(\boldsymbol{z}) - w_i(\widehat{\boldsymbol{z}}))(A_i\boldsymbol{x} + B_i\boldsymbol{u} + M_i\boldsymbol{d} + a_i)\|$

is also Lipschitz continuous in $\boldsymbol{e}_a$, with the same Lipschitz constant μ. In this case, combining Theorem 7.3 and Theorem 4.5, the following conditions can be formulated.

Corollary 7.2. *The error system* (7.19)*, under the assumption that*

$$\left\| \sum_{i=1}^{m} (w_i(\boldsymbol{z}) - w_i(\hat{\boldsymbol{z}}))(A_i\boldsymbol{x} + B_i\boldsymbol{u} + M_i\boldsymbol{d} + a_i) \right\| \leq \mu\|\boldsymbol{e}\|$$

where $\mu > 0$ *is a known constant, is asymptotically stable, if there exist* $P = P^T > 0$, $Q = Q^T > 0$, L_i*, and* Λ_i^j, $i = 1, 2, \ldots, m$, $j = 1, 2, \ldots, p$*, so that*

$$\begin{aligned} &\mathcal{H}\left(P \begin{pmatrix} A_i - L_iC & M_i & 0 & \cdots & 0 \\ -\Lambda_i^1 C & 0 & I & \cdots & 0 \\ \vdots & \vdots & \vdots & \ddots & \vdots \\ -\Lambda_i^{p-1} C & 0 & 0 & \cdots & I \\ -\Lambda_i^p C & 0 & 0 & \cdots & 0 \end{pmatrix} \right) < -Q \\ &\begin{pmatrix} Q - \mu^2 I & P \\ P & I \end{pmatrix} > 0 \end{aligned} \tag{7.20}$$

for $i = 1, 2, \ldots, m$.

The conditions (7.20) are not LMIs, but they can easily be transformed into LMIs using the change of variables $X_i = P\left(-L_i \; -\Lambda_i^1 \; \ldots \; -\Lambda_i^{p-1} \; -\Lambda_i^p\right)^T$. The design of an observer when the scheduling vector has to be estimated is illustrated on the following example.

Example 7.4. Consider the nonlinear dynamic system

$$\begin{aligned} \dot{\boldsymbol{x}} &= \begin{pmatrix} -x_3^2 & 1 & 2 \\ 1 & -2 & 1 \\ 0.1 & x_1 & -3 \end{pmatrix} \begin{pmatrix} x_1 \\ x_2 \\ x_3 \end{pmatrix} + \begin{pmatrix} 1 & 0 \\ 0 & 1 \\ 0 & 0 \end{pmatrix} \begin{pmatrix} d_1 \\ d_2 \end{pmatrix} \\ \boldsymbol{y} &= \begin{pmatrix} 1 & 0 & 0 \\ 0 & 1 & 0 \end{pmatrix} \begin{pmatrix} x_1 \\ x_2 \\ x_3 \end{pmatrix} \end{aligned} \tag{7.21}$$

with $x_1, x_2 \in [-1, 1]$, $x_3 \in [-0.2, 0.2]$, where $\boldsymbol{d} = \left(d_1 \; d_2\right)^T$ is an unknown input that is a first order polynomial function of time.

This system can be exactly represented (using the sector nonlinearity approach) by a 4-rule fuzzy system with the local matrices

$$A_1 = \begin{pmatrix} -0.04 & 1 & 2 \\ 1 & -2 & 1 \\ 0.1 & 1 & -3 \end{pmatrix} \qquad A_2 = \begin{pmatrix} -0.04 & 1 & 2 \\ 1 & -2 & 1 \\ 0.1 & -1 & -3 \end{pmatrix}$$

$$A_3 = \begin{pmatrix} 0 & 1 & 2 \\ 1 & -2 & 1 \\ 0.1 & 1 & -3 \end{pmatrix} \qquad A_4 = \begin{pmatrix} 0 & 1 & 2 \\ 1 & -2 & 1 \\ 0.1 & -1 & -3 \end{pmatrix}$$

$$M_i = \begin{pmatrix} 1 & 0 \\ 0 & 1 \\ 0 & 0 \end{pmatrix} \qquad C_i = \begin{pmatrix} 1 & 0 & 0 \\ 0 & 1 & 0 \end{pmatrix} \qquad i = 1,\, 2,\, 3,\, 4$$

The weighting functions (see Section 2.3.1) for the nonlinearities $-x_3^2$ and x_1 are $\eta_0^1 = \frac{x_3^2}{0.09}$, $\eta_1^1 = 1 - \eta_0^1$, and $\eta_0^2 = \frac{1-x(1)}{2}$, $\eta_1^2 = 1 - \eta_0^2$, respectively. Note that the weighting functions for the first nonlinearity depend on x_3, a state that has to be estimated. The condition $\| \sum_{i=1}^{m} (w_i(\boldsymbol{z}) - w_i(\widehat{\boldsymbol{z}}))(A_i \boldsymbol{x} + M_i \boldsymbol{d}) \| \leq \mu \|\boldsymbol{e}\|$ is satisfied with $\mu = 0.4$.

To design the observer, conditions (7.20) are transformed into LMIs using the change of variables $X_i = P \left(-L_i \ -\Lambda_i^1 \ \dots \ -\Lambda_i^{p-1} \ -\Lambda_i^p \right)$. Solving the LMIs, the observer gains are obtained as

$$L_1 = 10^2 \begin{pmatrix} 3.23 & 0.93 \\ 0.91 & 1.83 \\ -16.12 & -8.08 \\ 226.60 & 100.82 \\ 100.15 & 76.34 \\ 187.97 & 82.12 \\ 81.55 & 65.58 \end{pmatrix} \qquad L_2 = 10^2 \begin{pmatrix} 3.23 & 0.93 \\ 0.92 & 1.83 \\ -16.12 & -8.10 \\ 226.64 & 100.69 \\ 100.21 & 76.30 \\ 188.00 & 82.01 \\ 81.60 & 65.54 \end{pmatrix}$$

$$L_3 = 10^2 \begin{pmatrix} 3.23 & 0.92 \\ 0.92 & 1.83 \\ -16.12 & -8.07 \\ 226.68 & 100.57 \\ 100.27 & 76.25 \\ 188.03 & 81.91 \\ 81.66 & 65.51 \end{pmatrix} \qquad L_4 = 10^2 \begin{pmatrix} 3.23 & 0.92 \\ 0.92 & 1.83 \\ -16.13 & -8.08 \\ 226.70 & 100.50 \\ 100.30 & 76.23 \\ 188.05 & 81.86 \\ 81.69 & 65.49 \end{pmatrix}$$

A trajectory of the estimation error of the states using the observer gains above is presented in Figure 7.4(a). The corresponding true and estimated unknown inputs are given in Figure 7.4(b). The input vector was the first order polynomial $\boldsymbol{d} = \left(-0.3t + 0.2 \ \ 0.2t + 0.5 \right)^T$. The true initial states were $(0.5 \ 0.1 \ -0.1)^T$, and the estimated initial states were $(0 \ 0 \ 0)^T$. As can be seen, both the states and the input are correctly estimated. □

Note that the condition $\begin{pmatrix} Q - \mu^2 I & P \\ P & I \end{pmatrix} > 0$ is very conservative. In fact, if the estimated initial states are close enough to the true initial states, the estimates may converge even if this condition is not satisfied. Such an example is presented next.

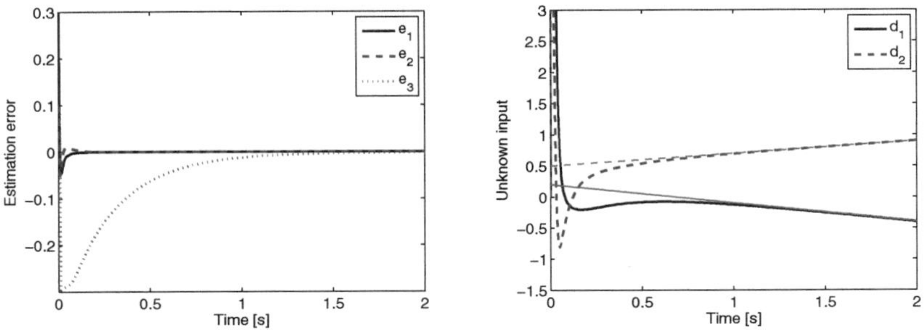

(a) Estimation errors for the states.

(b) The unknown inputs (thin line) and their estimates (thick line).

Fig. 7.4 Simulation results for Example 7.4.

Example 7.5. Consider the nonlinear dynamic system

$$\begin{aligned}\dot{\boldsymbol{x}} &= \begin{pmatrix} -x_3^2 & 1 & 2 \\ 1 & -2 & 1 \\ 0.1 & x_1 & -3 \end{pmatrix} \begin{pmatrix} x_1 \\ x_2 \\ x_3 \end{pmatrix} + \begin{pmatrix} 1 & 0 \\ 0 & 1 \\ 0 & 0 \end{pmatrix} \begin{pmatrix} d_1 \\ d_2 \end{pmatrix} \\ \boldsymbol{y} &= \begin{pmatrix} 1 & 0 & 0 \\ 0 & 1 & 0 \end{pmatrix} \begin{pmatrix} x_1 \\ x_2 \\ x_3 \end{pmatrix}\end{aligned} \tag{7.22}$$

with x_1, x_2, $x_3 \in [-1, 1]$, where $\boldsymbol{d} = \begin{pmatrix} d_1 & d_2 \end{pmatrix}^T$ is an unknown first order polynomial input.

Similarly to Example 7.4, this system can be exactly represented (using the sector nonlinearity approach) by a 4-rule fuzzy system with the local matrices

$$\begin{aligned} A_1 &= \begin{pmatrix} -1 & 1 & 2 \\ 1 & -2 & 1 \\ 0.1 & 1 & -3 \end{pmatrix} & A_2 &= \begin{pmatrix} -1 & 1 & 2 \\ 1 & -2 & 1 \\ 0.1 & -1 & -3 \end{pmatrix} \\ A_3 &= \begin{pmatrix} 0 & 1 & 2 \\ 1 & -2 & 1 \\ 0.1 & 1 & -3 \end{pmatrix} & A_4 &= \begin{pmatrix} 0 & 1 & 2 \\ 1 & -2 & 1 \\ 0.1 & -1 & -3 \end{pmatrix} \\ M_i &= \begin{pmatrix} 1 & 0 \\ 0 & 1 \\ 0 & 0 \end{pmatrix} & C_i &= \begin{pmatrix} 1 & 0 & 0 \\ 0 & 1 & 0 \end{pmatrix} \qquad i = 1,\, 2,\, 3,\, 4 \end{aligned}$$

The weighting functions (see Section 2.3.1) for the nonlinearities $-x_3^2$ and x_1 are $\eta_0^1 = x_3^2$, $\eta_1^1 = 1 - \eta_0^1$, and $\eta_0^2 = \frac{1-x(1)}{2}$, $\eta_1^2 = 1 - \eta_0^2$, respectively. Note that the weighting functions for the first nonlinearity depend on x_3, a state that has to be estimated.

The term $\sum_{i=1}^{m}(w_i(\boldsymbol{z}) - w_i(\widehat{\boldsymbol{z}}))(A_i\boldsymbol{x} + B_i\boldsymbol{u} + M_i\boldsymbol{d} + a_i)$, is bounded, and

$$\left\| \sum_{i=1}^{m}(w_i(\boldsymbol{z}) - w_i(\widehat{\boldsymbol{z}}))(A_i\boldsymbol{x} + B_i\boldsymbol{u} + M_i\boldsymbol{d} + a_i) \right\| \leq \|\boldsymbol{e}\|$$

i.e., $\mu = 1$. However, for this value of μ, the conditions (7.20) are not feasible, in particular, the condition $\begin{pmatrix} Q - \mu^2 I & P \\ P & I \end{pmatrix} > 0$ cannot be satisfied.

An observer (which does not guarantee that the estimation error will converge to zero) has been designed without using the condition $\begin{pmatrix} Q - \mu^2 I & P \\ P & I \end{pmatrix} > 0$, and the observer gains are

$$L_1 = \begin{pmatrix} 18.65 & 0.71 \\ 1.42 & 17.46 \\ 2.62 & 2.28 \\ 153.74 & 10.15 \\ 14.97 & 134.11 \\ 339.09 & 43.73 \\ 53.88 & 263.04 \end{pmatrix} \quad L_2 = \begin{pmatrix} 18.50 & 2.64 \\ -0.28 & 17.60 \\ 2.66 & 0.20 \\ 152.45 & 23.54 \\ 3.52 & 135.40 \\ 334.43 & 73.37 \\ 31.24 & 267.71 \end{pmatrix}$$

$$L_3 = \begin{pmatrix} 19.77 & -0.36 \\ 2.38 & 17.41 \\ 2.59 & 2.32 \\ 154.78 & 2.66 \\ 21.43 & 133.65 \\ 342.39 & 27.20 \\ 66.69 & 260.95 \end{pmatrix} \quad L_4 = \begin{pmatrix} 19.63 & 1.56 \\ 0.67 & 17.56 \\ 2.63 & 0.24 \\ 153.49 & 16.05 \\ 9.98 & 134.94 \\ 337.73 & 56.84 \\ 44.06 & 265.61 \end{pmatrix}$$

A trajectory of the estimation error of the states using the observer gains above is presented in Figure 7.5(a). The corresponding true and estimated unknown inputs are given in Figure 7.5(b). The input was the first order polynomial vector $\boldsymbol{d} = \left(-0.3t + 0.2 \;\; 0.2t + 0.5\right)^T$. The true initial states were $(0.5 \; 0.1 \; -0.1)^T$, and the estimated initial states were $(0 \; 0 \; 0)^T$. As can be seen, both the states and the input are correctly estimated. □

Consider now the case when the unknown input is not a polynomial function, but its derivative is bounded, that is, $\|\boldsymbol{d}^{(p)}\| \leq \mu_{\boldsymbol{d}}$. A bound similar to, although even more conservative than (7.16) can be computed in this case. For simplicity of notation, the computation is presented only for the case when the measurement matrices are common for all the rules. Then, the error dynamics are those in (7.19), and assume that

$$\left\| \sum_{i=1}^{m}(w_i(\boldsymbol{z}) - w_i(\widehat{\boldsymbol{z}}))(A_i\boldsymbol{x} + B_i\boldsymbol{u} + M_i\boldsymbol{d} + a_i) \right\| \leq \mu\|\boldsymbol{e}\|$$

with $\mu > 0$ a known constant. Let the condition

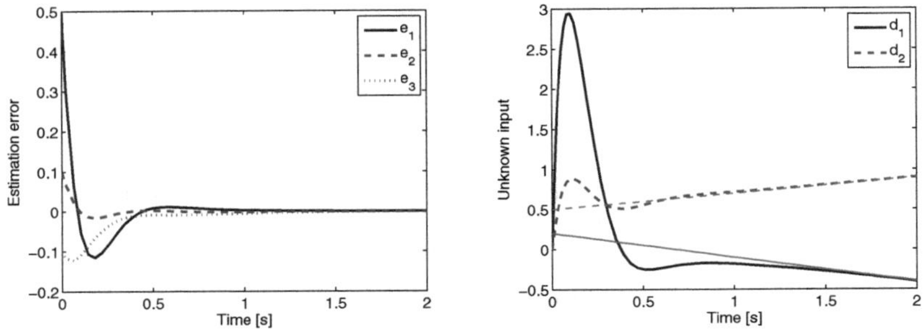

(a) Estimation errors for the states.

(b) The unknown inputs (thin line) and their estimates (thick line).

Fig. 7.5 Simulation results for Example 7.5.

$$
\begin{aligned}
&\mathcal{H}\left(P\begin{pmatrix} A_i - L_iC & M_i & 0 & \cdots & 0 \\ -\Lambda_i^1 C & 0 & I & \cdots & 0 \\ \vdots & \vdots & \vdots & \ddots & \vdots \\ -\Lambda_i^{p-1} C & 0 & 0 & \cdots & I \\ -\Lambda_i^{p} C & 0 & 0 & \cdots & 0 \end{pmatrix}\right) < -2Q \\
&P = P^T > 0 \\
&Q = Q^T > 0 \\
&\begin{pmatrix} Q - \mu^2 I & P \\ P & I \end{pmatrix} > 0
\end{aligned}
\tag{7.23}
$$

for $i = 1, 2, \ldots, m$, hold. Using a quadratic Lyapunov function $V = \boldsymbol{e}_a^T P \boldsymbol{e}_a$ for the extended error vector $\boldsymbol{e}_a$, we have

$$
\begin{aligned}
\dot{V} = &\sum_{i=1}^{m} w_i(\boldsymbol{z}) \boldsymbol{e}_a^T \mathcal{H}\left(P\begin{pmatrix} A_i - L_iC & M_i & 0 & \cdots & 0 \\ -\Lambda_i^1 C & 0 & I & \cdots & 0 \\ \vdots & \vdots & \vdots & \ddots & \vdots \\ -\Lambda_i^{p-1} C & 0 & 0 & \cdots & I \\ -\Lambda_i^{p} C & 0 & 0 & \cdots & 0 \end{pmatrix}\right)\boldsymbol{e}_a \\
&+ \sum_{i=1}^{m} w_i(\boldsymbol{z}) 2\boldsymbol{e}_a^T P \left(0\ 0\ 0\ \cdots\ {\boldsymbol{d}^{(p)}}^T\right)^T \\
&+ 2\boldsymbol{e}_a^T P \sum_{i=1}^{m} ((w_i(\boldsymbol{z}) - w_i(\widehat{\boldsymbol{z}})))(I\ 0\ \cdots\ 0)^T (A_i\boldsymbol{x} + B_i\boldsymbol{u} + M_i\boldsymbol{d}) \\
&\le -2\lambda_{\min}(Q)\|\boldsymbol{e}_a\|^2 + 2\lambda_{\max}(P)\mu\|\boldsymbol{e}_a\|^2 + 2\lambda_{\max}(P)\|\boldsymbol{e}_a\|\mu_{\boldsymbol{d}} \\
&\le -2(1-\sigma)(\lambda_{\min}(Q) - \mu\lambda_{\max}(P))\|\boldsymbol{e}_a\|^2 \\
&\quad - 2(\sigma(\lambda_{\min}(Q) - \mu\lambda_{\max}(P))\|\boldsymbol{e}_a\|^2 - \lambda_{\max}(P)\|\boldsymbol{e}_a\|\mu_{\boldsymbol{d}})
\end{aligned}
$$

where $\sigma \in (0,\ 1)$ is arbitrarily chosen and $Q = Q^T$ is a positive definite matrix such that (7.23) is satisfied. Then, $\dot{V}$ is negative definite if

$$\sigma(\lambda_{\min}(Q) - \mu\lambda_{\max}(P))\|\boldsymbol{e}_a\|^2 - \lambda_{\max}(P)\|\boldsymbol{e}_a\|\mu_{\boldsymbol{d}} > 0$$
$$\|\boldsymbol{e}_a\| > \frac{\lambda_{\max}(P)\mu_{\boldsymbol{d}}}{\sigma(\lambda_{\min}(Q) - \mu\lambda_{\max}(P))}$$

Since $\lambda_{\min}(P)\|\boldsymbol{e}_a\|^2 \leq V \leq \lambda_{\max}(P)\|\boldsymbol{e}_a\|^2$, using Theorem 4.18 of (Khalil, 2002) it can be concluded that $\|\boldsymbol{e}_a\|$ converges exponentially to stay within a ball with radius

$$\gamma = \sqrt{\frac{\lambda_{\max}(P)}{\lambda_{\min}(P)}} \frac{\lambda_{\max}(P)\mu_{\boldsymbol{d}}}{\sigma(\lambda_{\min}(Q) - \mu\lambda_{\max}(P))} \tag{7.24}$$

This bound can also be minimized using the conditions (7.17), together with the condition $\lambda_{\min}(Q) > \mu\lambda_{\max}(P)$.

Example 7.6. Consider the nonlinear system in Example 7.4, with the fuzzy representation developed in Example 7.4. For the current example, the true unknown input is $\boldsymbol{d} = (-\sin(0.1t)\ \cos(0.1t))^T$. This input is approximated by a second order polynomial, i.e., in the observer, a second order polynomial function was assumed. The second order derivative of $\boldsymbol{d}$ is bounded, with $\|\boldsymbol{d}^{(2)}\| \leq 0.14$.

To design the observer, conditions (7.23) were solved and the bound on the estimation error was obtained as 369.94. Naturally, this bound is extremely conservative, as it is confirmed by the simulation results. A trajectory of the estimation error of the states using the observer gains above is presented in Figure 7.6(a). The corresponding true and estimated unknown inputs are given in Figure 7.6(b). The true initial states were $(0.5\ 0.1\ -0.1)^T$, and the estimated initial states were $(0\ 0\ 0)^T$. As can be seen, both the estimated states and the estimated input converge close to the true values. □

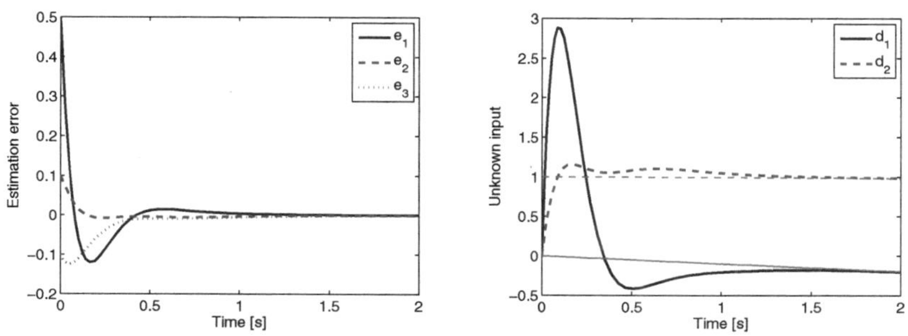

(a) Estimation errors for the states.

(b) The unknown inputs (thin line) and their estimates (thick line).

Fig. 7.6 Simulation results for Example 7.6.

7.4 Estimation of Unmodelled Dynamics

In this section, we consider unknown inputs that are not polynomial, but that are due to unmodelled dynamics. This means that the fuzzy system is of the form

$$\begin{aligned}\dot{\boldsymbol{x}} &= \sum_{i=1}^{m} w_i(\boldsymbol{z})(A_i\boldsymbol{x} + B_i\boldsymbol{u} + M_i(A_{\delta i}\boldsymbol{x} + B_{\delta i}\boldsymbol{u} + \theta_i)) \\ \boldsymbol{y} &= C\boldsymbol{x}\end{aligned} \tag{7.25}$$

where A_i, B_i, $i = 1, 2, \ldots, m$ are the known local models, and the matrices $A_{\delta i}$, $B_{\delta i}$ and the vectors θ_i, $i = 1, 2, \ldots, m$ are unknown, but $A_{\delta i}$, $i = 1, 2, \ldots, m$ are bounded by a known bound $\mu_{\max}$, $\max \|A_{\delta i}\| \leq \mu_{\max}$. This corresponds to the situation when part of the true dynamics is unmodeled. The goal is to determine sufficient conditions and to design an observer that estimates $\boldsymbol{x}$ and also the constant matrices $A_{\delta i}$, $B_{\delta i}$ and the vector θ_i, $i = 1, 2, \ldots, m$. Therefore, our goal is to estimate the unknown dynamics.

For the simplicity of the derivations, we present only the case when the measurement matrix is common for all rules of the model. If the measurement matrices are different, similar results can be derived. In what follows, two cases are distinguished: 1) the scheduling vector is known and 2) the scheduling vector depends on states that have to be estimated.

7.4.1 Measured Scheduling Vector

Consider first the case when the scheduling vector does not depend on states to be estimated. For system (7.25), the following observer is considered:

$$\begin{aligned}\dot{\widehat{\boldsymbol{x}}} &= \sum_{i=1}^{m} w_i(\boldsymbol{z})(A_i\widehat{\boldsymbol{x}} + B_i\boldsymbol{u} + L_i(\boldsymbol{y} - \widehat{\boldsymbol{y}}) + M_i(\widehat{A}_{\delta i}\widehat{\boldsymbol{x}} + \widehat{B}_{\delta i}\boldsymbol{u} + \widehat{\theta}_i)) \\ \widehat{\boldsymbol{y}} &= C\widehat{\boldsymbol{x}} \\ \dot{\widehat{A}}_{\delta i} &= \boldsymbol{f}_i^{\mathrm{u}}(\widehat{A}_{\delta i}, \boldsymbol{w}(\boldsymbol{z}), \widehat{\boldsymbol{x}}, \boldsymbol{y}) \\ \dot{\widehat{B}}_{\delta i} &= \boldsymbol{g}_i^{\mathrm{u}}(\widehat{B}_{\delta i}, \boldsymbol{w}(\boldsymbol{z}), \widehat{\boldsymbol{x}}, \boldsymbol{y}, \boldsymbol{u}) \\ \dot{\widehat{\theta}}_i &= \boldsymbol{h}_i^{\mathrm{u}}(\widehat{\theta}_i, \boldsymbol{w}(\boldsymbol{z}), \widehat{\boldsymbol{x}}, \boldsymbol{y})\end{aligned} \tag{7.26}$$

where L_i, $i = 1, 2, \ldots, m$ are the gain matrices for each rule, and the update laws f_i^{u}, g_i^{u}, h_i^{u}, $i = 1, 2, \ldots, m$ should be determined so that the estimation errors $\boldsymbol{x} - \widehat{\boldsymbol{x}}$, $A_{\delta i} - \widehat{A}_{\delta i}$, $B_{\delta i} - \widehat{B}_{\delta i}$, and $\theta_i - \widehat{\theta}_i$ converge asymptotically to zero.

The error dynamics when using the observer (7.26) can be expressed as:

$$\begin{aligned}\dot{\boldsymbol{e}} &= \sum_{i=1}^{m} w_i(\boldsymbol{z})[(A_i - L_iC + M_iA_{\delta i})\boldsymbol{e} + M_i(\bar{A}_{\delta i}\widehat{\boldsymbol{x}} + \bar{B}_{\delta i}\boldsymbol{u} + \bar{\theta}_i)] \\ \boldsymbol{e}_y &= C\boldsymbol{e}\end{aligned} \tag{7.27}$$

with $\bar{A}_{\delta i} = A_{\delta i} - \widehat{A}_{\delta i}$, $\bar{B}_{\delta i} = B_{\delta i} - \widehat{B}_{\delta i}$, $\bar{\theta}_i = \theta_i - \widehat{\theta}_i$.

Consider first the following part of the error expressed in (7.27):

$$\dot{\tilde{e}} = \sum_{i=1}^{m} w_i(z)(A_i - L_iC + M_iA_{\delta i})\tilde{e} \tag{7.28}$$

Since a bound on $A_{\delta i}$, $i = 1, 2, \ldots, m$ is known, i.e., $\max \|A_{\delta i}\| \leq \mu_{\max}$, stability conditions for perturbed fuzzy systems (Bergsten, 2001) can be used to render (7.28) stable and to design the gain matrices L_i, $i = 1, 2, \ldots, m$:

find $P = P^T > 0$, $Q = Q^T > 0$, *and* L_i, $i = 1, 2, \ldots, m$, *so that*

$$\begin{aligned} \|M_i\|\mu_{\max} &\leq \frac{\lambda_{\min}(Q)}{\lambda_{\max}(P)} \\ \mathcal{H}(P(A_i - L_iC)) &\leq -2Q \end{aligned} \tag{7.29}$$

for $i = 1, 2, \ldots, m$. Note that this means that the error dynamics is robustly stable.

Consider now a Lyapunov function of the form

$$V = e^T P e + \sum_{i=1}^{m} \mathrm{tr}(\bar{A}_{\delta i}^T \bar{A}_{\delta i}) + \sum_{i=1}^{m} \mathrm{tr}(\bar{B}_{\delta i}^T \bar{B}_{\delta i}) + \sum_{i=1}^{m} (\bar{\theta}_i^T \bar{\theta}_i)$$

for the error system (7.27), where P satisfies (7.29). We have

$$\begin{aligned} \dot{V} =& \sum_{i=1}^{n} w_i(z) e^T [(A_i - L_iC + M_iA_{\delta i})^T P + P(A_i - L_iC + M_iA_{\delta i})] e \\ &+ 2e^T P \sum_{i=1}^{m} w_i(z) M_i \bar{A}_{\delta i} \hat{x} + 2e^T P \sum_{i=1}^{m} w_i(z) M_i \bar{B}_{\delta i} u \\ &+ 2e^T P \sum_{i=1}^{m} w_i(z) M_i \bar{\theta}_i - 2\sum_{i=1}^{m} \mathrm{tr}(\dot{\hat{A}}_{\delta i}^T \bar{A}_{\delta i}) \\ &- 2\sum_{i=1}^{m} \mathrm{tr}(\dot{\hat{B}}_{\delta i}^T \bar{B}_{\delta i}) - 2\sum_{i=1}^{m} (\dot{\hat{\theta}}_i^T \bar{\theta}_i) \\ =& \sum_{i=1}^{n} w_i(z) e^T G_i e + 2\sum_{i=1}^{m} (\mathrm{tr}(\hat{x} e^T P M_i w_i(z) \bar{A}_{\delta i}) - \mathrm{tr}(\dot{\hat{A}}_{\delta i}^T \bar{A}_{\delta i})) \\ &+ 2\sum_{i=1}^{m} (\mathrm{tr}(u e^T P M_i w_i(z) \bar{B}_{\delta i}) - \mathrm{tr}(\dot{\hat{B}}_{\delta i}^T \bar{B}_{\delta i})) \\ &+ 2\sum_{i=1}^{m} (e^T P M_i w_i(z) \bar{\theta}_i - \dot{\hat{\theta}}_i^T \bar{\theta}_i) \\ =& \sum_{i=1}^{n} w_i(z) e^T G_i e + 2\sum_{i=1}^{m} \mathrm{tr}((\hat{x} e^T P M_i w_i(z) - \dot{\hat{A}}_{\delta i}^T) \bar{A}_{\delta i}) \\ &+ 2\sum_{i=1}^{m} \mathrm{tr}((u e^T P M_i w_i(z) - \dot{\hat{B}}_{\delta i}^T) \bar{B}_{\delta i}) + 2\sum_{i=1}^{m} (e^T P w_i(z) - \dot{\hat{\theta}}_i^T) \bar{\theta}_i \end{aligned}$$

with $G_i = \mathcal{H}(P(A_i - L_iC + M_iA_{\delta i}))$, $i = 1, 2, \ldots, m$.

Since $V > 0$ and from (7.29) $G_i < 0$, for $i = 1, 2, \ldots, m$, $\dot{V} < 0$ is rendered negative definite if $\mathrm{tr}((\widehat{\boldsymbol{x}}\boldsymbol{e}^T PM_i w_i(\boldsymbol{z}) - \dot{\widehat{A}}_{\delta i}^T)\bar{A}_{\delta i}) = 0$, $\mathrm{tr}((\boldsymbol{u}\boldsymbol{e}^T PM_i w_i(\boldsymbol{z}) - \dot{\widehat{B}}_{\delta i}^T)\bar{B}_{\delta i})$, and $\boldsymbol{e}^T PM_i w_i(\boldsymbol{z}) - \dot{\widehat{\theta}}_i^T = 0$, for $i = 1, 2, \ldots, m$. These conditions lead to the update laws:

$$\begin{aligned} \dot{\widehat{A}}_{\delta i} &= w_i(\boldsymbol{z}) M_i^T P \boldsymbol{e} \widehat{\boldsymbol{x}}^T \\ \dot{\widehat{B}}_{\delta i} &= w_i(\boldsymbol{z}) M_i^T P \boldsymbol{e} \boldsymbol{u}^T \\ \dot{\widehat{\theta}}_i &= w_i(\boldsymbol{z}) M_i^T P \boldsymbol{e} \end{aligned} \tag{7.30}$$

In general $\boldsymbol{e}$ is not directly available. However, given Assumption 7.1, there exist matrices Λ_i, $i = 1, 2, \ldots, m$, so that $\Lambda_i C = M_i^T P$, $\Lambda_i = M_i^T P C^\dagger$, where $C^\dagger$ denotes the Moore-Penrose pseudoinverse of C.

Therefore, the update laws can be expressed as:

$$\begin{aligned} \dot{\widehat{A}}_{\delta i} &= w_i(\boldsymbol{z}) M_i^T P C^\dagger \boldsymbol{e}_y \widehat{\boldsymbol{x}}^T \\ \dot{\widehat{B}}_{\delta i} &= w_i(\boldsymbol{z}) M_i^T P C^\dagger \boldsymbol{e}_y \boldsymbol{u}^T \\ \dot{\widehat{\theta}}_i &= w_i(\boldsymbol{z}) M_i^T P C^\dagger \boldsymbol{e}_y \end{aligned} \tag{7.31}$$

If all the rules are sufficiently excited, both the error system and the estimation error of the unknown matrices are asymptotically stable. It can easily be seen that, assuming nonzero and varying $\boldsymbol{x}$, $\boldsymbol{u}$, the only invariant set of the error system (7.27) is $\boldsymbol{e} = 0$, $\bar{A}_{\delta i} = 0$, $\bar{B}_{\delta i} = 0$ and $\bar{\theta}_i = 0$. If $w_i(\boldsymbol{z})$, $i = 1, 2, \ldots, m$,are sufficiently smooth and the fuzzy model is defined on a compact set of variables, then based on Barbalat's lemma and LaSalle's invariance principle – see Lemma 8.2 and Theorem 4.4 of (Khalil, 2002) –, the dynamics (7.27), together with the update laws above are asymptotically stable.

The results can be summarized as follows:

Theorem 7.5. *The error dynamics* (7.27) *are asymptotically stable, if the update laws* (7.31) *are used, the membership functions* w_i, $i = 1, 2, \ldots, m$ *are smooth, all rules are sufficiently excited, and, furthermore, there exist* $P = P^T > 0$, $Q = Q^T > 0$, *and* L_i, $i = 1, 2, \ldots, m$ *so that*

$$\begin{aligned} & P > 0 \\ & \mathcal{H}(P(A_i - L_i C)) < -Q \\ & \|M_i\| \mu_{\max} \leq \frac{\lambda_{\min}(Q)}{\lambda_{\max}(P)} \end{aligned} \tag{7.32}$$

for $i = 1, 2, \ldots, m$.

Example 7.7. Consider the four-rule TS fuzzy system

$$\dot{\boldsymbol{x}} = \sum_{i=1}^{4} w_i(\boldsymbol{z}) A_i \boldsymbol{x}$$
$$\boldsymbol{y} = C\boldsymbol{x}$$

with the local matrices

$$A_1 = \begin{pmatrix} -2 & -2 & 1 \\ -1 & -4 & 1 \\ 1 & 1 & -2 \end{pmatrix} \qquad A_2 = \begin{pmatrix} -2 & -2 & 1 \\ 1 & -4 & 1 \\ 1 & 1 & -2 \end{pmatrix}$$
$$A_3 = \begin{pmatrix} 2 & -2 & 1 \\ -1 & -4 & 1 \\ 1 & 1 & -2 \end{pmatrix} \qquad A_4 = \begin{pmatrix} 2 & -2 & 1 \\ 1 & -4 & 1 \\ 1 & 1 & -2 \end{pmatrix}$$
$$C = \begin{pmatrix} 1 & 0 & 0 \\ 0 & 1 & 0 \end{pmatrix}$$

with $\boldsymbol{z} = (z_1,\, z_2)^T$ measured scheduling vector and membership functions

$$w_1(\boldsymbol{z}) = (2 + z_1 - z_2)(1 - z_1)/8 \qquad w_2(\boldsymbol{z}) = (2 + z_1 - z_2)(1 + z_1)/8$$
$$w_3(\boldsymbol{z}) = (-1 - z_1 + z_2)(1 - z_1)/8 \qquad w_4(\boldsymbol{z}) = (-1 - z_1 + z_2)(1 + z_1)/8$$

All the variables are assumed to be bounded, $z_i \in [-1,\, 1]$, $i = 1,\, 2$, $x_i \in [-1,\, 1]$, $i = 1,\, 2,\, 3$.

For this system, a four-rule approximate TS model is available, with the same scheduling vector and membership functions, but with the local models being

$$A_1 = \begin{pmatrix} -3 & -2. & 1 \\ -1 & -3.5 & 1 \\ 1 & 1 & -2 \end{pmatrix} \qquad A_2 = \begin{pmatrix} -3 & -2 & 1 \\ 1 & -3.5 & 1 \\ 1 & 1 & -2 \end{pmatrix}$$
$$A_3 = \begin{pmatrix} 1 & -2 & 1 \\ -1 & -3.5 & 1 \\ 1 & 1 & -2 \end{pmatrix} \qquad A_4 = \begin{pmatrix} 1 & -2 & 1 \\ 1 & -3.5 & 1 \\ 1 & 1 & -2 \end{pmatrix}$$
$$C = \begin{pmatrix} 1 & 0 & 0 \\ 0 & 1 & 0 \end{pmatrix}$$

that is, the dynamics of the first two states are not modeled correctly. The distribution matrix of the unmodeled dynamics is $M_i = \begin{pmatrix} 1 & 0 \\ 0 & 1 \\ 0 & 0 \end{pmatrix}$, $i = 1,\, 2,\, 3,\, 4$, and the matrix of the unmodeled dynamics is bounded, with $\mu_{\max} = 1$.

For the approximate system, an observer has been designed by solving the conditions (7.32). The observer gains are

$$L_1 = \begin{pmatrix} 7.65 & 0.36 \\ 0.38 & 7.30 \\ 3.22 & 3.36 \end{pmatrix} \qquad L_2 = \begin{pmatrix} 7.48 & 1.42 \\ 1.33 & 7.47 \\ 3.07 & 3.51 \end{pmatrix}$$

$$L_3 = \begin{pmatrix} 10.90 & -0.03 \\ 0.53 & 6.47 \\ 3.15 & 3.21 \end{pmatrix} \qquad L_4 = \begin{pmatrix} 10.73 & 1.02 \\ 1.47 & 6.63 \\ 3.00 & 3.36 \end{pmatrix}$$

A trajectory of the estimation error using the observer gains above is presented in Figure 7.7. The true initial states were $(0.5\ 0.1\ \ -0.1)^T$, and the estimated initial states were $(0\ 0\ 0)^T$. The scheduling variables were randomly generated. As can be seen, the estimated states converge to the true values. □

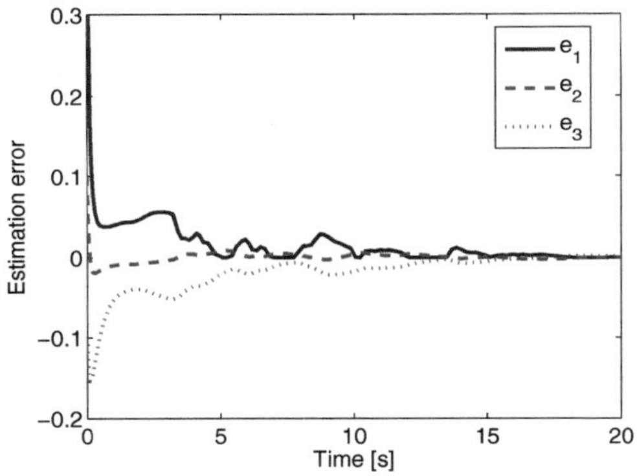

Fig. 7.7 Simulation results for Example 7.7.

Remark: Note that if the measurement matrix is different for each rule of the fuzzy model, the update laws for the matrices of the unknown dynamics can still be expressed as (7.30). Update laws similar to (7.31) can be derived if $(\sum_{i=1}^{m} w_i(\boldsymbol{z})C_i)^\dagger$, is defined $\forall \boldsymbol{z}$. In this case, the update laws are

$$\begin{aligned} \dot{\hat{A}}_{\delta i} &= w_i(\boldsymbol{z}) M_i^T P \Big(\sum_{i=1}^{m} w_i(\boldsymbol{z}) C_i \Big)^\dagger \boldsymbol{e}_y \hat{\boldsymbol{x}}^T \\ \dot{\hat{B}}_{\delta i} &= w_i(\boldsymbol{z}) M_i^T P \Big(\sum_{i=1}^{m} w_i(\boldsymbol{z}) C_i \Big)^\dagger \boldsymbol{e}_y \boldsymbol{u}^T \\ \dot{\hat{\theta}}_i &= w_i(\boldsymbol{z}) M_i^T P \Big(\sum_{i=1}^{m} w_i(\boldsymbol{z}) C_i \Big)^\dagger \boldsymbol{e}_y \end{aligned} \tag{7.33}$$

and the observer gains are given by (7.32). However, to have a unique solution, Assumption 7.1 has to be modified to $\mathrm{rank}(\sum_{j=1}^{m} w_j(\boldsymbol{z})C_j M_i) = \mathrm{rank}(M_i)$, $i = 1, 2, \ldots, m, \forall \boldsymbol{z}$.

Remark: If the unknown matrices are not constant, but slowly varying, such that $\dot{A_{\delta i}} \simeq 0$, etc., the results above can still be applied.

7.4.2 Estimated Scheduling Vector

Consider now the case when the scheduling vector depends on state variables to be estimated, that is, in the observer, the estimated state vector has to be used. Then, instead of (7.26), the observer

$$\begin{aligned}
\dot{\widehat{\boldsymbol{x}}} &= \sum_{i=1}^{m} w_i(\widehat{\boldsymbol{z}})(A_i\widehat{\boldsymbol{x}} + B_i\boldsymbol{u} + L_i(\boldsymbol{y} - \widehat{\boldsymbol{y}}) + M_i(\widehat{A}_{\delta i}\widehat{\boldsymbol{x}} + \widehat{B}_{\delta i}\boldsymbol{u} + \widehat{\theta}_i)) \\
\widehat{\boldsymbol{y}} &= C\widehat{\boldsymbol{x}} \\
\dot{\widehat{A}}_{\delta i} &= \boldsymbol{f}_i^{\mathrm{u}}(\widehat{A}_{\delta i}, w(\widehat{\boldsymbol{z}}), \widehat{\boldsymbol{x}}, \boldsymbol{y}) \\
\dot{\widehat{B}}_{\delta i} &= \boldsymbol{g}_i^{\mathrm{u}}(\widehat{B}_{\delta i}, w(\widehat{\boldsymbol{z}}), \widehat{\boldsymbol{x}}, \boldsymbol{y}, \boldsymbol{u}) \\
\dot{\widehat{\theta}}_i &= \boldsymbol{h}_i^{\mathrm{u}}(\widehat{\theta}_i, w(\widehat{\boldsymbol{z}}), \widehat{\boldsymbol{x}}, \boldsymbol{y})
\end{aligned} \tag{7.34}$$

has to be used, and the error system (7.27) becomes

$$\begin{aligned}
\dot{\boldsymbol{e}} &= \sum_{i=1}^{m} w_i(\widehat{\boldsymbol{z}})[(A_i - L_iC + M_iA_{\delta i})\boldsymbol{e} + M_i(\bar{A}_{\delta i}\widehat{\boldsymbol{x}} + \bar{B}_{\delta i}\boldsymbol{u} + \bar{\theta}_i)] \\
&\quad + \sum_{i=1}^{m}(w_i(\boldsymbol{z}) - w_i(\widehat{\boldsymbol{z}})) \cdot (A_i\boldsymbol{x} + B_i\boldsymbol{u} + M_i(A_{\delta i}\boldsymbol{x} + B_{\delta i}\boldsymbol{u} + \theta_i)) \\
\boldsymbol{e}_y &= C\boldsymbol{e}
\end{aligned} \tag{7.35}$$

Under the assumption that $\| \sum_{i=1}^{m}(w_i(\boldsymbol{z}) - w_i(\widehat{\boldsymbol{z}}))(A_i\boldsymbol{x} + B_i\boldsymbol{u} + M_i(A_{\delta i}\boldsymbol{x} + B_{\delta i}\boldsymbol{u} + \theta_i))\| \leq \mu\|\boldsymbol{e}\|$, and by combining the conditions in Theorems 4.5 and 7.5, the following result is obtained:

Corollary 7.3. *Assuming that*

$$\left\| \sum_{i=1}^{m}(w_i(\boldsymbol{z}) - w_i(\widehat{\boldsymbol{z}}))(A_i\boldsymbol{x} + B_i\boldsymbol{u} + M_i(A_{\delta i}\boldsymbol{x} + B_{\delta i}\boldsymbol{u} + \theta_i)) \right\| \leq \mu\|\boldsymbol{e}\|$$

holds with $\mu > 0$ *a known constant, the error system* (7.35)*, together with the update laws*

$$\begin{aligned}
\dot{\widehat{A}}_{\delta i} &= w_i(\widehat{\boldsymbol{z}})M_i^T PC^{\dagger}\boldsymbol{e}_y\widehat{\boldsymbol{x}}^T \\
\dot{\widehat{B}}_{\delta i} &= w_i(\widehat{\boldsymbol{z}})M_i^T PC^{\dagger}\boldsymbol{e}_y\boldsymbol{u}^T \\
\dot{\widehat{\theta}}_i &= w_i(\widehat{\boldsymbol{z}})M_i^T PC^{\dagger}\boldsymbol{e}_y
\end{aligned} \tag{7.36}$$

is asymptotically stable, if the membership functions are smooth, all rules are sufficiently excited, and, furthermore, if there exist $P = P^T > 0$, $Q = Q^T > 0$, L_i, $i = 1, 2, \ldots, m$ *so that*

$$\begin{aligned} & P > 0 \\ & \mathcal{H}(P(A_i - L_i C)) < -Q \\ & \begin{pmatrix} Q - (\mu^2 + \|M_i\|^2 \mu_{\max}^2) I & P \\ P & I \end{pmatrix} > 0 \end{aligned} \tag{7.37}$$

for $i = 1, 2, \ldots, m$.

Remark: Similarly to the case when the scheduling vector does not depend on the states, if the measurement matrix is different for each rule of the fuzzy model, update laws for the matrices of the unknown dynamics can be derived if $(\sum_{i=1}^m w_i(\widehat{\boldsymbol{z}})C_i)^\dagger$, is defined $\forall \widehat{\boldsymbol{z}}$. In this case, the update laws become

$$\begin{aligned} \dot{\widehat{A}}_{\delta i} &= w_i(\widehat{\boldsymbol{z}}) M_i^T P \Big(\sum_{i=1}^m w_i(\widehat{\boldsymbol{z}}) C_i \Big)^\dagger \boldsymbol{e}_y \widehat{\boldsymbol{x}}^T \\ \dot{\widehat{B}}_{\delta i} &= w_i(\widehat{\boldsymbol{z}}) M_i^T P \Big(\sum_{i=1}^m w_i(\widehat{\boldsymbol{z}}) C_i \Big)^\dagger \boldsymbol{e}_y \boldsymbol{u}^T \\ \dot{\widehat{\theta}}_i &= w_i(\widehat{\boldsymbol{z}}) M_i^T P \Big(\sum_{i=1}^m w_i(\widehat{\boldsymbol{z}}) C_i \Big)^\dagger \boldsymbol{e}_y \end{aligned} \tag{7.38}$$

and the observer gains are given by (7.37).

Example 7.8. Consider the four-rule TS fuzzy system

$$\begin{aligned} \dot{\boldsymbol{x}} &= \sum_{i=1}^4 w_i(\boldsymbol{z}) A_i \boldsymbol{x} \\ \boldsymbol{y} &= C\boldsymbol{x} \end{aligned}$$

with the local matrices

$$A_1 = \begin{pmatrix} 1 & -2 & 1 \\ -1 & -4 & 1 \\ 1 & 1 & -2 \end{pmatrix} \qquad A_2 = \begin{pmatrix} -1 & -2 & 1 \\ 1 & -4 & 1 \\ 1 & 1 & -2 \end{pmatrix}$$

$$A_3 = \begin{pmatrix} -1 & -2 & 1 \\ -1 & -4 & 1 \\ 1 & 1 & -2 \end{pmatrix} \qquad A_4 = \begin{pmatrix} -1 & -2 & 1 \\ 1 & -4 & 1 \\ 1 & 1 & -2 \end{pmatrix}$$

$$C = \begin{pmatrix} 1 & 0 & 0 \\ 0 & 1 & 0 \end{pmatrix}$$

with $\boldsymbol{z} = (x_1, x_3)^T$ the scheduling vector, with x_1 measured, and membership functions

$$w_1(\boldsymbol{z}) = (1 - x_3)(1 - x_1)/4 \qquad w_2(\boldsymbol{z}) = (1 - x_3)(1 + x_1)/4$$
$$w_3(\boldsymbol{z}) = (1 + x_3)(1 - x_1)/4 \qquad w_4(\boldsymbol{z}) = (1 + x_3)(1 + x_1)/8$$

all variables assumed to be bounded, $x_i \in [-1, 1]$, $i = 1, 2, 3$.

For this system, a four-rule approximate TS model is available, with the same scheduling vector and membership functions, but with the local models being

$$A_1 = \begin{pmatrix} 3 & -2. & 1 \\ -1 & -3.5 & 1 \\ 1 & 1 & -2 \end{pmatrix} \qquad A_2 = \begin{pmatrix} 3 & -2 & 1 \\ 1 & -3.5 & 1 \\ 1 & 1 & -2 \end{pmatrix}$$
$$A_3 = \begin{pmatrix} 1 & -2 & 1 \\ -1 & -3.5 & 1 \\ 1 & 1 & -2 \end{pmatrix} \qquad A_4 = \begin{pmatrix} 1 & -2 & 1 \\ 1 & -3.5 & 1 \\ 1 & 1 & -2 \end{pmatrix}$$
$$C = \begin{pmatrix} 1 & 0 & 0 \\ 0 & 1 & 0 \end{pmatrix}$$

that is, the dynamics of the first two states are not modeled correctly. The distribution matrix of the unmodeled dynamics is $M_i = \begin{pmatrix} 1 & 0 \\ 0 & 1 \\ 0 & 0 \end{pmatrix}$, $i = 1, 2, 3, 4$, and the matrix of the unmodeled dynamics is bounded, with $\mu_{\max} = 2$.

Note that the membership functions depend on x_3, a state that has to be estimated. The term $\sum_{i=1}^{m}(w_i(\boldsymbol{z}) - w_i(\widehat{\boldsymbol{z}})) \cdot (A_i\boldsymbol{x} + M_iA_{\delta i}\boldsymbol{x})$ is bounded, $\| \sum_{i=1}^{m}(w_i(\boldsymbol{z}) - w_i(\widehat{\boldsymbol{z}})) \cdot (A_i\boldsymbol{x} + M_iA_{\delta i}\boldsymbol{x})\| \leq \mu\|\boldsymbol{e}\|$, with $\mu = 2$. Moreover $\|M_iA_{\delta i}\| \leq 2$, $i = 1, 2, 3, 4$.

For the approximate system, an observer has been designed by solving the conditions (7.37). The observer gains are

$$L_1 = \begin{pmatrix} 8.05 & -1.37 \\ -1.30 & 1.49 \\ 2.19 & 2.23 \end{pmatrix} \qquad L_2 = \begin{pmatrix} 8.03 & -0.36 \\ -0.31 & 1.51 \\ 2.12 & 2.30 \end{pmatrix}$$
$$L_3 = \begin{pmatrix} 6.02 & -1.36 \\ -1.31 & 1.49 \\ 2.18 & 2.23 \end{pmatrix} \qquad L_4 = \begin{pmatrix} 6.01 & -0.35 \\ -0.32 & 1.51 \\ 2.12 & 2.30 \end{pmatrix}$$

A trajectory of the estimation error using the observer gains above is presented in Figure 7.8. The true initial states were $(0.5\ 0.1\ \ -0.1)^T$, and the estimated initial states were $(0\ 0\ 0)^T$. The scheduling variables were randomly generated. As can be seen, the estimated states converge to the true values. □

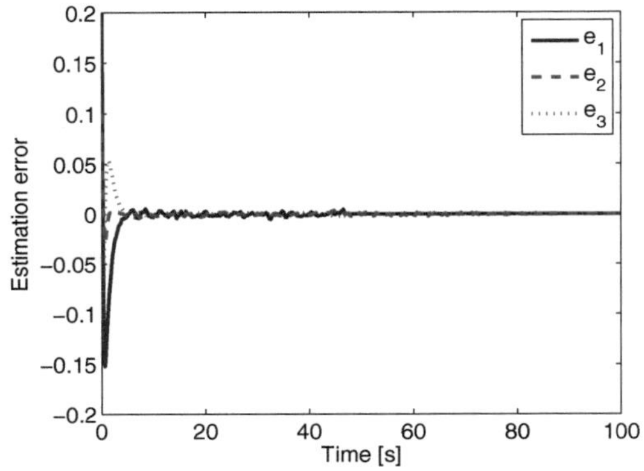

Fig. 7.8 Simulation results for Example 7.8.

7.5 Summary

In this chapter a method for designing observers that estimate the state and unknown inputs of TS fuzzy systems has been presented. The design of unknown input observers is important in practice, since in many cases not all the inputs are known. These unknown inputs can represent disturbances acting on the process, effects of unmodelled dynamics, or actuator faults. The observers presented in this chapter are designed based on a known part of the dynamic model, and the observer gains are computed by solving a system of LMIs. When the unknown inputs are represented or approximated by polynomial functions of time, sufficient conditions that guarantee the exponential convergence of the error and also an ultimate bound on the error signal have been presented. In the case of estimating unmodelled dynamics, sufficient conditions have been given for the asymptotic convergence of the observer.

Two shortcomings of the presented approach for the estimation of unmodelled dynamics has to be noted. First, a prerequisite of the approach is that an upper bound on the state matrix of the unmodelled dynamics is known, and that a robust observer can be designed. Second, the presented method only guarantees asymptotic stability, not exponential stability. Moreover, the methods rely on the existence of a common quadratic Lyapunov function, which introduces conservativeness in itself.

Glossary

Conventions

The following conventions are used throughout the book:

- The standard control-theoretic conventions are used. For instance, the state is denoted by $\boldsymbol{x}$, the control action by $\boldsymbol{u}$, the process dynamics by $\boldsymbol{f}$, the measurements by $\boldsymbol{y}$, and the measurement function by $\boldsymbol{h}$.
- All the vectors used in this thesis are column vectors. The transpose of a vector is denoted by the superscript T. For instance, the transpose of $\boldsymbol{x}$ is $\boldsymbol{x}^T$.
- Boldface notation is used for vector or matrix functions, e.g., $\boldsymbol{f}$ is a vector function.

List of Symbols and Notations

General Notations

I	Identity matrix.
0	Zero matrix.
$\mathcal{H}$	Hermitian of a matrix $\mathcal{H}(A) = A + A^T$.
$A > 0$	A is positive definite matrix.
$\widehat{s}$	Estimated value of the signal s.
$\dot{s}$	Derivative of the signal s.
$\mathcal{C}_{\boldsymbol{x}}$	Domain where the variable $\boldsymbol{x}$ is defined.
$\lVert \cdot \rVert$	Euclidian norm of a vector or induced norm of a matrix.
$\gamma,\ \mu$	Positive constants used as bounds.
$i,\ j,\ k,\ l$	Indices.

Fuzzy Sets and Systems

z_i	ith scheduling variable.
Z_j^i	Fuzzy set corresponding to the jth variable in the ith rule.

ω_{ij}	Membership value of z_j in the fuzzy set Z_j^i.
φ_i	(Non-normalized) membership function of rule i.
w_i	Normalized membership function of rule i.

Dynamic Systems

$\boldsymbol{x}$	State vector.
$\boldsymbol{u}$	Input vector.
$\boldsymbol{y}$	Output vector.
$\boldsymbol{d}$	Disturbance/unknown input.
n_x	Dimension of the state vector.
n_u	Dimension of the input vector.
n_y	Dimension of the output vector.
n_d	Dimension of the unknown input.
n_s	Number of subsystems.
$\boldsymbol{f}$	State transition function; general nonlinear vector function.
$\boldsymbol{h}$	Measurement function; general nonlinear vector function.
A	State transition matrix (linear systems).
B	Input matrix (linear systems).
C	Measurement matrix (linear systems).
a	Affine term in the state equation (linear systems).
c	Affine term in the output equation (linear systems).
t	Time.
$\boldsymbol{f}^{\mathrm{m}}$, $\boldsymbol{h}^{\mathrm{m}}$, $\boldsymbol{h}^{\mathrm{m}}$	Matrix functions.

TS Fuzzy Systems

η_i^j	Weighting function of the jth term obtained by the sector nonlinearity approach, $i \in \{0, 1\}$.
i, j	Indices for local linear models.
m	Number of rules.
$\boldsymbol{z}$	Vector of scheduling variables.
$\boldsymbol{e}$	Error vector.
A_i	State matrix of the ith local model.
B_i	Input matrix of the ith local model.
C_i	Measurement matrix of the ith local model.
a_i	Affine term in the ith state model.
c_i	Affine term in the ith measurement model.
P	Lyapunov matrix.
V	Lyapunov function.
L_i	Observer gain of the ith local model.
$\boldsymbol{A}(\boldsymbol{x})$	Matrix function.
$A_{\delta i}$	Uncertainty in the state matrix of the ith local model.
$B_{\delta i}$	Uncertainty in the input matrix of the ith local model.
$\boldsymbol{f}^{\mathrm{u}}$, $\boldsymbol{h}^{\mathrm{u}}$, $\boldsymbol{h}^{\mathrm{u}}$	Update laws (matrix functions).

References

Abonyi, J., Babuška, R., Szeifert, F.: Modified Gath-Geva fuzzy clustering for identification of Takagi-Sugeno fuzzy models. IEEE Transactions on Systems, Man, and Cybernetics, Part B: Cybernetics 32(5), 612–621 (2002)

Akar, M., Özgüner, U.: Decentralized techniques for the analysis and control of Takagi-Sugeno fuzzy systems. IEEE Transactions on Fuzzy Systems 8(6), 691–704 (2000)

Angelov, P., Filev, D.: On-line design of Takagi-Sugeno models. In: De Baets, B., Kaynak, O., Bilgiç, T. (eds.) IFSA 2003. LNCS, vol. 2715, pp. 92–165. Springer, Heidelberg (2003)

Angelov, P., Filev, D.: An approach to online identification of Takagi-Sugeno fuzzy models. IEEE Transactions on Systems, Man and Cybernetics – Part B: Cybernetics 34(1), 484–498 (2004a)

Angelov, P., Filev, D.: Flexible models with evolving structure. International Journal of Intelligent Systems 19(4), 327–340 (2004b)

Angelov, P., Kordon, A., Zhou, X.: Evolving fuzzy inferential sensors for process industry. In: Proceedings of the 3rd International Workshop on Genetic & Evolving Fuzzy Systems, pp. 41–46. Witten-Bommerholz, Germany (2008)

Angelov, P., Zhou, X.: On line learning fuzzy rulebased system structure from data streams. In: 2008 IEEE World Congress on Computational Intelligence, Hong-Kong, China, pp. 915–922 (2008)

Arcak, M., Angeli, D., Sontag, E.: A unifying integral ISS framework for stability of nonlinear cascades. SIAM Journal of Control and Optimization 40(6), 1888–1904 (2002)

Åström, K.J., Wittenmark, B.: Computer Controlled Systems. Prentice-Hall, Upper Saddle River (1990)

Babuška, R., van der Veen, P., Kaymak, U.: Improved covariance estimation for Gustafson-Kessel clustering. In: Proceedings of the IEEE International Conference on Fuzzy Systems, Honolulu, HI, USA, vol. 2, pp. 1081–1085 (2002)

Bacciotti, A., Boieri, P., Mazzi, L.: Linear stabilization of nonlinear cascade systems. Mathematics of Control, Signals and Systems 6(2), 146–165 (1993)

Baddou, A., Naib, M., Tadeo, F., Benzaouia, A.: On improving the convergence rate of linear continuous-time systems subject to asymmetrically constrained control. Mathematical Problems in Engineering, Article ID 89429, 14 Pages (2006)

Bakule, L.: Decentralized control: An overview. Annual Reviews in Control 32(1), 87–98 (2008)

Bavafa-Toosi, Y., Ohmori, H., Labibi, B.: A generic approach to the design of decentralized linear output-feedback controllers. Systems & Control Letters 55(4), 282–292 (2006)

Benigni, A., Ghisla, U., D'Antona, G., Monti, A., Ponci, F.: A decentralized observer for electrical power systems: Implementation and experimental validation. In: Proceedings of the IEEE Instrumentation and Measurement Technology Conference, Vancouver, Canada, pp. 859–864 (2008)

Bergsten, P.: Observers and Controllers for Takagi-Sugeno Fuzzy Systems. PhD thesis, Örebro University, Sweden (2001)

Bergsten, P., Palm, R.: Thau-Luenberger observers for TS fuzzy systems. In: The Ninth IEEE International Conference on Fuzzy Systems, San Antonio, TX, USA, pp. 671–676 (2000)

Bergsten, P., Palm, R., Driankov, D.: Fuzzy observers. In: Proceedings of the 10th IEEE International Conference on Fuzzy Systems, Melbourne, Australia, vol. 2, pp. 700–703 (2001)

Bernal, M., Guerra, T.M.: Generalized non-quadratic stability of continuous-time Takagi-Sugeno models. IEEE Transactions on Fuzzy Systems 18(4), 815–822 (2010)

Bernal, M., Guerra, T.M., Kruszewski, A.: A membership-function-dependent approach for stability analysis and controller synthesis of Takagi-Sugeno models. Fuzzy Sets and Systems 160(19), 2776–2795 (2009)

Besançon, G.: Observer design for nonlinear systems. In: Loria, A., et al. (eds.) Advanced Topics in Control Systems Theory, pp. 61–89. Springer, London (2006)

Blanco, Y., Perruquetti, W., Borne, P.: Stability and stabilization of nonlinear systems and Takagi-Sugeno's fuzzy models. Mathematical Problems in Engineering 7(3), 221–240 (2001)

Bolic, M., Djuric, P.M., Hong, S.: Resampling algorithms and architectures for distributed particle filters. IEEE Transactions on Signal Processing 53(7), 2442–2450 (2004)

Boyd, S., El Ghaoui, L., Féron, E., Balakrishnan, V.: Linear Matrix Inequalities in System and Control Theory. Society for Industrial and Applied Mathematics, Philadelphia, PA, USA (1994)

Chaillet, A., Loria, A.: Necessary and sufficient conditions for uniform semiglobal practical asymptotic stability: Application to cascaded systems. Automatica 42(11), 1899–1906 (2006)

Chiang, C.-C.: Decentralized robust fuzzy-model-based control of uncertain large-scale systems with input delay. In: Proceedings of the 2006 IEEE International Conference on Fuzzy Systems, Vancouver, Canada, pp. 498–505 (2006)

Chiang, C.-C., Kuo, Z.-H.: Decentralized adaptive fuzzy controller design of large-scale nonlinear systems with unmatched uncertainties. In: Proceedings of the 2002 IEEE International Conference on Fuzzy Systems, Honolulu, HI, USA, pp. 668–673 (2002)

Chiang, C.-C., Lu, W.-H.: Decentralized adaptive fuzzy controller design for uncertain large-scale systems with unknown dead-zone. In: Proceedings of the 2007 IEEE International Conference on Fuzzy Systems, London, UK, pp. 1–6 (2007)

Chiang, C.-C., Wang, W.-H.: Decentralized robust adaptive fuzzy controller for large-scale nonlinear uncertain systems. In: Proceedings of the 12th IEEE International Conference on Fuzzy Systems, St. Louis, MO, USA, pp. 436–440 (2003)

Chien, C.-J., Er, M.J.: Decentralized adaptive fuzzy iterative learning control for repeatable nonlinear interconnected systems. In: Proceedings of the 2006 IEEE International Conference on Systems, Man, and Cybernetics, Taipei, Taiwan, pp. 1710–1715 (2006)

Chilali, M., Gahinet, P.: H_∞ design with pole placement constraints: An LMI approach. IEEE Transactions on Automatic Control 41(3), 358–367 (1996)

Choi, D., Park, P.: Guaranteed cost controller design for discrete-time switching fuzzy systems. IEEE Transactions on Systems, Man, and Cybernetics, Part B 34(1), 110–119 (2004)

Coates, M.: Distributed particle filters for sensor networks. In: Proceedings of the 3rd International Symposium on Information Processing in Sensor Networks, Berkeley, CA, USA, pp. 99–107 (2004)

Corless, M., Tu, J.: State and input estimation for a class of uncertain systems. Automatica 34(6), 757–764 (1998)

Delmotte, F., Guerra, T.M., Ksontini, M.: Continuous Takagi-Sugeno's models: Reduction of the number of LMI conditions in various fuzzy control design technics. IEEE Transactions on Fuzzy Systems 15(3), 426–438 (2007)

Dhbaibi, S., Tlili, A.S., Elloumi, S., Braiek, N.B.: H_∞ decentralized observation and control of nonlinear interconnected systems. ISA Transactions 48(4), 458–467 (2009)

Ding, B.: Quadratic boundedness via dynamic output feedback for constrained nonlinear systems in Takagi–Sugeno's form. Automatica 45(9), 2093–2098 (2009)

Dong, J., Yang, G.: Dynamic output feedback H_∞ control synthesis for discrete-time T-S fuzzy systems via switching fuzzy controllers. Fuzzy Sets and Systems 160(19), 482–499 (2009)

Driankov, D., Hellendoorn, H., Reinfrank, M.: An Introduction to Fuzzy Control. Springer-Verlag New York, Inc., New York (1993)

Durrant-Whyte, H., Rao, B., Hu, H.: Toward a fully decentralized architecture for multi-sensor data fusion. In: Proceedings of the IEEE International Conference on Robotics and Automation, Cincinnati, OH, USA, vol. 2, pp. 1331–1336 (1990)

Fantuzzi, C., Rovatti, R.: On the approximation capabilities of the homogeneous Takagi-Sugeno model. In: Proceedings of the Fifth IEEE International Conference on Fuzzy Systems, New Orleans, LA, USA, pp. 1067–1072 (1996)

Feng, G.: Controller synthesis of fuzzy dynamic systems based on piecewise Lyapunov functions. IEEE Transactions on Fuzzy Systems 11(5), 605–612 (2003)

Feng, G.: A survey on analysis and design of model-based fuzzy control systems. IEEE Transactions on Fuzzy Systems 14(5), 676–697 (2006)

Frank, P.M.: Fault diagnosis in dynamic systems using analytical and knowledge-based redundancy: A survey and some new results. Automatica 26(3), 459–474 (1990)

Franklin, G.F., Powell, J.D., Workman, M.L.: Digital Control of Dynamic Systems. Addison-Wesley, Boston (1990)

Gahinet, P., Nemirovski, A., Laub, A.J., Chilali, M.: LMI control toolbox. The Mathworks Inc. (1995)

Gegov, A.E., Frank, P.M.: Decomposition of multivariable systems for distributed fuzzy control. Fuzzy Sets and Systems 73(3), 329–340 (1995)

Guelton, K., Bouarar, T., Manamanni, N.: Fuzzy Lyapunov LMI based output feedback stabilization of Takagi-Sugeno systems using descriptor redundancy. In: Proceedings of the 2008 IEEE International Conference on Fuzzy Systems, Hong Kong, China, pp. 1212–1218 (2008a)

Guelton, K., Bouarar, T., Manamanni, N.: Robust dynamic output feedback fuzzy Lyapunov stabilization of Takagi-Sugeno systems – a descriptor redundancy approach. Fuzzy Sets and Systems 160(19), 2796–2811 (2009)

Guelton, K., Delprat, S., Guerra, T.M.: An alternative to inverse dynamics joint torques estimation in human stance based on a Takagi-Sugeno unknown-inputs observer in the descriptor form. Control Engineering Practice 16(12), 1414–1426 (2008b)

Guerra, T.M., Bernal, M.: A way to escape from the quadratic framework. In: Proceedings of the IEEE International Conference on Fuzzy Systems, Jeju, Korea, pp. 784–789 (2009)

Guerra, T.M., Kruszewski, A., Vermeiren, L., Tirmant, H.: Conditions of output stabilization for nonlinear models in the Takagi-Sugeno's form. Fuzzy Sets and Systems 159, 1248–1259 (2006)

Guerra, T.M., Vermeiren, L.: LMI-based relaxed nonquadratic stabilization conditions for nonlinear systems in the Takagi-Sugeno's form. Automatica 40(5), 823–829 (2004)

Ha, Q.P., Trinh, H.: State and input simultaneous estimation for a class of nonlinear systems. Automatica 40(10), 1779–1785 (2004)

Haijun, G., Tianping, Z., Qikun, S.: Decentralized model reference adaptive sliding mode control based on fuzzy model. Journal of Systems Engineering and Electronics 17(1), 182–186 (2006)

Ho, H., Wong, Y., Rad, A., Lo, W.: State observer based indirect adaptive fuzzy tracking control. Simulation Modelling Practice and Theory 7(13), 646–663 (2005)

Hostetter, G.H., Meditch, J.S.: On the generalization of observers to systems with unmeasurable unknown inputs. Automatica 9(6), 721–724 (1973)

Hou, M., Müller, P.C.: Design of decentralized linear state function observers. Automatica 30(11), 1801–1805 (1994)

Hou, M., Patton, R.J.: Input observability and input reconstruction. Automatica 34(6), 789–794 (1998)

Hovakimyan, N., Calise, A., Madyastha, V.: An adaptive observer design methodology for bounded nonlinear processes. In: Proceedings of the 41st IEEE Conference on Decision and Control, Las Vegas, NV, USA, vol. 4, pp. 4700–4705 (2002)

Hsiao, F.-H., Chen, C.-W., Liang, Y.-W., Xu, S.-D., Chiang, W.-L.: T-S fuzzy controllers for nonlinear interconnected systems with multiple time delays. IEEE Transactions on Circuits and Systems I: Regular Papers 52(9), 1883–1893 (2005a)

Hsiao, F.-H., Hwang, J.-D.: Stability analysis of fuzzy large-scale systems. IEEE Transactions on Systems, Man, and Cybernetics, Part B 32(1), 122–126 (2002)

Hsiao, F.-H., Hwang, J.-D., Chen, C.-W., Tsai, Z.-R.: Robust stabilization of nonlinear multiple time-delay large-scale systems via decentralized fuzzy control. IEEE Transactions on Fuzzy Systems 13(1), 152–163 (2005b)

Hsieh, C.-S.: Extension of unbiased minimum-variance input and state estimation for systems with unknown inputs. Automatica 45(9), 2149–2153 (2009)

Hua, C., Guan, X., Shi, P.: Adaptive fuzzy control for uncertain interconnected time-delay systems. Fuzzy Sets and Systems 153(3), 447–458 (2005)

Huang, H., Ho, D.: Delay-dependent robust control of uncertain stochastic fuzzy systems with time-varying delay. IET Control Theory and Applications 1(4), 1075–1085 (2007)

Huang, M., Dey, S.: Distributed state estimation for hidden Markov models with dynamic quantization and rate allocation. In: Proceedings of the 16th IFAC World Congress, vol. 16. Czech Republic, Prague (2005)

Huang, Y.-S., Zhou, D.-Q., Xiao, S.-P., Lin, D.: Coordinated decentralised hybrid adaptive output feedback fuzzy control for a class of large-scale non-linear systems with strong interconnections. IET Control Theory & Applications 3(9), 1261–1274 (2009)

Hyun, C.-H., Kim, J.-H., Kim, E., Park, M.: Adaptive fuzzy observer based synchronization design and secure communications of chaotic systems. Chaos, Solitons & Fractals 27(4), 930–940 (2006)

Jankovic, M., Sepulchre, R., Kokotovic, P.: Constructive Lyapunov stabilization of nonlinear cascade systems. IEEE Transactions on Automatic Control 41(12), 1723–1735 (1996)

Jiang, Z.-P.: Decentralized and adaptive nonlinear tracking of large-scale systems via output feedback. IEEE Transactions on Automatic Control 45(11), 2122–2128 (2000)

Jiang, Z.-P., Repperger, D., Hill, D.: Decentralized nonlinear output-feedback stabilization with disturbance attenuation. IEEE Transactions on Automatic Control 46(10), 1623–1629 (2001)

Johansen, T., Babuška, R.: Multiobjective identification of Takagi-Sugeno fuzzy models. IEEE Transactions on Fuzzy Systems 11(6), 847–860 (2003)

Johansen, T.A., Shorten, R., Murray-Smith, R.: On the interpretation and identification of dynamic Takagi-Sugeno fuzzy models. IEEE Transactions on Fuzzy Systems 8(3), 297–313 (2000)

Johansson, M., Rantzer, A.: Computation of piecewise quadratic Lyapunov functions for hybrid systems. IEEE Transactions on Automatic Control 43(4), 555–559 (1998)

Johansson, M., Rantzer, A., Arzen, K.: Piecewise quadratic stability of fuzzy systems. IEEE Transactions on Fuzzy Systems 7(6), 713–722 (1999)

Kailath, T.: Linear Systems. Prentice Hall, Upper Saddle River (1980)

Kailath, T., Sayed, A., Hassibi, B.: Linear Estimation. Prentice Hall, Upper Saddle River (2000)

Kau, S.-W., Lee, H.-J., Yang, C.-M., Lee, C.-H., Hong, L., Fang, C.-H.: Robust H_∞ fuzzy static output feedback control of T-S fuzzy systems with parametric uncertainties. Fuzzy Sets and Systems 158(2), 135–146 (2007)

Kaymak, U., van den Berg, J.: On constructing probabilistic fuzzy classifiers from weighted fuzzy clustering. In: Proceedings of the International Conference on Fuzzy Systems, Budapest, Hungary, vol. 1, pp. 395–400 (2004)

Khalil, H.K.: Nonlinear Systems. Prentice-Hall, Upper Saddle River (2002)

Kim, E., Lee, H.: New approaches to relaxed quadratic stability condition of fuzzy control systems. IEEE Transactions on Fuzzy Systems 8(5), 523–534 (2000)

Kiriakidis, K.: Nonlinear modeling by interpolation between linear dynamics and its application in control. Journal of Dynamic Systems, Measurement, and Control 129(6), 813–824 (2007)

Kosko, B.: Fuzzy systems as universal approximators. IEEE Transactions on Computers 43(11), 1329–1333 (1994)

Kočvara, M., Stingl, M.: Penbmi, version 2.1. (2008), `http://www.penopt.com`

Krishnamurthy, P., Khorrami, F.: Decentralized control and disturbance attenuation for large-scale nonlinear systems in generalized output-feedback canonical form. Automatica 39(11), 1923–1933 (2003)

Kruse, R., Gebhardt, J.E., Klowon, F.: Foundations of Fuzzy Systems. John Wiley & Sons, Inc., New York (1994)

Kruszewski, A., Guerra, T.M.: Stabilization of a class of nonlinear model with periodic parameters in the Takagi-Sugeno form. In: Proceedings of the IFAC Workshop Periodic Control Systems, Saint Petersburg, Russia, pp. 1–6 (2007)

Kruszewski, A., Sala, A., Guerra, T.M., Arino, C.: A triangulation approach to asymptotically exact conditions for fuzzy summations. IEEE Transactions on Fuzzy Systems 17(5), 985–994 (2009)

Kruszewski, A., Wang, R., Guerra, T.M.: Nonquadratic stabilization conditions for a class of uncertain nonlinear discrete time TS fuzzy models: A new approach. IEEE Transactions on Automatic Control 53(2), 606–611 (2008)

Kukolj, D., Levi, E.: Identification of complex systems based on neural and Takagi-Sugeno fuzzy model. IEEE Transactions on Systems, Man, and Cybernetics, Part B: Cybernetics 34(1), 272–282 (2004)

Labiod, S., Guerra, T.M.: Adaptive fuzzy control of a class of SISO nonaffine nonlinear systems. Fuzzy Sets and Systems 158(10), 1126–1137 (2007)

Landau, Y.D.: Adaptive Control: The Model Reference Approach. Marcel Dekker, Inc., New York (1979)
Lendek, Zs., Babuška, R., De Schutter, B.: Stability analysis and observer design for decentralized TS fuzzy systems. In: Proceedings of the IEEE International Conference on Fuzzy Systems, Hong Kong, China, pp. 631–636 (2008)
Lendek, Zs., Babuška, R., De Schutter, B.: Stability bounds for fuzzy estimation and control — Part I: State estimation. In: Proceedings of the 2010 IEEE International Conference on Automation, Quality and Testing, Robotics, Cluj, Romania, pp. 1–6 (2010a)
Lendek, Zs., Guerra, T.M., Babuška, R.: On non-PDC local observers for TS fuzzy systems. In: Proceedings of the IEEE World Congress on Computational Intelligence, Barcelona, Spain, pp. 2436–2442 (2010b)
Li, D.D., Chen, C.: A novel approach to estimate load factor of variable-speed wind turbines. IEEE Transactions on Power Systems 20(2), 1186–1188 (2005)
Liao, T.L., Huang, N.S.: An observer based approach for chaotic synchronization with applications to secure communications. IEEE Transactions on Circuits and Systems 46(9), 1144–1150 (1999)
Liu, F., Wu, M., He, Y., Yokoyama, R.: New delay-dependent stability criteria for T–S fuzzy systems with time-varying delay. Fuzzy Sets and Systems 161(15), 2033–2042 (2010)
Liu, S.-J., Zhang, J.-F., Jiang, Z.-P.: Decentralized adaptive output-feedback stabilization for large-scale stochastic nonlinear systems. Automatica 43(2), 238–251 (2007)
Liu, X., Zhang, H.: Stability analysis of uncertain fuzzy large-scale system. Chaos, Solitons & Fractals 25(5), 1107–1122 (2005)
Liu, X., Zhang, H., Liu, G.: Robust and non-fragile H_∞ control for affine fuzzy large-scale systems. In: Proceedings of the 7th World Congress on Intelligent Control and Automation, Chongquing, China, vol. 1, pp. 6107–6112 (2008)
Liu, X., Zhang, Q.: New approaches to H_∞ controller designs based on fuzzy observers for T-S fuzzy systems via LMI. Automatica 39(9), 1571–1582 (2003)
Lo, J.-C., Lin, M.-L.: Robust H_∞ nonlinear control via fuzzy static output feedback. IEEE Transactions on Circuits and Systems I: Fundamental Theory and Applications 50(11), 1494–1502 (2003)
Löfberg, J.: YALMIP: a toolbox for modeling and optimization in MATLAB. In: Proceedings of the CACSD Conference, Taipei, Taiwan, pp. 284–289 (2004)
López-Orozco, J.A., de la Cruz, J.M., Besada, E., Ruipérez, P.: An asynchronous, robust, and distributed multisensor fusion system for mobile robots. International Journal of Robotics Research 19(12), 914–932 (2000)
Loria, A., Panteley, E.: Cascaded nonlinear time-varying systems: Analysis and design. In: Advanced Topics in Control Systems Theory, pp. 23–64. Springer, Heidelberg (2005)
Luenberger, D.G.: Observers for multivariable systems. IEEE Transactions on Automatic Control 11(2), 190–197 (1966)
Ma, X.-J., Sun, Z.-Q., He, Y.-Y.: Analysis and design of fuzzy controller and fuzzy observer. IEEE Transactions on Fuzzy Systems 6(1), 41–51 (1998)
Marino, R.: Adaptive observers for single output nonlinear systems. IEEE Transactions on Automatic Control 35(9), 1054–1058 (1990)
Marino, R., Santosuosso, G., Tomei, P.: Robust adaptive observers for nonlinear systems with bounded disturbances. IEEE Transactions on Automatic Control 46(6), 967–972 (2001)
Marino, R., Tomei, P.: Adaptive observers with arbitrary exponential rate of convergence for nonlinear systems. IEEE Transactions on Automatic Control 40(7), 1300–1304 (1995)

Marx, B., Koenig, D., Ragot, J.: Design of observers for Takagi-Sugeno descriptor systems with unknown inputs and application to fault diagnosis. IET Control Theory & Applications 1(5), 1487–1495 (2007)

Michel, A.N., Miller, R.K., Tang, W.: Lyapunov stability of interconnected systems: Decompositions into strongly connected subsystems. IEEE Transactions on Circuits and Systems 25(9), 799–809 (1978)

Mozelli, L.A., Palhares, R.M., Souza, F.O., Mendes, E.M.A.M.: Reducing conservativeness in recent stability conditions of TS fuzzy systems. Automatica 45(6), 1580–1583 (2009)

Narendra, K.S., Annaswamy, A.M.: Stable Adaptive Systems. Prentice-Hall, Inc., Upper Saddle River (1989)

Ohtake, H., Tanaka, K., Wang, H.: Fuzzy modeling via sector nonlinearity concept. In: Proceedings of the Joint 9th IFSA World Congress and 20th NAFIPS International Conference, Vancouver, Canada, vol. 1, pp. 127–132 (2001)

Oudghiri, M., Chadli, M., El Hajjaji, A.: Lateral vehicle velocity estimation using fuzzy sliding mode observer. In: Proceedings of the Mediterranean Conference on Control and Automation, Athens, Greece, pp. 1–6 (2007)

Palm, R., Bergsten, P.: Sliding mode observer for a Takagi-Sugeno fuzzy system. In: Proceedings of the 9th IEEE International Conference on Fuzzy Systems, San Antonio, TX, USA, vol. 2, pp. 665–670 (2000)

Palm, R., Driankov, D.: Towards a systematic analysis of fuzzy observers. In: Proceedings of the 18th International Conference of the North American Fuzzy Information Processing Society, New York, NY, USA, pp. 179–183 (1999)

Park, J.-H., Park, G.-T.: Adaptive fuzzy observer with minimal dynamic order for uncertain nonlinear systems. IEE Proceedings–Control Theory and Applications 150(2), 189–197 (2003)

Park, J.-H., Park, G.-T., Kim, S.-H., Moon, C.-J.: Output-feedback control of uncertain nonlinear systems using a self-structuring adaptive fuzzy observer. Fuzzy Sets and Systems 151(1), 21–42 (2005)

Park, J.-H., Yoon, P.-S., Park, G.-T.: Robust adaptive observer using fuzzy systems for uncertain nonlinear systems. In: Proceedings of the 10th IEEE International Conference on Fuzzy Systems, Melbourne, Australia, vol. 2, pp. 749–752 (2001)

Pertew, A., Marquez, H., Zhao, Q.: H_∞ synthesis of unknown input observers for non-linear Lipschitz systems. International Journal of Control 78(15), 1155–1165 (2005)

Pertew, A., Marquez, H., Zhao, Q.: H_∞ observer design for Lipschitz nonlinear systems. IEEE Transactions on Automatic Control 51(7), 1211–1216 (2006)

Priscoli, F.D., Marconi, L., Isidori, A.: Adaptive observers as nonlinear internal models. Systems & Control Letters 55(8), 640–649 (2006)

Rajamani, R., Hedrick, J.: Adaptive observers for active automotive suspensions: theory and experiment. IEEE Transactions on Control Systems Technology 3(1), 86–93 (1995)

Rhee, B., Won, S.: A new fuzzy Lyapunov function approach for a Takagi-Sugeno fuzzy control system design. Fuzzy Sets and Systems 157(9), 1211–1228 (2006)

Roebenack, K., Lynch, A.: Observer design using a partial nonlinear observer canonical form. International Journal of Applied Mathematics and Computer Science 16(3), 333–343 (2006)

Roumeliotis, S., Bekey, G.: Distributed multirobot localization. IEEE Transactions on Robotics and Automation 18(5), 781–795 (2002)

Ruiz Vargas, J., Hemerly, E.: Adaptive observers for unknown general nonlinear systems. IEEE Transactions on Systems, Man and Cybernetics, Part B 31(5), 683–690 (2001)

Saif, M., Guan, Y.: Decentralized state estimation in large-scale interconnected dynamical systems. Automatica 28(1), 215–219 (1992)

Sala, A., Ariño, C.: Asymptotically necessary and sufficient conditions for stability and performance in fuzzy control: Applications of Polya's theorem. Fuzzy Sets and Systems 158(24), 2671–2686 (2007)

Sala, A., Guerra, T.M., Babuška, R.: Perspectives of fuzzy systems and control. Fuzzy Sets and Systems 156(3), 432–444 (2005)

Sandell, N., Varaiya, P., Athans, M., Safonov, M.: Survey of decentralized control methods for large scale systems. IEEE Transactions on Automatic Control 23(2), 108–128 (1978)

Sastry, S., Bodson, M.: Adaptive Control: Stability, Convergence, and Robustness. Prentice Hall, Englewood Cliffs (1989)

Scherer, C., Weiland, S.: Linear Matrix Inequalities in Control. Delft University, The Netherlands (2005)

Schmitt, T., Hanek, R., Beetz, M., Buck, S., Radig, B.: Cooperative probabilistic state estimation for vision-based autonomous mobile robots. IEEE Transactions on Robotics and Automation 18(5), 670–684 (2002)

Seibert, P., Suarez, R.: Global stabilization of nonlinear cascade systems. Systems & Control Letters 14(4), 347–352 (1990)

Sheikholeslam, S.: Observer-based parameter identifiers for nonlinear systems with parameter dependencies. IEEE Transactions on Automatic Control 40, 382–387 (1995)

Sheldrake, A.L.: Estimation of Plant Electrical Load, ch. 1, pp. 1–18. John Wiley & Sons, New York (2005)

Sontag, E.: Remarks on stabilization and input-to-state stability. In: Proceedings of the 28th IEEE Conference on Decision and Control, Tampa, FL, USA, pp. 1376–1378 (1989a)

Sontag, E.: On the input-to-state stability property. European Journal of Control 1(1), 24–36 (1995)

Sontag, E.D.: Smooth stabilization implies coprime factorization. IEEE Transactions on Automatic Control 34(4), 435–443 (1989b)

Sontag, E.D., Wang, Y.: On characterizations of the input-to-state stability property. Systems & Control Letters 24(5), 351–359 (1995)

Sundareshan, M.K., Elbanna, R.M.: Design of decentralized observation schemes for large-scale interconnected systems: Some new results. Automatica 26(4), 789–796 (1990)

Takagi, T., Sugeno, M.: Fuzzy identification of systems and its applications to modeling and control. IEEE Transactions on Systems, Man, and Cybernetics 15(1), 116–132 (1985)

Tanaka, K., Hori, T., Wang, H.: A multiple Lyapunov function approach to stabilization of fuzzy control systems. IEEE Transactions on Fuzzy Systems 11(4), 582–589 (2003)

Tanaka, K., Ikeda, T., Wang, H.: Fuzzy regulators and fuzzy observers: relaxed stability conditions and LMI-based designs. IEEE Transactions on Fuzzy Systems 6(2), 250–265 (1998)

Tanaka, K., Sano, M.: On the concepts of regulator and observer of fuzzy control systems. In: Proceedings of the Third IEEE International Conference on Fuzzy Systems, Orlando, FL, USA, vol. 2, pp. 767–772 (1994)

Tanaka, K., Wang, H.: Fuzzy regulators and fuzzy observers: a linear matrix inequality approach. In: Proceedings of the 36th IEEE Conference on Decision and Control, San Diego, CA, USA, vol. 2, pp. 1315–1320 (1997)

Tanaka, K., Wang, H.O.: Fuzzy Control System Design and Analysis: A Linear Matrix Inequality Approach. John Wiley & Sons, New York (2001)

Taniguchi, T., Tanaka, K., Ohtake, H., Wang, H.: Model construction, rule reduction, and robust compensation for generalized form of Takagi-Sugeno fuzzy systems. IEEE Transactions on Fuzzy Systems 9(4), 525–538 (2001)

Taniguchi, T., Tanaka, K., Yamafuji, K., Wang, H.: A new PDC for fuzzy reference models. In: Proceedings of the IEEE International Conference on Fuzzy Systems, Seoul, Korea, vol. 2, pp. 898–903 (1999a)

Taniguchi, T., Tanaka, K., Yamafuji, K., Wang, H.: Nonlinear model following control via Takagi-Sugeno fuzzy model. In: Proceedings of the, American Control Conference, San Diego, CA, USA, vol. 3, pp. 1837–1841 (1999b)

Tong, S., Li, H.-X., Wang, W.: Observer-based adaptive fuzzy control for SISO nonlinear systems. Fuzzy Sets and Systems 148(3), 355–376 (2004)

Tong, S., Wang, W., Qu, L.: Decentralized robust control for uncertain T-S fuzzy large-scale systems with time-delay. International Journal of Innovative Computing, Information and Control 3(3), 657–672 (2007)

Tong, S., Zhang, Q.: Robust control and stability analysis of fuzzy large-scale systems. In: Proceedings of the 3rd International Conference on Innovative Computing Information and Control, Dalian, China, vol. 1, p. 116 (2008)

Tseng, C.-S.: A novel approach to H_∞ decentralized fuzzy-observer-based fuzzy control design for nonlinear interconnected systems. IEEE Transactions on Fuzzy Systems 16(5), 1337–1350 (2008)

Tseng, C.-S., Chen, B.-S.: H_∞ decentralized fuzzy model reference tracking control design for nonlinear interconnected systems. IEEE Transactions on Fuzzy Systems 9(6), 795–809 (2001)

Tuan, H., Apkarian, P., Narikiyo, T., Yamamoto, Y.: Parameterized linear matrix inequality techniques in fuzzy control system design. IEEE Transactions on Fuzzy Systems 9(2), 324–332 (2001)

Uang, H.-J., Chen, B.-S.: Fuzzy decentralized controller and observer design for nonlinear interconnected systems. In: Proceedings of the Ninth IEEE International Conference on Fuzzy Systems, San Antonio, TX, USA, vol. 2, pp. 945–948 (2000)

Wang, H., Tanaka, K., Griffin, M.: An approach to fuzzy control of nonlinear systems: stability and design issues. IEEE Transactions on Fuzzy Systems 4(1), 14–23 (1996)

Wang, L.-X., Mendel, J.: Fuzzy basis functions, universal approximation, and orthogonal least-squares learning. IEEE Transactions on Neural Networks 3(5), 807–814 (1992)

Wang, T., Qu, L.-J.: Robust decentralized control for T-S uncertain fuzzy interconnected systems with time-delay. In: Proceedings of the Sixth International Conference on Machine Learning and Cybernetics, Hong Kong, China, pp. 631–636 (2007)

Wang, T., Tong, S.: Decentralized fuzzy model reference H_∞ tracking control for nonlinear large-scale systems. In: Proceedings of the Sixth World Congress on Intelligent Control and Automation, Dalian, China, vol. 1, pp. 75–79 (2006)

Wang, W.-J., Lin, W.-W.: PDC synthesis for TS fuzzy large-scale systems. In: Proceedings of the IEEE International Conference on Systems, Man and Cybernetics, Washington, DC, USA, pp. 4956–4961 (2003)

Wang, W.-J., Lin, W.-W.: Decentralized PDC for large-scale T-S fuzzy systems. IEEE Transactions on Fuzzy Systems 13(6), 779–786 (2005)

Wang, W.-J., Luoh, L.: Stability and stabilization of fuzzy large-scale systems. IEEE Transactions on Fuzzy Systems 12(3), 309–315 (2004)

Wang, Y., Chai, T.: Output-feedback control of uncertain nonlinear systems using adaptive fuzzy observer. In: Proceedings of the American Control Conference, Portland, OR, USA, vol. 4, pp. 2613–2618 (2005)

Wang, Y., Papageorgiou, M.: Real-time freeway traffic state estimation based on extended Kalman filter: A general approach. Transportation Research B 39(2), 141–167 (2005)

Wang, Y., Papageorgiou, M., Messmer, A., Coppola, P., Tzimitsi, A., Nuzzolo, A.: An adaptive freeway traffic state estimator. Automatica 45(1), 10–24 (2009)

Wang, Z., Ho, D., Liu, X.: State estimation for delayed neural networks. IEEE Transactions on Neural Networks 16(1), 279–284 (2005)

Waurajitti, N., Ngamwiwit, J., Prempraneerach, Y.: Adaptive fuzzy sliding mode controller for two cascaded tanks level control. In: Proceedings of TENCON 2000, Kuala Lumpur, Malaysia, vol. 2, pp. 592–597 (2000)

Welch, G., Bishop, G.: An introduction to the Kalman filter. Technical Report TR 95-041, Department of Computer Science, University of North Carolina, NC, USA (2002)

Wu, H.-N.: Robust H_2 fuzzy output feedback control for discrete-time nonlinear systems with parametric uncertainties. International Journal of Approximate Reasoning 46(1), 151–165 (2007)

Xiong, Y., Saif, M.: Unknown disturbance inputs estimation based on a state functional observer design. Automatica 39(8), 1389–1398 (2003)

Xu, D., Li, Y., Wu, T.: Stability analysis of large-scale nonlinear systems with hybrid models. In: Proceedings of the Sixth World Congress on Intelligent Control and Automation, Dalian, China, vol. 1, pp. 1196–1200 (2006)

Xu, S., Song, B., Lu, J., Lam, J.: Robust stability of uncertain discrete-time singular fuzzy systems. Fuzzy Sets and Systems 158(20), 2306–2316 (2007)

Ying, H.: Sufficient conditions on general fuzzy systems as function approximators. Automatica 30(3), 521–525 (1994)

Yoneyama, J.: Robust stability and stabilization for uncertain Takagi-Sugeno fuzzy time-delay systems. Fuzzy Sets and Systems 158(2), 115–134 (2007)

Yoneyama, J., Nishikawa, M., Katayama, H., Ichikawa, A.: Output stabilization of Takagi–Sugeno fuzzy systems. Fuzzy Sets and Systems 111(2), 253–266 (2000)

Yoneyama, J., Nishikawa, M., Katayama, H., Ichikawa, A.: Design of output feedback controllers for Takagi–Sugeno fuzzy systems. Fuzzy Sets and Systems 121(1), 127–148 (2001)

Zhang, H., Li, C., Liao, X.: Stability analysis and H_∞ controller design of fuzzy large-scale systems based on piecewise Lyapunov functions. IEEE Transactions on Systems, Man and Cybernetics, Part B 36(3), 685–698 (2006)

Zhang, Q., Xu, A.: Implicit adaptive observers for a class of nonlinear systems. In: Proceedings of the American Control Conference, Arlington, VA, USA, vol. 2, pp. 1551–1556 (2001)

Zhang, X., Parisini, T., Polycarpou, M.M.: Sensor bias fault isolation in a class of nonlinear systems. IEEE Transactions on Automatic Control 50(3), 370–376 (2005)

Zhou, S., Feng, G.: Generalised H_2 controller synthesis for uncertain discrete-time fuzzy systems via basis-dependent Lyapunov function. IEE Control Theory and Applications 153(1), 74–80 (2006)

Zhu, Y., Pagilla, P.: Adaptive controller and observer design for a class of nonlinear systems. In: Proceedings of the 42nd IEEE Conference on Decision and Control, Maui, HI, USA, vol. 2, pp. 1687–1692 (2003)

Index

Adaptive observer, 154, 156

Boundedness, 80, 135, 163, 164, 171

Cascade
  Nonlinear system, 74, 80
  Observer, 90
  Stability, 80, 84
  TS system, 82
Controllability, 50

Definite, 25
Distributed
  Analysis, 106
  Linear system, 126
  Observer design, 125, 127
  Stability analysis, 106, 114
  TS model, 104
Double sum, 28
  Relaxation, 28, 59, 160
Dynamic system, 7

Eigenvalue problem, *see* EVP
Estimation error, 50, 52, 62, 92, 97, 130, 138, 156, 158, 167
Evolving fuzzy models, 150
EVP, 27

Global asymptotic stability, *see* GAS

Hurwitz matrix, 31

Input-to-state stability, *see* ISS

Linear matrix inequality, *see* LMI
Lipschitz constant, 64, 112, 150, 168
LMI, 26, 37, 53, 60, 119, 132, 136, 160, 168
Lyapunov function
  Non-quadratic, 34, 35
  Quadratic, 30, 107, 159

Membership function, 6, 19

Observability, 50

Parallel distributed compensation, *see* PDC
Partitioning, 75, 76
PDC, 36
Performance
  H_∞ attenuation, 38
  Decay rate, 31, 60, 88, 93, 162
  LMI regions, 32
Perturbation method, 104

Scheduling variables, 5, 12, 18
Schur complement, 27, 118, 141
Sector nonlinearity, 11
Stability
  GAS, 31, 39, 80
  ISS, 46, 80
  Non-quadratic, 33
  Quadratic, 30
Stabilization
  H_∞ attenuation, 38

Observer-based, 68
Output feedback, 45
Robust, 41
State feedback, 36
Subsystems
Cascaded, 74
Distributed, 104

Taylor series, 17
TS model
Affine, 9, 86, 110
Linear, 9
Uncertain, 38, 112
TS observer, 50, 129, 157
Estimated scheduling, 62, 96, 138, 166, 179
Measured scheduling, 52, 92, 130, 158, 174

Unknown input, 150, 152
Derivative, 152, 157
Polynomial, 158, 166
Unmodelled dynamics, 174

Vanishing disturbance, 85

Weighting function, 12

MIX
Papier aus verantwortungsvollen Quellen
Paper from responsible sources
FSC® C105338

If you have any concerns about our products,
you can contact us on
ProductSafety@springernature.com

In case Publisher is established outside the EU,
the EU authorized representative is:
Springer Nature Customer Service Center GmbH
Europaplatz 3, 69115 Heidelberg, Germany

Printed by Libri Plureos GmbH
in Hamburg, Germany